Worldwide Telecommunications Guide for the Business Manager

WILEY SERIES IN TELECOMMUNICATIONS

Donald L. Schilling, Editor
City College of New York

Worldwide Telecommunications Guide for the Business Manager
Walter L. Vignault

Expert Systems Applications to Telecommunications
Jay Liebowitz (in preparation)

Introduction to Telecommunications Engineering, 2nd Edition
Robert Gagliardi (in preparation)

Digital Signal Estimation
Robert J. Mammone (in preparation)

Worldwide Telecommunications Guide for the Business Manager

WALTER L. VIGNAULT

A Wiley-Interscience Publication
JOHN WILEY & SONS
New York • Chichester • Brisbane • Toronto • Singapore

Library of Congress Cataloging in Publication Data:

Vignault, Walter L.
Worldwide telecommunications guide for the business manager.

(Wiley series in telecommunications)
"Wiley-Interscience."
Bibliography: p. xiv
1. Business—Communication systems. 2. Telecommunication systems.
3. Telephone in business.
I. Title. II. Series.
HF5541.T4V54 1987 384 87-2002
ISBN 0-471-85828-5

Printed in the United States of America

87 88 10 9 8 7 6 5 4 3 2 1

To Jane, Mark, John, and Michelle

PREFACE

After numerous lectures and briefings to worldwide visitors from the United States, Europe, the Middle East, South America, and Japan, I have concluded that there is a need for information on international telecommunications. However, there is no central group or department, either private or public, that pulls together international services that are available throughout the world. Although the material is available in bits and pieces, usually consultants or very large firms with international telecommunications departments acquire ongoing information on several countries. The process of gathering information is difficult, time-consuming, and very expensive and requires contact with many persons within each telecommunications administration. The reports of consultants cover several telecommunications topics, but by nature of their consultants' business, most reports are in response to a client's specific problem within a geographical area.

An expanding enterprise that is adding new work groups or locations will generate communications problems. As a professional, you will find that there are continuing telecommunications problems that disrupt your work or that you may be asked to solve. How do you solve them?

My intention is to provide information on several alternatives for U.S. and international telecommunications. Telecommunications solutions for networks and services will be presented on a country basis.

In general, you focus on immediate local problems on a day-to-day basis. This reaction is normal, and to increase your knowledge of the U.S. environment and international telecommunications, several subjects are included that will provide you with a base for further study.

Although voice communications will be discussed, most of the material will cover data communications.

One advantage of a four-year assignment in the IBM Telecommunications Center for Europe, Middle East, and Africa was access to telecommunications workers in this region and throughout the world. People working in telecommunications relations from over forty countries participated in the final review process for this book. My thanks to over 100 IBMers throughout the world that have made contributions to this material. Several of my colleagues at IBM have also made important contributions: Michel Humbert, Dr. Jacques Besseyre, Marc Boisseau, William G. Burke, Claude Laurens, and Robert Ure, all at La Gaude, France, and Tatsuya Sato, in Tokyo. Contributions were also made by Anthony Parish, London; Mary Anne Angell of Paris, France; Tom Kullmann, Atlanta; Heinz Mertin, International Education Center, La Hulpe, Belgium; Karen Boudreau, Purchase, NY; and my colleagues at La Gaude: Walter Bozyczko, Mike Hamill-Stewart, and Cuong Ngo-Mai. In addition, I sincerely appreciate the assistance of Gary Stout of American Telephone and Telegraph Company, Morristown, New Jersey. Several people from Cable and Wireless, London, MCI Communications, Telenet (U.S. Sprint), Tymnet McDonnell Douglas, and the Western Union Telegraph Company were very helpful in providing material.

Thanks to Publishers Network, whose careful editing and excellent art and layout work are responsible for the form of the final product.

The subject material has been verified as of this writing. However, information could change as a result of new technology, regulations, public policies, and tariffs. Although most common carriers will change their tariffs for telecommunications services, the methodology should remain constant for many years. Tariff update services should be subscribed to for current information.

The material in this book is the author's and in no way represents the opinion or position of the IBM Corporation.

AT&T is a trademark of the American Telephone & Telegraph Company, Morristown, New Jersey. GE is a trademark of the General Electric Company, Fairfield, Connecticut. References to GE are to the U.S.-based corporation. IBM is a trademark of the International Business Machines Corporation, Armonk, New York.

Walter L. Vignault

Atlanta, Georgia
July 1987

CONTENTS

Worldwide Telecommunications Guide for the Business Manager

1

INTRODUCTION

The Boston Converter Company, an east coast manufacturer with revenues of $3 million a year, would like to expand their marketing coverage on the west coast. They have an employee, Fred Fox, who has worked in the applications engineering section of the marketing department and is quite familiar with the product line. Boston Converter has been selling products through manufacturer representatives and would like Fred to establish a district office in Los Angeles and market directly to clients. After Fred has selected an office location and furnishings, one of his first tasks is to order a telephone. During the initial period, the telephone will satisfy his business needs and liaison with the home office. As his business increases, communications will increase to the home office. Since there is a 3-hour difference between Los Angeles and the home office, he may consider some alternatives to reduce his communications costs.

One alternative is a telex machine that can run unattended and allow Fred to send and receive messages while he is out of the office. Depending on usage, a telex system could cost about $200 to $350 dollars per month, including rental. Another choice is a facsimile or Fax machine that can send exact copies of customer purchase orders, including letterhead and signature if the company requires documentation before shipping goods. A Fax Group 3 type of machine that can send a page in less than 1 minute over a telephone line would cost about $2500.

Another alternative is a private (or leased) line. After checking with the telephone company, Fred will find that a leased line offers 24-hour service with unlimited calls to the home office, but that his office will be charged over $1600 per month for a direct line.

Another possibility is a packet data network offering from Telenet or Tymnet. Fred would have to invest in a terminal or personal computer (PC). A terminal would permit him to send messages or data to a home office computer. The connection for an interactive session with the computer is about $5 to $6 dollars per hour and could be attractive if he used the system about 20–30 hours per month. A summary of the options available to Fred Fox is listed in Table 1.1. Each type of service supports different types of data transfer, and although all services cannot be compared directly, Boston Converter may require a combination of these services to support their data communications requirements.

Later, Boston Converter Company has expanded, and Fred has been promoted to regional manager and has added an office in San Francisco. The company now has other offices in New York, Atlanta, Miami, Chicago, and Dallas.

Revenues are now approaching $10 million, and Boston Converter would like Fred to start a European office in Paris to give closer support to the European manufacturer representatives.

Fred's secretary will first arrange for a telephone from the Post, Telephone, and Telegraph company (PTT) since it could take from 2 weeks to 3 months to acquire one. Fred will have communications problems similar to those in the United States, but they will be compounded due to the poorer quality of international switched

TABLE 1.1. Alternative Telecom Services for District Office

Service	Advantages	Disadvantages
Telephone	Direct personal contact	Telephone tag is costly; monthly bills do go over budget
Telex	Adequate message transfer	Data must be keyed in by hand, distributed by internal mail at the home office
Leased line	Unlimited data and voice to home office	Point-to-point to one telephone number; cost is high and does not apply to other business calls
Packet data network	Good for messages; company prepares data once, input directly to the company computer	Requires terminal at district office and home office computer support
Fax	Originals sent directly to home office; no typing required; Fax is used more and more for message transfers	Requires compatible units at both ends; home office may require investment

telephone service. Telephone conversations will often have noise during part of the conversation, other voices may be heard, and sometimes the line will drop out for a few seconds.

Telex offers a reasonable message alternative due to a 6-hour time difference from Boston; however, answers do not come quickly enough. Is Fax a possibility? Yes; originals can be sent without modification, reformatting, or retyping. Another solution is a connection to an international packet carrier. Fred should ask his company to contact either Telenet, Tymnet, or another packet network carrier in the United States, and he will contact the French counterpart, TRANSPAC, the national packet network in France. Two contracts will be required to establish communications, one in France and the second in the United States.

Fred may request a private (or leased) line, but when checking with the PTT, he will find that the monthly line cost is $3775 for the U.S. side and 18,000 FF ($2825) for the French side, for a total of $6600. If constant contact with the home office is crucial, he may be forced to use a leased line. As Boston Converter grows, a small computer connected by a dial-up or switched line in the off hours is possible if summary data or small files are transferred. A distributed processing system may be a good solution.

1.1. HOW THIS BOOK WILL BENEFIT YOU

Telecommunications is the transmission, emission, or reception of signals, writings, images, sounds, or information of any nature whatsoever through wire, fiber optics, satellite links, broadcast or other electromagnetic systems between locations. Today, telecommunications includes a terminal connection to a computer located within a building or building complex, across a state, across a country, or internationally through telephone company or private facilities. A dialog is established between a terminal and the computer in order to send a message, transfer text, a file, or a program. Although this book will discuss voice and image, data communications will be covered in greater depth.

This book provides quick access to computer networking functions, telephone company services, new enhanced data services, and alternative network solutions.

The book is intended for regular use by professionals, managers, and executives and those in marketing and telecommunications departments. Marketing, financial, and manufacturing managers will find this book invaluable in planning new international office locations, expanding existing markets, establishing communications to new plant facilities, and merely improving communications to international locations. If you plan to communicate by telephone, send messages, or move data around the world, this book is for you.

The book discusses various types of telecommunications alternatives for communications between persons and for dialogs between a terminal and a remote computer or remote terminal. Examples will be provided for calculating telephone company charges and for evaluating alternatives for voice and data communications, message, text, electronic mail, Teletex, and Videotex. Although voice communications will appear throughout the book, as we have noted, there is more emphasis on data communications. Chapters that include telephone company facilities will cover many of the major countries of the world.

1.2. WHY THIS BOOK WAS WRITTEN

Why a guide to international telecommunications? Because each telephone company is unique and has separate procedures and regulations. Tariff prices and structures are different in each country and are different for both national and international traffic.

Canada, Japan, the United Kingdom, and the United States have the most liberal telecommunications markets. In other countries, the situation is quite different in terms of liberalization of services and availability of competitive equipment. However, there are trends towards liberalization in Denmark, Finland, France, the Netherlands and Norway.

International information is difficult to obtain for a number of reasons, the most basic being that countries print information in their national language, for example, Japanese in Japan, Flemish and French in Belgium, Finnish in Finland, and Swedish in Sweden. Most importantly, telecommunications are regulated by and are dependent on the national telephone companies, whose following characteristics are pertinent:

- Almost without exception, they are monopolies.
- Generally they are one of the largest employers in the country and want to maintain employment and their national markets.
- They are tied into suppliers that use the national market as a base for exports.
- Telecommunications revenue, mainly from telephone and telex services, may be used to finance the post office or fund social programs within the country. This leads to restrictions or higher tariffs on telecommunications lines and in particular leased lines. Users have only one choice, and that is to use telephone company products and services.

A restricted environment can result in lower or less efficient operating speeds and delays in delivering a message, data file, or program.

New services such as electronic data interchange, document distribution, and electronic mail are taken for granted in the United States but may be restricted, not allowed, or at best negotiable in other countries. Many countries have not been able to integrate the proliferation of information technology into their telecommunications facilities. As the technology gap has widened, restrictions have been added to protect customer revenue.

There are many ways of handling restrictions and coping with regulations. This book will explain them to you.

1.3. WHAT THIS BOOK IS ABOUT

This book explains telecommunications alternatives, new public and private vendor offerings that will allow you and/or your company to move data, voice, message, text, and image around the world.

To assist you in your decision making process, a brief overview of each chapter follows.

Chapter 2: Information Center Environment. An information system network consists of computers, programs, a group of terminals, modems, multiplexers, switches, controllers, and common carrier telecommunication facilities. The International Organization for Standardization (ISO) defines an information system network as a configuration of information processing equipment, such as processors, controllers and terminals, for the purpose of information processing and information exchange, which may use transmission facilities. Information can move between computer and terminal, from terminal to terminal, or between terminal and host. Today, terminals cover a range of products from a simple display to more sophisticated personal computers or word processors. Normally, a centralized information system network starts with the main computer and works outward to a local or remote terminal. Decentralized and distributed data processing approaches are discussed in this chapter.

Chapter 3: Worldwide Environment. Can you connect both ends of a telephone line with compatible equipment to any country? No. Some line attachment products may not be allowed in some countries due to telephone company restrictions. How do you get around telephone company restrictions? The key is negotiating. Suggestions will be offered to improve your chances of obtaining the facilities you need.

Chapter 4: International Information Flow. Data protection and security of data were in effect in most European countries at year end 1986. You will be required to comply with the new laws to do business because lack of compliance could result in fines or even prison. Trade barriers in the form of excessive import duties, value added taxes, or lack of copyright protection also restrict operation in many countries. There are several international organizations that can offer assistance. When satellites and microwave are used in telephone networks, security of data becomes a major concern. Anyone with a satellite receiving station can pick off the information at the same time you receive the data. Thus, the use of cryptography to protect information is becoming increasingly important.

Chapter 5: Network Attachment Products. Attaching your terminal to the telephone company lines will require a device that will modify and condition the computer's output signal to telephone company or carrier facilities. To communicate between two locations, compatible products are required. However, U.S. and Canadian telecommunications are based on one set of standards that are different from the standards in the remainder of the world. Products that meet international standards are required at each end to have connectivity. New telecommunications products such as multiplexers and concentrators are now available to reduce line costs, permit future growth, and improve system performance. On a worldwide basis, open systems interconnection (OSI) is a requirement for communications among different architectures supplied by multiple vendors.

Chapter 6: Office Systems. The office is the major growth area for computerized systems, which are replacing the traditional typewriter and message systems. Automation is eliminating intermediate work steps, resulting in improved productivity and information flow with new techniques for word processing, electronic messaging, and resource sharing. Local area networks (LANs) can provide connectivity within an establishment between incompatible office products and resource sharing (disk, printer, application programs).

Chapter 7: Digital Voice and Data Networks. Future integrated voice and data networks or Integrated Services Digital Networks (ISDNs) will allow new and different services not available or cost effective today. New digital data networks in the form of circuit switching and packet switching are making it possible to add new applications and services. Circuit switching data networks are ideal for short transactions, while packet networks are distance independent. Messages sent on a packet network cost the same across town or across the nation. New integrated voice and data networks or ISDN will allow new and different services that are not available or cost effective today.

Chapter 8: U.S. Network Services. There are many types of services that are necessary for voice and data transmission in all businesses. Private networks, analog and digital services, and public switched facilities will be discussed. Monthly cost calculations are included for public and private leased lines and packet switched data networks. Examples and calculations are also provided for point-to-point leased lines and multipoint lines.

Chapter 9: U.S. Network Alternatives. What are the relative costs of network alternatives? Tariffs alone do not tell the whole story. Leased telephone lines may cost more but can save time and frustration and may be the only alternative for sending messages or data files. What are the new digital network alternatives? Voice and data can be mixed on new high-speed T1 digital networks. Comparisons are provided for point-to-point star networks, multipoint networks and high speed digital data networks.

Chapter 10: International Traffic. Consistent services are important for world-wide communications to simplify the management and control of moving information. New networks and services are available for communicating to other countries. Telex gets your message to more countries than any other means. What are the other alternatives? Private lines and data and packet switched data networks may offer an alternative solution for international data and message exchange. How do you get there from here? Countries, by nature of geography and historical background, play a key role in international traffic.

Chapter 11: Value Added Network Services (VANS). International services vary by country. In some countries, voice and data transmissions may not be mixed. What are the trade-offs for each service? Your choices for equipment may be limited. A VANS provider may be a simple cost effective approach to meeting international needs. VANS could offer a lower-cost networking alternative. They can put you in business around the world within weeks, handle network management, and simplify negotiations with local and foreign telephone companies. International service companies offer an effective solution for start-up situations in new company locations around the world.

Chapter 12: Telematics. There are many electronic mail services available nationally and internationally. A new alternative to telex is Teletex (a telex replacement). In addition, there are a number of new electronic mail services from VANS and telephone companies. Teletex started in Europe. What are the latest trends? Will the new X.400 Message Handling System replace existing electronic mail services? Telefax usage is increasing as image processing gains acceptance. For some applications, Telefax costs less than telex. In Europe, the national telephone companies are competing with private industry for new data processing services. What will be the impact of these new services on your business? Will Videotex open up a new marketing channel for your company? How can these new services help you?

Chapter 13: Satellite Communications. Satellite transmission is key to moving international voice, video information, and large volumes of data. Can you design your own network, set up your own ground station equipment, and bypass national telephone companies? What are the restrictions in some countries?

Chapter 14: Trends through the Year 2000. Standardization is now becoming mandatory in most countries. Will this result in better products and services at lower cost? New technology is being implemented by all major world telephone companies to reduce operational costs and increase service offerings. A new telephone company network, called Integrated Services Digital Network (ISDN), will offer high-speed voice and data over the same line. New information services will be offered by telephone companies. How will these services affect you in the next 10 years or into the twenty-first century? Are teleports the new information centers of the future?

1.4. WORLD TELECOMMUNICATIONS—AN OVERVIEW

There are a few basic facts that you must know before reading the chapters that follow. When discussing the United States, we will use the term *Telcos* to represent the Bell Operating Companies, GTE Operating Companies, and independent telephone companies. When discussing the rest of the world, we will use the term *PTTs*. PTTs represent the post, telephone and telegraph companies. However, many governments are making the post office a separate entity entity as a result of a public mandate for new and additional telecommunications services.

In many European countries, the profits generated by the telephone and telegraph activities subsidize the post office branch. In some countries, such as France, the PTT revenue is used to support social programs (officials believe that raising a telephone bill is more palatable to voters than raising taxes). By separating the post office, which is labor intensive, profits from telephone and telegraph services can be used for new services and upgrading existing telecommunications facilities.

In the United Kingdom, there are significant changes in the total structure due to the Telecommunications Act of 1984. Telecommunications is now separate from the post office and includes two competing carriers, British Telecom and Mercury Communications. Both British Telecom and Mercury are licensed by the Director General of Telecommunications to compete as common carriers for voice and data traffic.

Telephone service and line quality vary throughout the world. After living in France for 4 years and placing numerous telephone calls to the United States, I have found that there are usually one or more telephone line problems during a conversation. There are problems with noise during part or all of the conversation, or another conversation can be heard, or there are periods where nothing is heard for a few seconds. You can put up with this on a conversational basis. However, if you intend to send data to a host system over the same dial-up line, you will experience loss of data and log-off from the host, which requires that you call again, restart, or log on the system again and, in some cases, input your data from the beginning or until you give up and mail the information. The mail service is another problem; it can take from 1 week to 3 months for a first class letter to cross the Atlantic. One answer could be a leased telephone line or private line for data, messages, and text transfer. The information may be crucial enough to force you to use a leased telephone line. However, obtaining a leased telephone can be a problem in some countries.

Can you send data across the ocean or around the world? Not unless you have compatible modems (required for sending and receiving data) at each end of a telephone line. In some countries you will be restricted from using your usual brand of modem and be required to use a worldwide International Telegraph and Telephone Consultative Committee (CCITT) standard modem. Once you solve this problem, you may be faced with message switching restrictions in some countries since message and text are considered to be in competition with PTT services. In some countries, these incompatibilities and restrictions may be negotiated with the PTT.

For companies with high traffic volumes to foreign subsidiaries, why not buy space on a satellite transponder and install your own satellite earth station on site? It's not a matter of a make or buy decision on a leased line versus satellite cost based on the volume of data traffic. Your decision may be OK for the United States, but in many countries, it's a different story. For example, in many European countries, the PTT owns the antenna and support equipment. So, if you want extra line capacity, you will have to rent the antenna and support equipment from the PTT.

Will satellite transmission open up to new competition? Possibly for TV reception in the near future. Direct broadcasting satellites (DBSs) will be launched that can beam signals to an antenna receiving station (an antenna of less than 2 meters, or 6.6 feet) on your rooftop. Luxembourg and Ireland have assigned orbital space over the equator and intend to launch a satellite into orbit.

Fax is growing in use and offers the ability to send exact copies of the original without rekeying information. If signatures, letterheads, or diagrams are required at the other end, Fax is a quicker alternative than mail. Depending on the number of pages transmitted, Fax can cost less than a telex message.

Another alternative for message transfer is communicating word processors, which are used primarily to replace telex messages. In the past, a telex was considered a formal document, and telex messages within companies or between divisions would be sent to a president or vice-president. As the message passed down the management chain, it could take weeks before it reached the person in the company who was to solve the problem. In most cases, the response would then go up the chain with reviews and comments along the way. Now, the telex has been replaced by communicating word processors, electronic document distribution systems, and Fax for most communications within an enterprise.

Today, you can send a message, text, or document directly to the individual that will handle your problem. Telecommunications has increased the human touch; you get to know who is saying yes or no to your request. It's now an individual response versus a group response. What has happened to management? Previously, managers would review every document that left the department. Now that the volumes are greater, they review only those documents that require their specific recommendations. The manager of the future must be more selective in the information he or she reviews, retains, and answers.

2

INFORMATION CENTER ENVIRONMENT

In this chapter we will

- Compare computer systems applications from PCs to mainframes
- Examine the information center and its role in teleprocessing
- Define teleprocessing applications
- Discuss networking products and networking functions

There are three main approaches to satisfying data processing needs: centralized, decentralized, and a combination of the two, distributed data processing.

In the *centralized* approach a data processing facility is set up to allow a user to share a computer's resources such as main storage, disks, and printers and to simplify the organization's control of operations. Originally, many companies set up data processing in one location based around a central processor complex. As their computing work load increased, systems were replaced with larger machines. The early computers were mostly batch oriented; that is, jobs were run consecutively, one at a time. Access to the mainframe or large computer by an engineering group or laboratory was second in priority to access by the accounting department. Therefore, it was difficult for a laboratory to tie up the central computer for an experimental project that could last days or weeks. In addition, access to the computers input/output logic was usually forbidden. Figure 2.1 is a layout of a centralized data processing system. Local and remote terminals are connected directly to the host for processing.

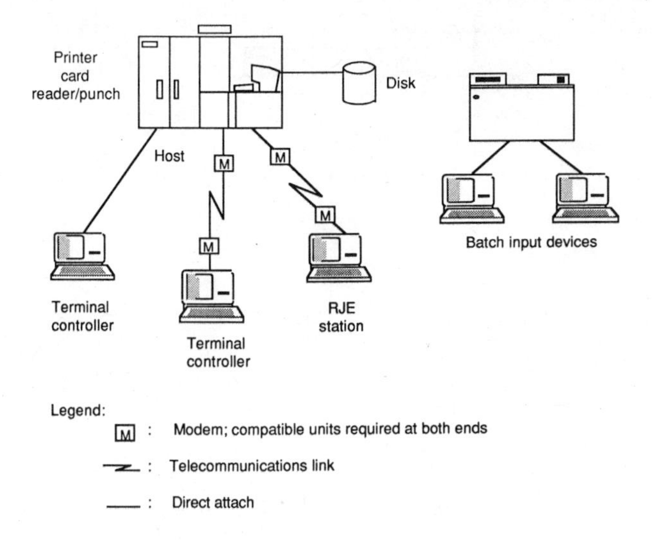

Figure 2.1. Centralized data processing.

The term *decentralized* suggests that the computer facilities should be under the control of the user—spread out geographically or conveniently located near the user. For example, a user or group of users require a stand-alone computer to satisfy local applications requirements but in addition require a simple telecommunications transfer to upload data to a host, download data from the host, or print a file. The user should have enough computer power to do the job, but access to the host system is also important to be able to share in the host's central resources. Local control coupled with economies of scale between remote processing and host resources is the key to a successfully balanced approach.

The trend to decentralized systems was pioneered by Digital Equipment Corporation (DEC) and others in the early 1960s. The introduction of the PDP-8 (Programmed Data Processor) with a flexible computer input and output structure, personal software such as an editor and a debugging routine, as well as an assembler, all at an $18,000 price, offered a relatively low cost computer solution for the laboratory. Some of the other new features that were added at the beginning were a cathode ray tube (CRT) display, an analog-to-digital converter, and digital outputs. These features coupled with a flexible line of digital building block modules made a cost effective computer solution for many laboratory experiments. The layout of a decentralized data processing system is shown in Figure 2.2.

Providing solutions to both centralized and decentralized systems is called *distributed* data processing. As an example, a user or group of users require a stand-alone computer to perform local applications. In addition, they require constant interaction with a central host system to update files, access the host database, perform calculations, store or archive files, or print large reports. Distributed processing was pioneered by Digital Equipment Corporation (DEC), starting with its introduction of the popular PDP-8 minicomputer. Minicomputers gained in popu-

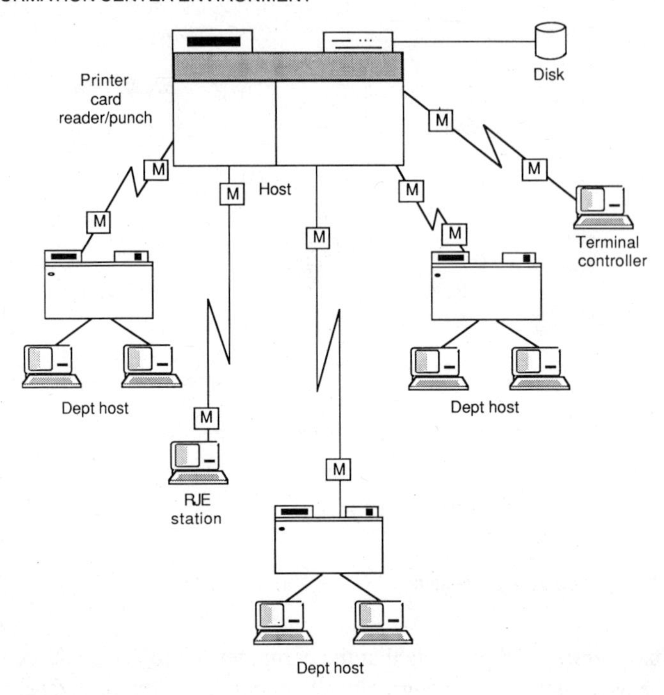

Figure 2.2. Decentralized data processing.

larity by off-loading the host, providing processing power at the source, and offering economic computer solutions for many scientific and engineering applications. Over time, commercial applications were added to minicomputers, resulting in many different types of systems within an organization. Distributed processing also evolved because of the geographical and organizational distribution of enterprises. Each distributed system is tailored to the needs of an establishment within an enterprise.

Although decentralized systems provide independent solutions, there is a need to tie many diverse systems and applications together to avoid redundant design and allow systems to communicate. Distributed processing should be a balance between centralized and decentralized processing: the power of larger centralized resources coupled with the need for computer power at the source, whether at a desk or in a lab or department. Therefore, distributed processing should offer a planned system approach to handling growth situations and developing common solutions for a variety of applications.

2.1. DISTRIBUTED PROCESSING TRENDS

Distributed processing will be accepted in most companies in the late 1980s because small systems and personal computers now offer a number of features previously found only on larger systems. In a distributed processing environment, computer resources are located as dictated by the user's workload or organizational requirements. Advantages of distributed data processing are as follows:

- Computer power is in the department or at the user's desk.
- The department or user has local control over programs, data, storage, and printing.
- There is better response time than with an interactive session with a remote host.
- Personal computers or terminals attached locally to a department computer usually offer faster terminal response time.
- Local computing requires fewer communications facilities than an on-line interactive terminal on a leased line. Communication costs are lower for dial-up calls or public data networks due to smaller sessions and data transfers over telephone lines.
- The simplicity of the system improves reliability and availability.

However, there is more to distributed data processing than local computing. Distributed processing includes local processing coupled with communications between systems. Applications, data, and messages are sent between systems in a network to many locations. Users are able to communicate information to other users at all levels of the organization. Therefore, distributed processing should be an extension of the central system. Otherwise, you have a random number of different types of computers throughout the organization with no capability of integrating them into the information network. More importantly, separate unattached remote computing isolates workers and departments and cuts them off from the information flow and mainstream of the company.

Personal computers, or small computers, normally complement the central database and off-load the central host site of some application development, processing, storage, and communications lines. A coordinated, loosely coupled control system that does not bog down the user provides an effective complementary system. The following scenarios show typical uses of distributed processing. See Figure 2.3 for a distributed data processing system topology.

For example, a U.S. insurance company with 50 agencies has a new flexible life program with adjustable premiums and wants to maintain control over the program in order to have consistent rates and standard policy conditions. The agencies, which are independently owned, pay for their small computer or personal computer and would like to use the computer for other applications within their office.

Another example of a country where many companies have gone to distributed processing to solve problems is Sweden. Sweden has one of the highest standards of living and the highest cost per capita for business employees. In addition, Sweden has the highest number of communicating terminals and work stations, 20 per 1000 working population. This figure compares with about 11 for Switzerland and 8.5 for Europe. According to a management study by the Eurodata Foundation; by 1991, Sweden's level will increase to 48, while Europe will increase to 22. The numbers are

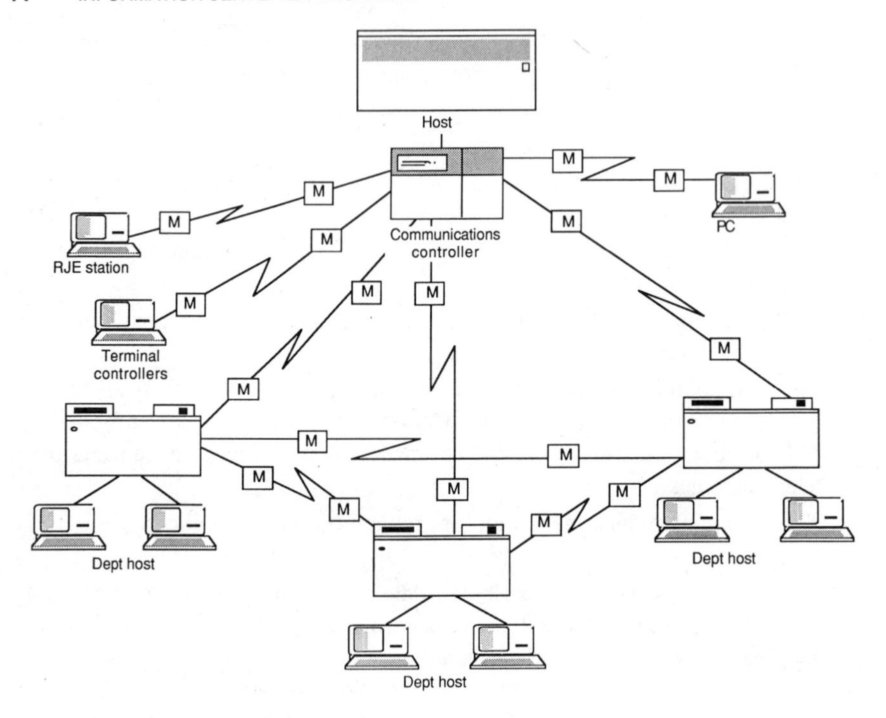

Figure 2.3. Distributed data processing.

based on the number of network termination points published by the PTTs. One explanation for these figures is that distributed terminals and work stations are required to improve the productivity of the average worker. Without the increase in productivity, Sweden might not be able to compete in world markets.

2.2. COMPUTER SYSTEMS—PCs TO MAINFRAMES

In many countries where local processing is implemented, the countries mandate certain requirements for local processing within the country. For example, you cannot transmit data unless it is processed within the country. However, there is also a need for information at the headquarters location from remote country offices or manufacturing facilities; marketing, financial, production, and personnel data need to be sent within a reasonable time.

There are a number of computer product alternatives to match business requirements for different communications applications. Computer systems range from microcomputers for a few hundred dollars to personal computers for a few thousand dollars to large computer mainframes for a few million dollars. At the high end, large computer mainframes handle multiprogramming and multitasking operations for hundreds of users.

Most consultants place the various systems types into four categories: the personal computer (PC), the small computer, the medium computer, and the large computer or mainframe systems. International Data Corporation (IDC), Framingham, Massachusetts reported that the 1986 world market for the four classes of computer systems was $66.4 billion. The market is segmented as follows:

	1985	**1986**
Personal computers (PC)	$19.1	$20.8
Small computers	$12.3	$13.3
Medium computers	$14.6	$15.0
Large computers or mainframes	$16.5	$17.3

The PC category is predominantly single-user applications and intelligent work stations configured for a multiuser environment. The value for each category typically includes peripherals like storage devices and printers. PCs range from a few hundred dollars to a top price of about $15,000. The main difference between a high-end PC and a low-end small computer is the systems engineering (e.g., input/output subsystem) that facilitates multiuser configurations. Small computers or minicomputers offer more computer power than a PC. In addition, they have the capacity and power to handle more devices and can be programmed to respond to external events. They range in price from $15,000 to $100,000. Companies use small computers to fill in a major gap where there is difficulty in tying PCs together with the mainframe into a cohesive information system, particularly at the departmental level. The demand for small systems in departments or work groups in 1985 was growing at 30% per year or at about twice the industry growth rate. There are about 4 million departments and work groups in companies and government agencies in the United States alone.

Medium-sized computers or superminicomputers range in price from $100,000 to $1 million and form the data processing center for medium-sized companies and divisional centers in large corporations. Compared to the large mainframes, medium-sized computers offer less computing power, less teleprocessing capability, and less concurrent support for large numbers of end user terminals. Although medium-sized computer systems usually offer software operating systems with less function, some medium-sized computers use the same operating systems as the large computers. Large systems or mainframes are required to simultaneously handle multiple programming tasks from hundreds of users from many locations. Mainframes, priced over $1 million, manage companywide communications and teleprocessing applications.

Table 2.1 includes a list of computer systems for the 10 leading vendors in each category based on IDC's 1985 worldwide dollar value of system shipments. Each system price includes a central processing unit (CPU), an operator's console, and typically a storage unit and printer.

TABLE 2.1. Leading Computer Systems Suppliers and Product Offerings

Personal Computers	Small Computers	Medium Computers	Large or Mainframe Computers
Apple Computer: Apple, Macintosh	Burroughs (Unisys): A 3 Series	Burroughs (Unisys): A 9, A 10 Series	Amdahl Computer: 580 Series, 3090-400
AT&T: 3 B2 Systems, PC 6300	Data General: Nova, Eclipse Series	Data General: Eclipse Series	Burroughs (Unisys): A 15 Systems
Atari: 520 ST, 1040 ST	Digital Equipment Corp.: PDP-11, VAX Series	Digital Equipment Corp.: VAX Series	Control Data Corp.: Cyber Series
Commodore Business Machines: Amiga, C 128	Hewlett Packard: HP 3000 Series	Hewlett Packard: HP 9000 Series	Cray: CRAY-1S, X-MP Series
Compaq Computer Corp.: Compaq Computer Systems	IBM Corp.: System/36, System/38	Honeywell (HIS): DPS Series	Digital Equipment Corp.: DECSYSTEM-20
Hewlett Packard: Series 80, Touchscreen	NCR: 9300 Series	IBM Corp.: System/38, 4300	Honeywell (HIS): DPS 90 Series
IBM Corp.: PC Series	Prime Computer: 2250, 2550	Prime Computer: 9650, 9750, 9955	IBM Corp.: 3080 Series, 3090
Tandy/Radio Shack: TRS & TANDY Series	Sperry (Unisys): Mapper Series	Sperry (Unisys): Series 1100	NAS: AS Series
Wang: Professional Computer	Texas Instruments: 100, 300, 600, 800	Tandem: Non-Stop	NCR: 8000 Series
Zenith: Z Series	Wang Laboratories: VS Series	Wang Laboratories: VS Series	Sperry (Unisys): Series 1100

2.3. THE INFORMATION CENTER—TELEPROCESSING CONTROL

Coordination of distributed processing and communications over national and international telephone company facilities is usually a function of the information center, which contains a large computer mainframe or computer complex or host system. The information center is normally an organization within the data processing department and provides nontechnical end users with programs, application packages, software tools and assistance that will allow them to access data; process information; send files, messages, and text; and use the resources of the computer. These resources include files, disk storage for their programs, printers, and communication facilities and can include databases.

The information center collects and then distributes the information throughout the system. In most multinational companies, there is a mix between the home office information centers and individual country or area information centers. In some cases, this mix is due to national requirements for local processing.

Information centers continue to grow for the following reasons:

- Corporate databases continue to expand with new applications and yearly accumulation of data.
- Large scale transaction processing increases as new users and uses are added.
- Mass archival storage continues to grow.
- Installed corporate applications keep changing and growing; existing applications do not go away.
- High-speed printing requires large systems to coordinate the printing of multiple documents.
- Network management continues to expand due to new user requirements and increasing choices of data networks and teleprocessing applications.

The information center's requirement for providing communications facilities for end users in the late 1980s and early 1990s will be an any-to-any connection—that is, by simple request, to be able to provide the end users working at their terminal with the ability to connect to any computer in the system over a choice of data networks. For the end user, communications should be transparent. That is, the user does not want to know how the system works but wants only to have access to local and remote applications. Figure 2.4 would allow a user to request services throughout the world. Terminal A could be attached to a local country information center (host A) by either public or private networks for local application support. In addition, terminal A could be passed through its information center A (host A) to another information center B (host B) in another country for remote application support. If terminal A would like to establish an on-line real-time dialog with terminal B, host B, which controls the attached resource in the remote country, would complete the connection. Figure 2.4 illustrates an any-to-any connection to host A in the same country and to host B or terminal B in another country.

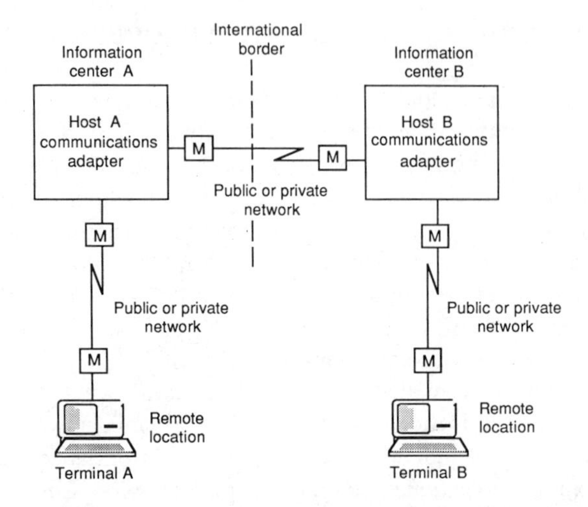

Figure 2.4. Any-to-any connection.

2.4. TELEPROCESSING APPLICATIONS—COMMON USAGE

There are three types of terminals that are used for most telecommunications applications to a host: nonintelligent terminals operating one data line at a time or with full-screen support, intelligent terminals such as a PC, and remote job entry terminals that submit an entire job.

1. *Nonintelligent terminals.* One type is a low-function asynchronous or start-stop terminal that uses a line mode procedure. That is, after a line of data is typed on a terminal and return, enter, or the carriage return key is pressed, the data is sent down the line to a computer that may be local or remote. The computer, after receiving the line of data, sends back a return or line feed that means that the data is accepted. Data transfer of a file occurs one line at a time. Line mode is time-consuming but uses less computer overhead and allows many interactive users due to the slow speeds, which range from 10 to 1200 characters per second. This character rate would correspond to line transfer rates of 110–2400 bits per second (bps). Another type of nonintelligent terminal offers full-screen support and block mode transfer. One type of full-screen support is the IBM 3270 Information Display System. There are several controllers and display stations in the 3270 series. A new

system in the series is the IBM 3174 Cluster Controller with IBM Model 3191 and 3193 display stations and a 3194 color display station. A 3194 with full-screen support has a working area of 24 vertical lines of characters by 80 rows of characters for data or message input. This results in 1920 characters if the full screen is used. On newer 3270 protocols, once you input your message or data on the screen and press enter, the contents of the data fields that have been changed are sent to the computer by the 3174 cluster controller as one block of data.

2. *Intelligent terminals.* Intelligent terminals offer flexibility and can operate as a stand-alone system, interact as a conventional on-line display, act as an on-line remote batch terminal, or emulate a 3270 display station, e.g., the IBM 3270 PC. Normally, if these systems include a multitasking operating system, operations are performed simultaneously.
3. *Remote job entry.* Data is input to a controller or terminal controller by a card reader, tape, disk, or keyboard and stored temporarily on a disk, diskette, or magnetic tape. When all data files are complete, usually once a day for remote locations, the data files or job along with a job procedure for processing the job are transmitted to a host computer for processing.

The three terminal types make up the majority of input devices to some of the more common teleprocessing applications such as data entry and data collection, remote job entry (RJE), and batch processing where transactions arrive in random order. Other applications are real-time data acquisition,which is handled the instant the event occurs, and time sharing, which accommodates multiple terminal interruptions and job processing. Time sharing is an optimum method by which end users share in the large processing capability or extensive database capabilities of a more expensive and resourceful mainframe, usually from a relatively inexpensive terminal. Most of these common teleprocessing applications grew because more computing power was needed to handle the processing of data, applications software support on the host, or large computational programs. The types of teleprocessing that can be accomplished on various computer systems follow.

2.4.1. Typical Teleprocessing Capability

Personal computers usually allow one user at a time to transfer a file or program to another system. Small computers have multiple users with file or program transfer and perform protocol conversion or message handling to another system within the department. Medium-sized systems manage multiple users, perform batch processing for local or remote jobs, control interactive processing of multiple transactions, and perform as a node controller in large networks. Large systems or mainframes perform medium-sized system applications but for more tasks, transactions, and users.

2.4.2. Data Entry and Data Collection

A terminal or data entry device may include a program or set of instructions for inputting information from a keyboard or machine-readable device into a set of data files. Data entry devices can be a simple nonintelligent terminal or include data storage in the form of diskette, disk, or magnetic tape for temporary storage before sending the file to the host for processing.

2.4.3. Remote Job Processing and Batch Processing

Remote job processing work stations, personal computers, and small computers normally perform the collection of information, provide temporary storage, and later, usually once a day, transfer the information and processing instructions to a remote larger system for processing. As a large company option, some RJE host systems are on line all day. They are polled frequently for file transfer in a batch mode. Medium- and large-sized computers poll and receive these data files from remote RJE work stations, PCs, or small computer systems and schedule the data files into their job stream processor. The medium and large systems process the files sequentially (one at a time) in a batch mode and return the results or output results to system facilities.

2.4.4. Word Processing

Word processing is a system of people, procedures, and equipment for entering, revising, storing, and printing simple text documents (e.g., memoranda and reports) and for processing paperwork and producing documents. Word processing involves typing procedures that would be done on an automated typewriter using simple commands. Word processors are used to transform handwritten drafts or revisions of typed material into typewritten or printed pages or documents. Word processor systems range from an intelligent typewriter to computer-based systems. Intelligent typewriters represent the low end followed by stand-alone nondisplay units with a keyboard and storage console; stand-alone display units that include a display, keyboard, and printer; and shared-logic systems, which consist of a central processor containing logic for text editing and printing, storage devices, and several input and editing terminals and printers. On larger systems or mainframes, word processing is offered as an applications software package.

2.4.5. Message Switching

Electronic mail and message switching programs usually offer memo creation and editing, delivery to another person's mailbox, or the sending of mail directly to the user's terminal. By the early 1990s, most company terminals and computer systems will be tied into an intracompany electronic mail system.

2.4.6. Conversational Time Sharing

Time sharing usually applies to general problem solving, programming and testing, calculations, engineering design, text editing, or automated banking applications. Time sharing options are usually available on small, medium, and large systems. The flexibility, computational power, and number of users normally increase with larger systems.

For each user, data imput, processing, and output are performed by the computer system at a slice of time on a round-robin or one-at-a-time basis. For example, if the system has 24 users logged on, each user is given a time slot, and the monitor progresses from user 1 to 2 to · · · to 24 and back to 1. Depending on the computer's operating system and computational power, the time slice for each application is usually measured in milliseconds. Since the computer operates on each user's application in microseconds or less, there are enough time slots available to make each user think that the computer is dedicated to his or her job or task.

2.4.7. Real-Time or Data Acquisition Systems

Most small computers or minicomputers are designed to respond to external real-time events at the instant they happen. Minis were developed with input/output structures to provide instantaneous reaction to external events, direct access to memory to collect information on the event, and instructions to minimize the time delays needed to process the incoming information. In general, medium and large systems perform processing and analysis on data files received from the small computers or minicomputers.

Medium computer or superminicomputer systems are used to handle many inputs that require high computational processing such as flight simulation programs or air traffic control. Computer-aided design (CAD) and computer-aided manufacturing (CAM) are areas where greater computational power can justify the higher system cost. In addition, many CAD/CAM applications as well as flight simulator programs are now available on PCs. Table 2.2 provides a list of various teleprocessing applications by computer size.

2.5. TELEPROCESSING APPLICATIONS—EMERGING AREAS

In the mid-to-late 1980s, the new growth areas for in-house teleprocessing will be in transaction processing, eletronic mail and message systems, document interchange distribution, information exchange, resource sharing, and new international OSI implementation. These new products will allow the user to have a virtual or any-to-any connection to distribute their messages or documents. That is, a terminal user will be able to connect to a host or remote host or send messages to another terminal around the world.

TABLE 2.2. Common Teleprocessing Usage

Usage	Personal Computers	Small Computers	Medium Computers	Large Computers
Teleprocessing capability	Transfer a file or program from one user to another	Manage transmissions of multiple users	Manage transmissions of multiple users	Manage control and process user sessions
	Perform protocol conversion	Perform protocol conversion	Could have a front-end processor to handle transmission	Usually have a front-end processor to handle transmission
Data entry and data collection	One user per terminal	Two to 16 users	17 to 128 users	Over 128 users
	Source of information, send results to larger host, preprocessing	Source of information, send results to larger host, some processing	Process data collected at remote locations	Process data collected at remote locations
RJE & batch processing	Accumulate data, send to host for processing, collect results	Accumulate data, send to host for processing, collect results	Receive remote job, process data collected, and return results	Receive remote job, perform batch processing, return results
Message switching	Send and receive messages through host or service bureau	Message switching for department or small company	Perform message switching for company	Manage, control, and process for worldwide locations
Conversational time sharing	No	Up to 16 users, multiple languages	Up to 64 users, multiple languages	Upwards of 100 users, multiple languages
		General problem solving	Scientific and matrix problems	Large matrices, scientific problems, seismographic and weather trends, simulation problems
Real time data acquisition	Medium usage	Heavy usage, processor designed to handle external events	Medium usage, depends on data processing work load	Specific unique applications that are cost-justified
Word processing	Heavy usage, one terminal per user	Shared logic systems, multiple users	Applications package, multiple users	Applications package, multiple users

2.5.1. Transaction Processing

Transaction processing is normally a short-term highly interactive procedure. One example is a credit card application where the user inputs a card into a reader and information is read off the card and sent to a computer system containing card numbers, credit limits, and user balance. The result is sent back to the originating reader as accept a charge or credit limit exceeded. Processing the card and the answerback usually takes less than a minute.

Another example is a remote banking or electronic banking terminal where the user

1. Keys data into the terminal,
2. Initiates inputs that are sent to a bank control system where the user file is accessed and changes are made based on the request, and
3. Reads/looks at a control device that informs the user of his or her account status and, if requested, drops money into a receptacle and then terminates the transaction. The entire process should take less than a few minutes.

2.5.2. Electronic Mail and Mailbox

Electronic mail is an outgrowth of word processing and communicating word processors. Today, electronic mail may be generated from a variety of remote terminals, work stations, personal computers, or departmental small computer systems. The message is created on a terminal and may include a distribution list. When completed, the message is stored in the computer, and each person on the distribution list is notified that he or she has a message waiting in the mailbox (a storage area or file in the computer's storage media). In addition to a local mailbox with store and forward capability, most electronic mail systems may be attached to remote locations through telephone company facilities: a dial-up or public switched line, a private leased line, a packet switched data network (X.25), a circuit switched data network (X.21), or a specific country digital data network.

2.5.3. Document Distribution

Document distribution provides for the creation, distribution, storage, retrieval, processing, archiving, and printing of documents and messages. In addition, an interchange software program should offer display, printing, and redistribution of received documents and document access protection.

2.5.4. Information Interchange or Any-to-Any Connection

Information interchange or an any-to-any connection should provide at least three remote options from a user terminal: (1) a dialog between the terminal and a host computer system in the same geographical area or country, (2) a dialog between the terminal and a remote computer system passing through one or more computer systems, nodes, and public or private networks in one or more countries, or (3) a dialog through that remote computer (through multiple systems, nodes and countries) to a local or remote attached terminal in the same country as the remote host computer. Refer to Figure 2.1 for a layout. An any-to-any connection means that a terminal will be able to use the resources of a local or remote host processor or terminal, independently of the type of network.

Information interchange offers computer-to-computer, computer-to-terminal, or terminal-to-terminal communications among company locations, such as headquarters, regions, or foreign offices, or between different enterprises, such as a retailer and manufacturing company and other suppliers to the retailer. A key feature is that it will handle a wide variety of architecturally dissimilar machines and provide internetworking attachment.

2.5.5. New International X.400 Message System

A new International Telegraph and Telephone Consultative Committee (CCITT) X.400 Message Handling System Recommendation has been introduced for independent message transfer. Terminals that implement the standard will offer the user local editing and preparation on a terminal, temporary storage of the message, and archiving. The system will accept messages from users and route the messages to their final destination. One main advantage of the X.400 will be the adoption of the standard by private and public electronic mail services. A common standard will allow electronic mail between end users independently of the networks used to transfer the message.

Table 2.3 lists the emerging teleprocessing applications in the preceding discussion.

2.6. NETWORKING PRODUCTS

Company data processing systems have progressed from distributed systems to coordinated interworking networks of hosts and terminals. Today, they can be connected internationally by public telephone or private data networks. As more terminals are added to the network, there must be minimal disruption to the organization while increasing the capability of interconnecting with other networks.

In a changing environment, a comprehensive set of standard specifications, for both hardware and software, is required to select the optimum data communications offerings. The solution is a set of specifications and products designed to provide maximum control and productivity. Over the past 10 years, networking architectures have been improved and products have been introduced to enable international teleprocessing. To support architectures, network products provide a set of functional hardware and software components that allows the system to share resources at a remote location. The number of functions implemented on each manufacturer's system is based on that system's computational power, storage capacity, software capabilities, and available application programming support.

TABLE 2.3. Emerging Teleprocessing Usage

Emerging	Personal Computers	Small Computers	Medium Computers	Large Computers
Electronic mail and mailbox	Input terminal, create message, send, receive, and display own message	Handle multiple departmental message switching system	Companywide Integrate into management information system (MIS)	International management and control Integrate into MIS
Document distribution	Input terminal, text editing for file, search, processing, and printing	Manage multiple inputs at departmental level Text editing for multiple users	Companywide Integrate into database applications Integrate into MIS	Manage international distribution and control document flow Integrate into database applications Integrate into MIS
Information exchange or any-to-any connection	Send and receive to larger host or another terminal	Process multiple inputs for department Protocol conversion: asynchronous to 3270 BSC, a synchronous to 3270 SNA/SDLC, 3270 to X.25 packet network	Process inputs for small- to medium-sized company Manage, control, and maintain directories	Process for intercompany, international operations Manage, control, and maintain directories
International OSI standard	Initally work through host system Later, implement all OSI layers	Initially implement first three layers of OSI Later, implement all OSI layers Cost-effective product to implement gateway	Provide gateway to match international standards Provide message formats at location to optimize network design Later, provide all OSI layers	Provide gateway for incompatibilities between private and public networks Later, provide all OSI layers
International X.400 standard	Provide external attachment for single interface	Provide departmental attachment to external network facilities	Provide small to medium company interface to optimize network charges Manage department traffic	Provide gateways to international network carriers Manage company network

Networking is a combination of interconnected equipment and programs for moving information between points where it can be written, processed, stored, and used. Interconnections can be over telephone lines, microwave links, and satellite links. Other facilities available for international networking will be discussed in later chapters.

2.6.1. Host Control

Host-controlled systems maintain control over terminals and smaller computer systems within their geographical working area. Usually, the host is brought up to a working state. Next, the system operator establishes a table with terminals that will have access to the host system and its resources. The terminals cannot operate independently and must be attached to the host. All traffic between the host and terminals is controlled by the host.

Host control offers central management and a concentration of personnel expertise in one location. However, when there are line problems to remote locations, the remote application is down. Line problems and time delays through the telephone company facilities may discourage remote interactive usage. Another alternative is peer-to-peer connection.

2.6.2. Peer-to-Peer Connection

A peer-to-peer or equal-to-equal connection allows any system to initiate a session with another local or remote system. Unlike a host-controlled system where only the host can initiate a session, a peer can initiate a session with any other peer. Depending on the application and network control required, a peer system could be a PC or larger computer system. In effect, the initiating system (source) becomes the session controller and controls the transaction to the remote destination. In this type of arrangement, there is no designated host system. Each system is on the same level from a communications and applications standpoint. If there is a line problem, each system operates independently, and users continue working.

A peer-to-peer connection is critical for international telecommunications. Due to difference in time zones and public line facilities, each system must operate independently within a country or geographical area. In addition, the system must be able to initiate a transfer of data back to headquarters or other locations when the system is ready and not at the request of a central host.

There are at least three applications: (1) A terminal accesses data from its attached computer system in the same country, (2) the same terminal connects through its host to a remote host computer for files or application programs or to a remote database in another country, and (3) the same terminal connects to a remote terminal through its host and the remote host to establish a two-way dialog between the two terminals.

A list of vendor networking products from various computer suppliers is given in Table 2.4. Most vendors' networking products are based on an architecture that includes a peer-to-peer capability. Naturally, each vendor should be contacted prior to designing a network.

TABLE 2.4. Computer Networking Products by Major Vendors

Supplier	Networking Product Name
Apollo Computers	Aegis
Computer Designed Systems	Avos
Control Data	Cyphernet
Data General	Xodiac
Datapoint	Attached Resource Computer (ARC)
Digital Equipment	DECnet
Formation	TMS
Hewlett Packard	AdvanceNet
Honeywell/Bull	Distributed Systems Architecture (DSA)
IBM	Systems Network Architecture (SNA)
Modcomp	Max IV
NCR	Distributed Network Architecture (DNA)
Perkin Elmer	Xelos
Philips	Sopho-Net
Prime Computer	Primenet
Proteon	80 Mbps LAN
Tandem	NonStop Network
Unisys	
Burroughs	Burroughs Network Architecture (BNA)
Sperry	Distributed Communications Architecture (DCA)
Wang Laboratories	Wangnet

2.7. NETWORKING FUNCTIONS

Networking products consist of a set of functional software components implemented on a system (host and terminal) that allows the system to share resources at a remote connected system. The number of functions implemented on each system is based on that system's computational power, storage capacity, software capabilities, or available application programming support and the network itself. Remote data sharing facilities allow PCs and small computers to access larger disks, printers, or application programs or files on larger systems.

To understand networking functions, note that the connected systems are designated as either a *source* or a *destination* system. The *source* system is the system that requests a resource, and the *destination* system has the requested resources (data files, peripheral devices, or system facilities). Communication support is transparent from a terminal.

Figure 2.5 illustrates the functions for remote data sharing and remote device sharing.

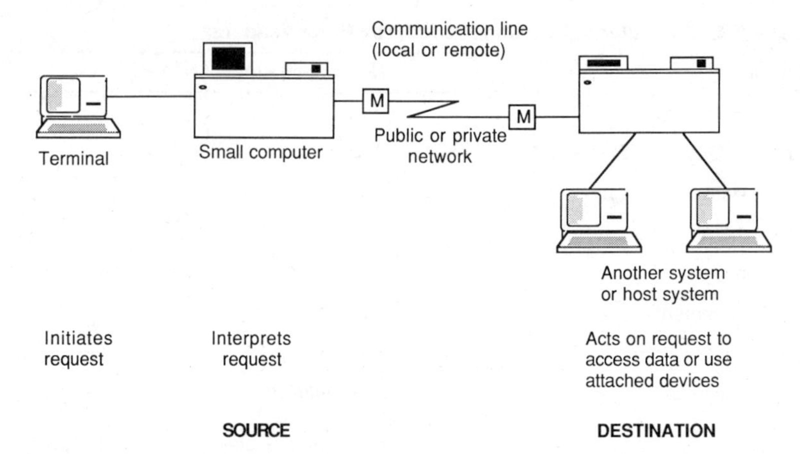

Figure 2.5. Network functions (remote data and device sharing).

2.8. PEER RELATIONSHIP FUNCTIONS

A peer relationship allows any two systems with comparable functions to interact with each other. In a networking hierarchy, peer systems are treated as equals, and interaction is independent of architecture, communications, and computer size. Any peer system can initiate a program or a data transfer or request resources from another peer system. Once communication is established between peer systems, resource sharing functions facilitate request handling.

Remote data sharing provides access to data files located in another system as if that system were locally attached. This allows, for example, a COBOL or FORTRAN program on a *source* system to use the same interface and file access method as if the user were executing on the *source* system even when data is being accessed on the *destination* system. Examples of data sharing include the following functions.

2.8.1. Copy File or File Transfer

Copy file allows the requesting *source* system to cause the transfer of a disk file to or from a specified *destination* system. That is, the request may be to send the disk file from the *source* to the *destination* system or to send the disk file from the *destination* to the *source* system.

The file to be copied may be consecutive, indexed, or random organization. The resulting file must be predefined on the receiving system. Space allocation, attribute definition, key field and location, and other attributes unique by system must be previously established by the *source* system.

A copy file feature is a basic requirement for transferring files from a PC or small system to a host for processing, printing or storage. Copy file is also necessary in the host to transfer files back to a PC or small system for updating, local processing, or

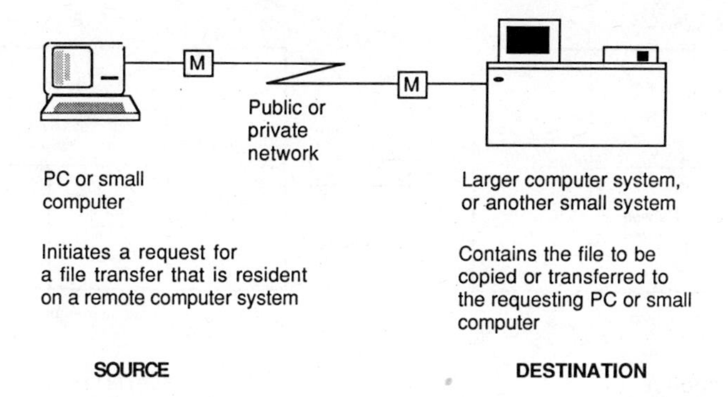

Figure 2.6. Example of copy file or file transfer.

printing. In electronic mail or electronic document distribution application programs, a copy file routine is embedded into the software program.

Figure 2.6 is an illustration of copy file or file transfer.

2.8.2. Program-to-Program Communication

The program-to-program function allows the user at a *source* system (at the console or by a program) to initiate and execute a program on a *destination* or remote system.

Interaction between PCs, word processors or small systems and larger systems is possible if programs in each system are compatible. In most cases, programs may be written by the same applications software group. A program to program feature on both a *source* and *destination* system would simplify the coordination between incompatible computer architectures and operating systems. For example, a program written in COBOL on a smaller *source* system could interact with a COBOL program on a larger incompatible *destination* system.

The programs are designed to provide information to or from each other independently of communications, architecture, and system size. Software application programs are usually written in the same language, such as COBOL, FORTRAN, or RPG. If both the *source* and *destination* programs are written in the same language, debugging is simpler.

2.8.3. Remote Program Execute

Remote program execute offers the ability to initiate a program on a remote computer system from a terminal. This program is very practical when a remote system has more or better resources such as a database, high quality printer, storage device, or graphics printer. Remote program execute would appear as Figure 2.5 except that the *source* would simply request to initiate a program in the *destination* system.

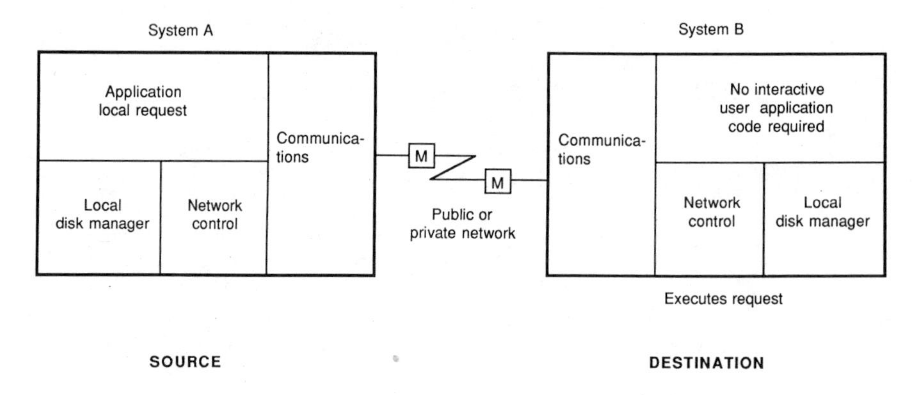

Figure 2.7. Example of direct record access.

2.8.4. Direct Record Access

Direct record access allows the *source* system to read or write to a *destination* disk file with local and remote transparency. An operator or executing program can request the READ, WRITE, ADD, DELETE, and UPDATE functions of a data file on the *destination* system. An example of a READ request is shown in Figure 2.7 followed by the steps in a READ request sequence.

The five steps in a READ request sequence are the following:

1. An application on system A requests support of the local disk manager.
2. If the file is not available locally, the disk manager interfaces the network control software, which translates the request to a communications protocol that will be understood by the remote system.
3. The common communications mechanism carries this request from system A to system B.
4. The network control software in system B (destination) translates from the communcations protocol to the system B format and turns over the functional request to its disk manager for execution.
5. The system B local disk manager passes the requested data back through the network software product to the system A application.

2.8.5. Remote Database Access and Inquiry

Remote database access and inquiry are similar to direct record access except for the database system. Database management facilities offer greater flexibility for multiple updates, sorting, and output formatting.

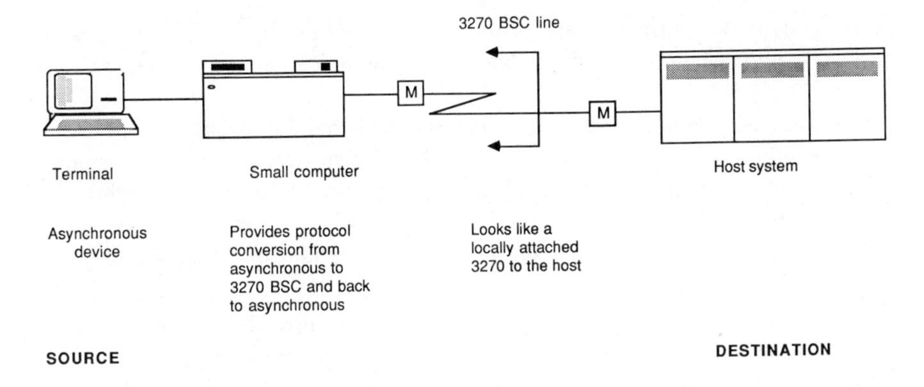

Figure 2.8. Work station passthrough.

2.9. WORK STATION PASSTHROUGH

Work station passthrough allows the terminal to go through the system to which it is attached to a *destination* system. The terminal is connected through a *source* system that performs no operation on the incoming or outgoing information except to make the terminal appear to the *destination* system as a locally attached terminal. Emulation programs on small computer systems provide passthrough; e.g., an asynchronous terminal is attached to a small computer. The small computer provides a protocol conversion on the incoming data stream and converts the data stream from asynchronous to 3270 binary synchronous communications (BSC). After this conversion, the terminal data is passed to the host system, which recognizes 3270 BSC. After processing in the host, the outgoing data stream to the small computer is 3270 BSC. The small computer then converts the data stream back to asynchronous characters to operate on the remote terminal. Figure 2.8 is an example of work station passthrough.

Reverse passthrough is a similar function in the opposite direction. For example, a terminal attached to a host system would like to access an application on the small system. The host converts the attached terminal's protocol to asynchronous, 3270 BSC, or other protocol necessary for communications with an application program on the small computer. In this case, the host is the *source,* and the small computer is the *destination* system.

2.10. SUMMARY

Networking products and networking functions contribute to satisfying user requests for greater resources from remote systems. Since the world environment continues to change and more restrictions are being added by PTTs, five different types of products would be helpful for international telecommunications:

1. *Electronic Mail and Mailbox.* This product is helpful for sending messages within a company to work around the different time zones. For example, London is normally 5 hours ahead of New York or Eastern time, while the continent of Europe is 6 hours ahead of Eastern time. Japan is 9 hours ahead of New York, and Hong Kong and Taiwan (Republic of China) are 13 hours ahead of New York.
2. *Document distribution.* Document distribution provides for the creation, distribution, storage, retrieval, processing, archiving (e.g., filing), and printing of documents and messages. Like electronic mail, document distribution allows for time differences and the sending of documents or text from a few pages to several hundred pages across international time zones.
3. *Public data network connections.* New data networks such as packet switched data networks are called Telenet or Tymnet in the United States and Datex-P, X.25 Packet Networks, or PSDNs in other countries. Circuit switched data networks, also called X.21 or Datex-L, are available in most European countries and Japan. Public data networks offer an alternative to private or leased line networks and are not restricted to applications. That is, electronic mail, data interchange, and electronic document distribution are allowed.
4. *Peer-to-peer connection.* Peer-to-peer connection is an important computer networking tool for data interchange between international systems. Due to the line quality on dial-up calls, restrictions on international leased lines for electronic mail, data interchange, and electronic document traffic, and national requirements for local processing, each major country or area location would require independent and separate information centers. The centers should be independent of a U.S. host to handle local data processing and local area traffic flow. Peer-to-peer networking would allow international entities to operate independently of a host-controlled system based in the United States and to transfer information as required. Companies would face fewer restrictions when applying for a leased line or registering a database to comply with data protection acts if they perform local processing and only exchange summary information between countries.
5. *Batch Transmission.* A remote job entry program is still the current method of transferring data files across international boundaries. Data files are usually transmitted once a day in off-peak hours. Amazingly, many large corporations still use this method for transmitting data across international boundaries. It's a time proven method, but file transfer may be a day later than actual data availability.

3

WORLDWIDE ENVIRONMENT

In this chapter we will

- Examine the worldwide regulations environment
- Evaluate customer premises equipment on a country-by-country basis
- Discuss volume charging and restrictions on international leased lines

Country regulations concerning PTTs affect the types of information an enterprise may exchange, the available network services, and international communications due to requirements for compatible equipment at both ends of an international dial-up (private) or leased line (private). Today, PTTs have set up regulations for services that are beyond the basic services within their original monopoly or charter. In addition to restrictive regulations on applications and services, data privacy can be an impediment to international data communications.

An analogy could be made with the highway department in a town, state, or country. Normally, the highway department plans, develops, and maintains roads. In addition, it would own and operate roadside restaurants, roadside service stations, and roadside retail stores. Furthermore, it would want to control the quality of its suppliers and would buy into suppliers of food, gasoline, oil, tires, clothes, etc. Suppliers to the highway department's retail stores would be required to meet the highway department's specifications, prices, and terms and conditions. Or in the case of some highway departments with licensed retail stores, the retailers would have to pay a percentage of sales for access from the highway department's right-of-way.

Sound far-reaching? It's normal practice by some PTTs for basic and enhanced telecommunications services. For example, one monopoly's charter states that all transmission media (radio, cable, satellites) and all telecommunications services (basic and enhanced) and all telecom equipment (including customer premises equipment) are legally a PTT monopoly. Enhanced services would include data processing and database applications. That is, if an enterprise wants to have a private network for electronic mail, electronic data interchange, or electronic document distribution, the applications may be restricted over the private leased line facilities.

The International Chamber of Commerce, a council set up to represent the interests of international businesses, favors a different approach. They would like basic and enhanced services defined into two categories. Basic services are within the PTT monopoly and cover the telephone dial-up service, telex, analog leased line networks, digital networks, and public data networks. Enhanced services are beyond the PTT monopolistic powers and should be open to competition. The PTT should not be restricted in providing enhanced services but there should be safeguards when PTTs compete with private business on an open market basis. Otherwise, the general public could be subsidizing a PTT's entry into unrelated businesses that have no return for the average consumer. In most cases, the new enhanced services are foreign to normal PTT operations, and the PTT is not equipped to handle hardware and software development and later market these new services and commercial applications. Therefore, the public would be subsidizing the PTT's entry into new markets without receiving a return on the investment. Enhanced services mainly fall into the data processing area and cover message switching, service bureau, remote job processing, time sharing, and data processing applications.

Today, the PTTs and Telcos are extending their business from basic voice and data services to include data processing services. A basic service is transmission of information, including switching, if required. There is no change in format, content, code, or protocol. Enhanced services are additions to basic services. They must provide alteration of format, content, code, or protocol. Note in Table 3.1 that enhanced services are mostly data processing applications that have been unregulated. Normally, the PTTs and Telcos are in the business of providing basic services.

TABLE 3.1. PTT Basic and Enhanced Services

Basic Services	Enhanced Services
Telephone dial-up network	Packet assembly/disassembly (PAD)
Telex data network (e.g., telex, telegram)	Protocol conversion
	Message storage and editing
	Electronic mail
Analog leased line network	Message handling systems
Digital leased line network	Document distribution
High-speed data networks	Remote time sharing
Satellite networks	Remote database applications
Packet switched data networks	Videotex applications—access to text and database
Circuit switched data networks	

3.1. POLICIES FOR CUSTOMER PREMISES EQUIPMENT

Customer premises equipment (CPE) is comprised of products that are purchased, leased, or rented by the customer for use with the PTT or Telco networks. Customer premises equipment includes telephone handsets, acoustical couplers, telephone answering systems, modems, multiplexers, telex terminals, facsimile machines, and private automatic branch exchanges (PABXs).

Although private vendors may supply CPE, the major domestic suppliers still supply most of the equipment to the PTTs. Today, there are some changes as a result of public pressure for new services and technology, and some governments are committed to competition. One example is the United Kingdom where there is stated open competition for customer premises equipment.

In general, national manufacturers resist opening up their markets because they are afraid that they will suffer in a free market. Governments are also anxious to protect their telecommunications industries and jobs. The danger is that European manufacturers, without a broad multinational base and the associated economies of scale, will not be as competitive as U.S. and Japanese manufacturers.

The user markets for customer premises equipment are being liberalized for the following reasons: It is now recognized that less expensive and better end user equipment stimulates the use of communications and, therefore, increases PTT revenue.

Certain types of CPE may or may not be allowed in a country depending on the amount of government protection required for local industry. In some countries, the PTT requires the customer to obtain the first telephone handset from the PTT. In some cases the customer will be allowed to purchase the second and subsequent telephone handsets from private companies. This rule also applies to telephone handsets attached to a PABX. Telephone answering equipment is normally allowed as a separate desk-top unit or as a component of a PABX. Answering systems that are not attached to PABX systems are allowed but may face import restrictions. A PABX is used by most public or private firms when there are several telephone handsets to be connected on the premises. The incoming trunk lines from the PTT are brought into the PABX to control and switch telephone lines in a building or building complex. PABXs range from several lines to more than 10,000 lines for a site facility.

Telephone handsets may be required from the PTTs for voice and data transmission. However, most of the world's PTT-supplied telephone handsets will not be a problem in communicating with U.S. equipment and Telco facilities. They are compatible because most PTTs in the world belong to the International Telecommunications Union (ITU) based in Geneva, Switzerland and adhere to the standards set forth by the organization. The ITU has 161 members, almost all of the countries of the world, with membership limited to national governments. The government delegations usually include telecommunications administration representatives. There are two permanent standards-making committees, the International Telegraph and Telephone Consultative Committee (CCITT) and the International Radio Consultative Committee (CCIR). They study technical, operating, and tariff questions and adopt recommendations in these areas. In addition, the ITU is a clearinghouse for information concerning telecommunications and sets standards through the CCITT recommendations.

CCITT members, in addition to the national delegations and their telecommunications administrations delegates, include recognized private operating agencies (RPOAs) such as the major telecommunications service providers (Telcos), scientific and industrial organizations, specialized UN agencies, and international organizations.

Every 4 years, study groups are formed to refine existing CCITT recommendations or formulate new recommendations:

Study Group	Area of Study
SG I	Quality of service aspects of telegraph, telex, and telematic services
SG II	Telephone network operation
SG III	General tariff principles
SG IV	Transmission maintenance of international lines and circuits; maintenance of automatic and semiautomatic networks
SG V	Protection against dangers and disturbances of electromagnetic origin
SG VI	Outside plant
SG VII	Data communications networks
SG VIII	Terminal equipment for telematic services (e.g., Telefax, Teletex, and Videotex)
SG IX	Telegraph networks and terminal equipment
SG X	Languages and methods for telecommunications applications
SG XI	ISDN and telephone network switching and signaling
SG XII	Transmission performance of telephone networks and terminals
SG XV	Transmission systems
SG XVII	Data transmission over the telephone network
SG XVIII	Digital networks including ISDNs

Table 3.2 is a list of customer premises equipment for North America and Europe, while Table 3.3 lists equipment for Central and South America, the Middle East, Africa, and the Asia/Pacific area. A designation *local* is a requirement that equipment must be supplied by manufacturers with plant facilities within the country. *Yes* in any column means that the user may either purchase, lease, or rent equipment from the PTT or from a private vendor. The designation *PTT* means that equipment is mandatory from the PTT and cannot be supplied by private or other sources without consent from the PTT. In terms of compatibility for telephone services, although the user will be required to purchase, rent, or lease local telephone handsets, the equipment will in almost all cases conform to international telephone standards. Ensuing tables in this chapter will further compare equipment allowed in many countries of the world.

TABLE 3.2. Telephone Equipment Status by Country

	First Telephone Handset in Home or Business	Additional Telephone Handset in Home or Business	Telephone Extension to PABX, Usually for a Business	Telephone Answering System in Home or Business	PABX, Usually for a Business
North America					
Canada	Yes	Yes	Yes	Yes	Yes
United States	Yes	Yes	Yes	Yes	Yes
Europe					
Austria	PTT	Yes	Yes	Yes	Yes
Belgium	PTT	PTT	Yes	Yes	Yes
Denmark	PTT	PTT	PTT	PTT	PTT
Finland	PTT	Yes	PTT	Yes	PTT
France	Yes	Yes	Yes	Yes	Yes
Germany, FRG	PTT	PTT	Yes	Yes	Yes
Greece	PTT	Yes	Yes	Yes	Yes
Ireland	PTT	Yes	Yes	Yes	Yes
Italy	PTT	Yes	Yes	Yes	Yes
Luxembourg	PTT	Yes	Yes	Yes	Yes
Netherlands	PTT	PTT	PTT	Yes	PTT
Norway	PTT	PTT	PTT	Yes	PTT
Portugal	PTT	Yes	Yes	Yes	PTT
Spain	PTT	PTT	PTT	PTT	Yes
Sweden	PTT	PTT	PTT	Yes	PTT
Switzerland	PTT	PTT	PTT	Yes	PTT
United Kingdom	PTT	PTT	PTT	Yes	Yes

Source: PTT published reports.

TABLE 3.3. Telephone Equipment Status by Country

	First Telephone Handset in Home or Business	Additional Telephone Handset in Home or Business	Telephone Extension to PABX, Usually for a Business	Telephone Answering System in Home or Business	PABX, Usually for a Business
Central & South America					
Argentina	Yes	Yes	Yes	Yes	Yes
Brazil	PTT	PTT	PTT	Local	Local
Columbia	Yes	Yes	Yes	Yes	Yes
Mexico	PTT	PTT	PTT	Yes	Local
Venezuela	Yes	Yes	Yes	Yes	Yes
Middle East & Africa					
Bahrain	PTT	PTT	PTT	Yes	PTT
Egypt	PTT	PTT	Yes	Yes	Yes
Israel	PTT	Yes	Yes	Yes	Yes
Kuwait	PTT	PTT	PTT	Yes	Yes
Saudi Arabia	PTT	PTT	PTT	Yes	Yes
South Africa	PTT	PTT	PTT	Yes	Yes
Asia/Pacific					
Australia	PTT	PTT	Yes	Yes	Yes
Hong Kong	PTT	PTT	Yes	Yes	Yes
Indonesia	PTT	PTT	Yes	Yes	Yes
Japan	PTT	Yes	Yes	Yes	PTT
Malaysia	PTT	PTT	Yes	Yes	Yes
New Zealand	PTT	PTT	Yes	Yes	Yes
Philippines	PTT	PTT	Yes	Yes	Yes
Singapore	PTT	PTT	PTT	Yes	Yes
South Korea	Yes	Yes	Yes	Yes	Yes
Taiwan (ROC)	PTT	PTT	PTT	Yes	Yes
Thailand	PTT	PTT	PTT	Yes	Yes

Source: PTT published reports.

Caution: In all cases where private telecommunications equipment is supplied, the user should check with the vendor for a PTT certification number or PTT acceptance before purchase. If a product does not conform to a PTT standard or is not PTT-type-approved, the equipment may not be attached to the PTT private leased or public switched network.

3.2. DATA CIRCUIT TERMINATING EQUIPMENT

Data circuit terminating equipment (DCE) is required for attachment and communications over Telco and PTT public and private data networks. DCE may be an acoustic coupler, modem, digital sending unit, multiplexer, or concentrator.

3.2.1. Acoustic Couplers

Acoustic couplers are used on dial-up or public switched telephone networks (PSTNs). An acoustic coupler allows the user to dial a remote computer or service bureau over an existing telephone line. After dialing the number and receiving a tone, the telephone handset is placed in two receptacles, and communications can begin. Acoustic couplers offer a simple low-cost connection for asynchronous communications between a terminal and host. They are flexible and portable and can be used at home, in a hotel room, or in an office. Some couplers are built into portable terminals. However, there may be restrictions on importing certain units.

3.2.2. Modems

Two modems (MOdulator-DEModulators), one at each end, are required for sending information over analog private or public telephone lines. They must be compatible at both ends and work at the same speed in order to establish communications. A modem converts the computer's digital output data into an analog signal for transporting data over telephone lines. At the receiving end, the modem converts the analog signal back to digital form to interface to a computer or work station.

Integrated modems are units that are embedded within a computer or communications controller. Normally, integrated modems face more PTT restrictions than stand-alone modems.

Stand-alone modems or DCEs may be supplied by either the PTTs or a private vendor. Integrated modems are usually allowed to be supplied by private vendors. Some PTTs allow the supply of modems for either public dial-up (switched) or private (leased) lines. Normally, DCEs can be supplied by a vendor for line speeds rated at 9600 bits per second (bps) or below. Higher-speed DTEs are PTT mandatory. Most international private leased line circuits operate at 9600 bps or lower. If the user has a special case, negotiations are possible with PTTs and Telcos for higher-speed data lines. In most countries, attachment to packet switched data networks (PSDNs), circuit switched data networks (CSDNs) or other data networks requires a DCE from the PTT.

TABLE 3.4. Data Communications Equipment – North America and Europe

	Acoustic Couplers	Integrated Modems	Modems	Autocall
North America				
Canada	Yes	Yes	Yes	PTT
United States	Yes	Yes	Yes	Yes
Europe				
Austria	Yes	Yes	Yes	Rest
Belgium	Yes	No	PTT (A)	Yes
Denmark	Yes	No	Yes	Rest
Finland	Yes	Yes	PTT	Yes
France	Yes	Yes	Yes	Rest
Germany, FRG	Yes	No	PTT (A)	Rest
Greece	Yes	Yes	Yes	Yes
Ireland	Yes	Yes	Yes	Yes
Italy	Yes	No	PTT	Rest
Luxembourg	Yes	No	Yes	Yes
Netherlands	Yes	Yes	Yes	Yes
Norway	Yes	No	PTT (A)	Rest
Portugal	Yes	Yes	Yes	Yes
Spain	Yes	No	PTT	Yes
Sweden	Yes	Yes	PTT (A)	Rest
Switzerland	Yes	Yes	PTT (A)	Rest
United Kingdom	Yes	Yes	Yes	Rest

Source: PTT published reports.

3.2.3. Autocall

Autocall is a modem feature or a separate autocall unit that fits between the computer and the telephone line and allows a computer connected to the public dial-up or switched telephone network to dial a number, retry the number if there is no response, and disconnect the call when completed. Autocall units, when allowed, operate differently in different countries. For example, a computer with an autocall feature could continue to call any number in the world repeatedly until the number is reached. The PTTs would view intermittent continuous calling as repetitive use of their switching equipment with no revenue. Therefore, countries differ in the number of calls allowed within a given period, the number of retries allowed on consecutive attempts, and the time period between retries.

A list of the common DCE attachments is shown in Table 3.4 for North America and Europe and in Table 3.5 for Central and South America, the Middle East, Africa, and the Asia/Pacific area. The following designations are used in Tables 3.4 and 3.5:

1. Yes: Supplied by PTT or private vendor.
2. PTT: Mandatory from the PTT.
3. PTT (A): Private vendors may supply either switched or leased line modems, but not both. They are usually limited to speeds of 9600 bps or less.

TABLE 3.5. Data Communications Equipment—Central and South America, the Middle East, Africa, and Asia/Pacific Area

	Acoustic Couplers	Integrated Modems	Modems	Autocall
Central & South America				
Argentina	Yes	Yes	Yes	Yes
Brazil	Negotiate	Local	Local	Yes
Columbia	Yes	Yes	Yes	Yes
Mexico	No	No	Local	No
Venezuela	Yes	Yes	Yes	Yes
Middle East & Africa				
Bahrain	Yes	Yes	Yes	Yes
Egypt	Yes	Yes	Yes	Yes
Israel	Yes	Yes	Yes	Negotiate
Kuwait	Yes	Yes	Yes	Yes
Saudi Arabia	Yes	Yes	Yes	Negotiate
South Africa	Yes	Yes	PTT	Yes
Asia/Pacific				
Australia	Yes	N/A	PTT	Yes
Hong Kong	Yes	Yes	Yes	Negotiate
Indonesia	Yes	Yes	Yes	Yes
Japan	Yes	Yes	PTT	PTT
Malaysia	No	Yes	Yes	Yes
New Zealand	Yes	PTT	PTT	Negotiate
Philippines	Yes	Yes	Yes	Yes
Singapore	No	No	PTT	No
South Korea	No	Yes	Yes	No
Taiwan (ROC)	No	No	PTT	No
Thailand	Yes	Yes	Yes	Yes

Source: PTT published reports.

4. Negotiate: Normally, the DCE is not allowed. However, the PTT is open to negotiations if there is a good business reason.
5. Rest: Restricted to number of attempts and retry procedure.
6. Local: Local manufacturer must supply equipment.
7. No: Not allowed for data calls.

Caution: Prior to purchasing equipment modems and autocall units, confirm that the product will conform to national PTT requirements and has been certified by the PTT.

3.3. DATA TERMINAL EQUIPMENT

Data terminal equipment (DTE) is comprised of terminals or controllers that attach to dial-up (public switched) or private leased lines or separate message networks for telex, Teletex, Telefax, Videotex, LANs, and public data networks.

3.3.1. Telex

Telex is a message exchange network with over 1,700,000 end points throughout the world. Telex provides a written record of the message transmitted to and received from one point to another or between the user terminal and the remote terminal the user dialed. The operator establishes a connection, similar to a dial-up call, and remains connected to the remote telex terminal until transmission of the message is complete.

Telex is the one means of communicating messages or data to most of the world. Telex terminals are relatively slow; messages are transmitted internationally at 66.67 words per minute based on 6 total characters per word including interword spacing, and the printout is in either upper- or lowercase, which can cause problems with interpretation. One advantage of telex is that a dialog or two-way correspondence is possible between two machines if the distant office is open at the time the transmission is received.

As of September 18, 1986, a method of transmitting both upper- and lowercase was approved by CCITT Study Group IX. Final approval of the draft recommendation, called S.2, will be decided at the IXth Plenary Assembly of the CCITT in 1988.

3.3.2. Teletex

Teletex is a set of interface and protocol standards that enables the communication of text between Teletex-compatible terminals on a memory-to-memory basis. Teletex is a new form of office-to-office telecommunications and uses a new type of terminal with an extended typewriter character set (over 300 characters) to prepare and transmit text. Teletex integrates with the office environment and is a logical supplement to existing telecommunications services such as telex (message exchange service) and Telefax (remote facsimile copying).

3.3.3. Videotex

Videotex is set of pages or frames consisting of alphanumeric characters or graphic images stored in a computer database. The pages are edited on a keyboard or generated from a computer-stored database. The database design is such that it permits rapid access and retrieval of specific items of information and the billing of customers using the system. The transmission lines between the user and the computer use the public telephone switched (dial-up) network. Modems are used to convert the analog signals to digital form for display on a TV set.

Most systems use a modified TV receiver or separate external box to translate the data and build up the images on the screen. Pages for transmission are selected by the user on a numeric keypad or an alphanumeric keyboard. There are many incentives for using Videotex. The French PTT has a program to replace the White Pages directory and later to replace the Yellow Pages by Videotex. There are over 2.5 million Minitel terminals in France whereby users have access to the French Videotex service called Teletel. Major advantages claimed are the savings in printing and distribution costs of the paper directories and the capability to update directories on a daily basis.

Tables 3.6 and 3.7 contain a list of communication terminals. A *yes* indicates that a terminal may be obtained from either the PTT or a private vendor. *PTT* means that terminals are PTT mandatory and can be supplied only by the PTT, while *N/A* indicates that the service is not available in the country.

3.4. VOLUME CHARGING ON INTERNATIONAL LEASED LINES

The public switched telephone network (PSTN), packet switched data network (PSDN), and circuit switched data network (CSDN) are switched facilities and are not restricted in the types of information flow in most countries of the world. The reasoning by the PTTs is that the user pays for the amount of traffic over the line by either time (PSTN and CSDN) or length of message or volume (PSDN).

International leased lines are vital for medium-sized and large companies dependent on quick, efficient flow of information. Banking, insurance, air travel, and computer-based service bureaus depend on international leased line circuits. Up to now, companies have depended on a flat rate structure with a fixed monthly charge.

TABLE 3.6. Data Terminal Equipment—North America and Europe

	Telex	Teletex	Videotex
North America			
Canada	Yes	Yes	Yes
United States	Yes	Yes	Yes
Europe			
Austria	Yes	Yes	Yes
Belgium	PTT	Yes	Yes
Denmark	PTT	Yes	Yes
Finland	PTT	Yes	Yes
France	Yes	Yes	Yes
Germany, FRG	Yes	Yes	Yes
Greece	Yes	N/A	N/A
Ireland	PTT	Yes	Yes
Italy	PTT	Yes	PTT
Luxembourg	Yes	Yes	Yes
Netherlands	PTT	Yes	Yes
Norway	PTT	Yes	Yes
Portugal	PTT	N/A	N/A
Spain	Yes	Yes	Yes
Sweden	Yes	Yes	Yes
Switzerland	PTT	Yes	Yes
United Kingdom	Yes	Yes	Yes

Source: PTT published reports.

TABLE 3.7. Data Terminal Equipment—Central and South America, the Middle East, Africa, and Asia/Pacific Area

	Telex	Teletex	Videotex
Central & South America			
Argentina	PTT	N/A	Yes
Brazil	PTT	N/A	Yes
Columbia	Yes	N/A	N/A
Mexico	PTT	N/A	N/A
Venezuela	Yes	N/A	N/A
Middle East & Africa			
Bahrain	PTT	N/A	N/A
Egypt	PTT	N/A	N/A
Israel	Yes	N/A	N/A
Kuwait	PTT	N/A	N/A
Saudi Arabia	PTT	N/A	N/A
South Africa	PTT	N/A	Yes
Asia/Pacific			
Australia	PTT	N/A	Yes
Hong Kong	Yes	PTT	PTT
Indonesia	PTT	N/A	N/A
Japan	PTT	Yes	PTT
Malaysia	Yes	N/A	PTT
New Zealand	PTT	N/A	PTT
Philippines	Yes	N/A	N/A
Singapore	PTT	PTT	PTT
South Korea	Yes	N/A	PTT
Taiwan (ROC)	Yes	N/A	PTT
Thailand	Yes	N/A	N/A

Source: PTT published reports.

Each PTT has a different set of transmission performance characteristics specified and guaranteed for international leased lines with appropriate penalty clauses under CCITT Recommendation D.1, Section 4, Allowances for Interruptions.

International leased circuits enable companies to develop private networks to match their business needs. Some of the reasons for choosing an international lease line are the following:

1. *Fixed rate.* A fixed rate allows businesses to plan communications and data processing needs. If the rate is changed to a volume charge, more local processing may be required with less bulk data transferred out for remote processing. New services such as video conferencing and international Videotex may lose their cost effectiveness.
2. *Privacy and data security.* Private circuits offer the possibility of tighter control.
3. *Reliability, availability, and performance.* Performance is guaranteed with penalty clauses based on the CCITT Recommendations for the Lease of International Telecommunication Circuits for Private Service, Recommen-

dation D.1, Section 4. There is no similar guarantee for performance on the public switched network. In some countries the switched telephone network and telex are subject to congestion and line problems. In cases where information transfer of voice, data, messages, and text is critical, international leased circuits are essential and the only viable choice.

As the need for international leased lines becomes more critical, the PTTs are attempting to add volume charging schemes to obtain more revenue. Volume charging was initiated by the Deutsche Bundespost (DBP) of the Federal Republic of Germany. They have introduced volume charging with rates up to nine times the current rate based on the network and usage. CEPT (Conference of European Postal & Telecommunication Administration) has also been studying this concept. CEPT membership includes PTT delegations from Austria, Belgium, Cyprus, Denmark, Germany, Finland, France, Greece, Iceland, Ireland, Italy, Liechtenstein, Luxembourg, Malta, Monaco, the Netherlands, Norway, Portugal, San Marino, Spain, Sweden, Switzerland, Turkey, the United Kingdom, Vatican City, and Yugoslavia. The CEPT issues recommendations and decisions after reaching a consensus.

The concept of volume charging can be seen as a strategy by PTTs to move companies to public data networks and new digital data networks. Most of the new data networks are underutilized, operate within the country, and are now priced artifically low to attract users. Once the PTTs establish a customer base and users invest in terminals, software, and computer interfacing equipment, the PTTs may increase tariffs in order to recover their investment.

3.5. NATIONAL LEASED LINES

National leased lines apply within a country and may be obtained by filing a request with the PTT. When filing, the user will be required to describe the application and potential uses of the line. A diagram usually accompanies the application. If the perception by the PTT is that the leased line will be used primarily for message switching and not data transfer between terminal and host, the application will not be approved.

Many PTTs have a requirement that both ends of a private leased line must terminate within the same legal entity. In some cases, a company subsidiary or another division may not be granted a leased line if it is in unrelated businesses. For example, the Deutsche Bundespost or PTT has viewed General Motors' Opel (automobile) Division and its Frigidaire (home appliances) Division as unrelated businesses and has declined a leased line between the two divisions.

In one or two countries, an upper limit of 20% of a computer's work load for message switching is allowed over a national leased line. This rule applies to one enterprise or one legal entity. A connection between two or more enterprises is a special case and requires a business need with related benefits for all users, for example, a group of banks or one bank providing service to a number of retail units within its banking area. In general, when applications for a national or international leased line are submitted, if there is any perception by the PTT that a company or a business group is setting up a network just to save tariffs, the application is automatically rejected.

3.6. INTERNATIONAL LEASED LINES

International leased lines are vital to medium-sized and large businesses. There is no question of an alternative choice; an international leased line is mandatory for most data communications. For example, credit card companies process millions of transactions on a daily basis. American Express Card services over 250,000 daily requests for card use authorization or transactions and provides an average response time of 5 seconds. This response is required and is not possible with any public switched network. Consider waiting in a checkout line at a retail store while the service person dials a number, waits for a response, and if the line is busy redials the number before your purchase is processed. Delays of this type would not be tolerated by most customers.

Submitting an application for an international leased line is similar to that for a national leased line. However, there are two monthly statements, one in the country where the line request was initiated and the other in the country where the line is terminated. Normally, the PTT will negotiate with its PTT or Telco counterpart in the other country to provide you with the international leased line. If intermediate countries are involved, negotiations are worked out by the PTT or Telco who initiates the request. Pricing and terms and conditions are based on published tariffs or are negotiated between PTTs. In many cases, the PTTs do not publish international leased line tariffs, and the monthly charge is subject to negotiation.

3.7. RESTRICTIONS ON INTERNATIONAL LEASED LINES

When filing an application for an international leased line, there are some restrictions the requester should consider. In most countries, the information provided on the application form may face restrictions that could result in no international leased line. Therefore, the requester should be open to negotiations, which translates to a higher price over normal tariffs. Since tariffs are lower in the U.S., you may want to apply for an international leased line in the United States.

Consider the following when preparing an application for an international leased line in a foreign country:

1. *Intracompany versus intercompany.* Leased line applications are less restrictive between intracompany (same) than intercompany (two or more) enterprises. A case must be established that the application for an intracompany line is between establishments that have related business needs and is not set up merely to reduce tariffs. Intercompany communications are more restricted, but lines can be obtained if there is a business requirement and the

PTTs cannot provide a comparable service or satisfy the customer's conditions otherwise. One example would be a travel service bureau connected to an automobile rental company whose databases are in another country.

2. *Basic service.* Competition with the PTT is not allowed for switched data networks or point-to-point message switching. Unlike private businesses, the PTTs have extended the definition of the telex message to include message store and forward services. If the requester's application states that message switching, message transfer, or document distribution is over 20% of the communication line's activity, the application may be rejected. In some cases, if the computer that holds the message, text, or document distribution application program is over 20% of the computer's work load, the application may be rejected or open to negotiation.
3. *Voice and data.* Normally, voice and data must be on separate private leased lines. Most European PTTs and many developing nations restrict or do not allow the combination of voice and data on the same line. Voice and data on the same line could be achieved by multiplexing equipment, to be discussed in Chapter 5. Many PTTs would view the combination as a loss of revenue because telephone calls are normally placed through the public switched telephone network.
4. *Access to switched or dial-up network.* In general, international leased lines must terminate in the computer system in the receiving country. The computer must not be used to switch data, messages, or text or distribute documents to a dial-up or switched telephone line. There must be a major change in the content, formatting, or repackaging of the incoming information before the computer may send out the information to the switched network. Normally, a leased line to switched line connection is considered a savings in line charges over a direct call from one end point to another and is disallowed.
5. *Third party traffic.* Third party traffic or resale of a leased line over a national or international leased line is not allowed. That is, you may not sublease a line to another company. In this sense, the enterprise is one party, the PTT is the second party, and another legal entity or enterprise would be the third party.

3.8. EXAMPLES OF INTERNATIONAL LEASED LINES

Although it is difficult to generalize about different applications for over 161 countries, five cases will be presented to provide an understanding of the alternatives for negotiation. By definition, if a company is doing business in a foreign country, the foreign company must be registered as a separate legal entity, and the enterprise is considered a multinational corporation.

Refer to Figure 3.1 for cases 1 and 2.

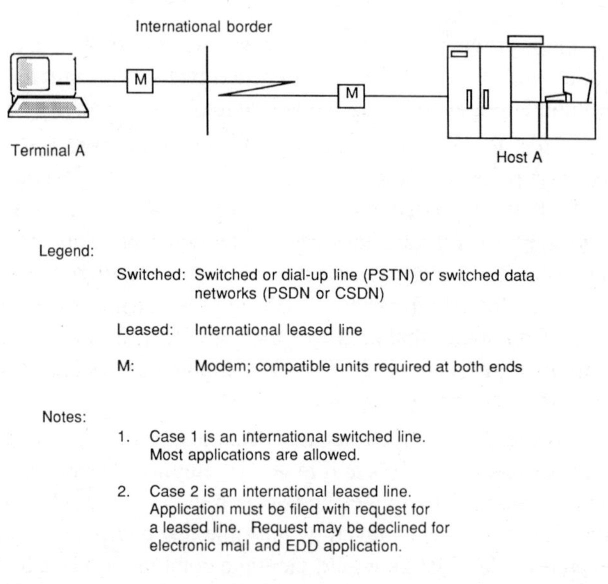

Figure 3.1. Terminal to host in another country.

Case 1 includes a terminal connected to a host over an international switched or private leased line within the same enterprise. By definition, the terminal and host are two legal entities because each location is registered as a separate entity in each country. Consider the switched line connection: Any combination of a switched line network—analog, digital, or switched data network—is allowed between terminal A and host A in the same enterprise or different enterprises. Message switching, text, or document distribution is allowed over all switched networks.

Case 2 is for an international leased line connection. Data transfer is allowed over international leased lines within the same enterprise. Message switching or document distribution may not be allowed or must be negotiated. In general the percentage of computer usage for message switching must be less than 20%. Message switching between different enterprises or interenterprises is not allowed.

Refer to Figure 3.2 for cases 3 to 5.

Case 3 is a switched line connection between two terminals, their respective hosts, and the two hosts. All equipment is within the same enterprise. A switched line connection would be allowed throughout the network. Message switching over switched lines would be allowed.

Case 4 is a switched line between terminal A and its host A, a leased line between host A and host B, and a switched line between host B and terminal B. This combination would not be allowed unless information flow from host A to host B terminates in host B. Before sending information to terminal B, the information must be reformatted or changed or some value added to the information. Otherwise, the PTT would perceive the combination as a savings on tariffs. Information must terminate in host B according to the CCITT D.1 Recommendation.

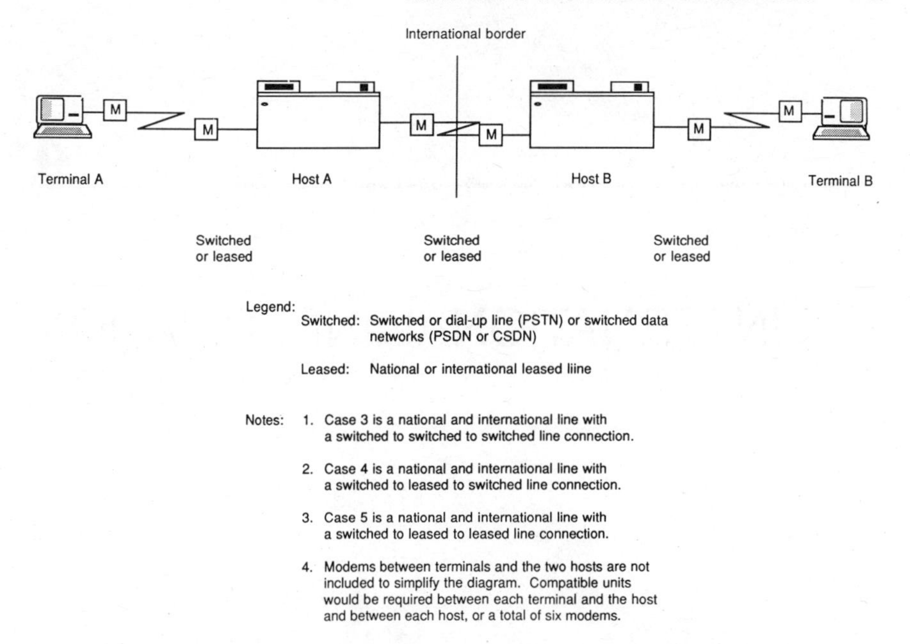

Figure 3.2. Terminal to remote host or remote terminal in another country.

Case 5 is a switched line connection from terminal A to host A, a leased line connection from host A to host B, and a leased line connection from host B to terminal B. This arrangement would be allowed under most regulations for data transfers. Message switching or document distribution would be allowed between terminal A and host A or host B and terminal B but not between host A and host B.

In cases 1 and 2, the terminal could be perceived as gathering information at a remote site for computation or processing on a large computer system because the remote terminal does not have resources capable of completing the job. For cases 3 to 5, the situation is open to discussion because there is a connection between two host systems in two countries. Switched lines between the countries do not present a problem. In general, if the information terminates in the remote host, either A or B, a good case can be made for processing on the remote host, e.g., remote application support or availability, remote database, or remote critical programming support. Message switching or document distribution would be more difficult depending on the country. In any case, plan your strategy and—negotiate—most PTTs will respond to a legitimate business need.

4

INTERNATIONAL INFORMATION FLOW

In this chapter we will

- Discuss data protection legislation, human rights, social identity, and economic independence.
- Determine the impediments to international information flow and trade barriers.
- Present data security and national laws, cryptography, and copyright laws applied to databases

One of the first main concerns about the consequences of data files and their processing was related to information about individual citizens. Sweden initiated the original personal privacy and data protection law in their "data act" of 1973. Since that time, the impact of *data protection* and the need to protect the individual from the rapid growth of accumulated information has spread to all countries. The basic rights of persons have been added to data protection acts and include economic and legal rights, national sovereignty, and the need to protect the character and personality of the nation itself.

Most of the national laws implement common principles with respect to the individual. The reasons for the accumulation of data on persons must be defined and proven to be valid to support the needs of a business. Personal data must be collected by legal and fair practices; it must be correct and not excessive. Every law in existence has a provision whereby the *data subject* has the right to review and correct the data pertinent to him or her, within a 30- to 90-day period. If another party is involved, personal data may be communicated if it is required for a business need and the recipient guarantees safekeeping. All information containing personal data must be classified and protected against unauthorized or accidental disclosure or modification. In addition, there is usually a time period during which the data must be kept. Personal data must not be accumulated over a long period, and irrelevant and obsolete data must be discarded. Existing laws carry a fine or prison term or both for non-compliance.

4.1. DATA PROTECTION ACTS

Each person, customer, and supplier has the right to expect that a *data controller* carefully manages the data related to them. A data controller is an organization or person who collects, stores, processes, and communicates personal data.

Protection of privacy is an example where society's concerns about the impact of information technology have resulted in widespread legislation and regulatory activity throughout the world. Table 4.1 is a compilation of existing data privacy Laws.

4.2. HUMAN RIGHTS

There is a concern for human rights or freedom of individuals in their relationships with others, their government, and other organizations. The individual should be able to receive information, use it, and transfer it to others. The protection of personal privacy must be maintained, and data about certain aspects of a person's private life must not be collected. There must be a guarantee that the personal information that is collected is correct and is used for relevant reasons. Human rights concerns were adopted by the United Nations in 1948 as stated in Article 19 of the UN Universal Declaration of Human Rights: "Everyone has the right to freedom of opinion and expression; this right includes freedom to hold opinion without interference and to seek, receive and impart information and ideas through any media regardless of frontiers."

The concept of data protection emerged based on the perception that there was a need to protect the individual's privacy by preventing the potential misuse of personal information held in computer files. A new technique called cryptography, or the encryption, scrambling, or transformation of data to obtain a secure means of protecting information transmitted over telephone lines, microwave, or satellite links, was developed to protect data.

In Europe, governments have defined data subjects' rights and data controller obligations (or mandatory requirements). European legislation covers both the public and private sector, as compared to the U.S. legislation where the law addresses only the public sector.

TABLE 4.1. Status of Data Protection Legislation—1986

Country	National Status
Australia	Draft legislation prepared
Austria	1978: Data Protection Act of Austria; law being revised
Belgium	Legislation in parliament
Canada	Privacy act passed July 1, 1983
Cyprus	Data protection bill proposed in 1985; may pass in 1987
Denmark	1978: law adopted, revision planned
Finland	New bill before parliament; a data protection board with a Privacy Commissioner will be established
France	1978: data protection law adopted
Germany (FRG)	Federal Data Protection Act (FDPA) of 1977, also called the Bundesdatenschultzgesetz (BDSG); revision planned
Greece	New bill proposed to parliament in 1985
Iceland	Act No. 63 passed by parliament and in effect from 1982 until December 31, 1985; a draft of a new act with no basic changes submitted to parliament for approval
Ireland	Government report prepared, private member bill prepared
Israel	The Protection of Privacy Act of 1981
Italy	Legislation in parliament
Japan	Government report prepared
Luxembourg	Data Protection Act of 1979
Netherlands	New data protection bill in 1987—rules for the protection of privacy in connection with personal data files
New Zeland	Data protection act adopted
Norway	1978: data protection act adopted; revision planned
Portugal	Constitutional provision, legislation in parliament
Spain	Data protection act proposed
Sweden	Data protection act adopted in 1973
Switzerland	Article 28 of the Civil Code; planning new data protection act
Turkey	Government report prepared, legislation in parliament
United Kingdom	Data protection act adopted in 1984
United States	Privacy Act of 1984
Yugoslavia	Government report prepared, legislation in parliament

Source: Transnational data and communications reports, Washington, D.C.

There have been other extensions that have an impact on business. Although European data protection legislation differs by country, five countries (Austria, Denmark, Iceland, Luxembourg, and Norway) have enacted laws that cover data related to a *legal person* (i.e., a company, trade association, or institution). Under the legal person provision, the corporation is treated as an individual. Applying the individual data subject to a business could result in a competitor demanding to review information being held by a competitor.

In addition, all European protection laws relate to computerized data, while some countries exclude manually handled data. For example, in countries where manual data is excluded, you are not allowed to transmit data electronically, but a briefcase full of confidential material is exempt from data protection laws.

All European data protection laws define *data subject rights* and *data controller obligations*. In addition, many European countries have extended the basic protection laws to include the establishment of national data inspectorates to implement transborder data flow (TDF) regulations. A data inspectorate supervises the implementation of legislation and in most countries the data inspectorate's prior approval is required before personal data may be transmitted across an international border by automatic means.

4.3. CULTURAL AND SOCIAL IDENTITY

Many governments are concerned that international information flow may affect national cultural structures and diversity. New communications technologies such as satellites allow reception of TV programs and advertising. Governments argue that advertising creates a demand for products provided by developed countries and influences the life-styles and cultural values of the receiving country. Another strong influence is the use of the English language in international databases and computer software. Overall, the perception is that it is a threat to national cultural structures.

There is an extensive amount of data in public and private information systems as a result of computerized tracking of personal profiles. The belief is that information technology developments are driven by what is possible technically and not by what is socially desirable. In many developing countries there is little understanding of information technology. Nations are vulnerable: The increase of international networks allows the accumulation of data within a country and the sending of this information abroad for compilation, summary, and decision making. In many cases, the information is collected without government knowledge.

4.4. ECONOMIC INDEPENDENCE

Some countries believe that data pertaining to economic potential and structure is a national resource and that their governments should be able to exercise control over its collection, use, and distribution. There is concern that international businesses can operate within a country while maintaining their decision-making centers abroad. High-speed information transfers over international networks and international networking services facilitate and encourage these transfers. The Swedish data protection act is based on the principle that technological applications should have socially desirable ends and not be driven by economic profit. Computer systems should be adapted to a user's capability and knowledge, and international information flow must be supervised and regulated so as not to undermine sovereignty. A Latin American view is that legal measures must be taken to restore national security. There is an imbalance of international information flow, and it affects national sovereignty and cultural identity.

4.5. NATIONAL INFORMATION SECURITY

In 1986, the U.S. National Security Administration initiated a 5-year $40 billion program to encode most of the millions of electronic messages sent each year by the U.S. government and defense contractors. The reason behind the program was the belief by the government security personnel that the Soviet Union has an electronic surveillance program directed against the United States. One of the missions of the National Security Agency is to protect the sensitive information of the United States. Surveillance is easier since most of the messages are transmitted at some point by satellites or microwave towers. The interception of messages in the air is far easier to detect than by underground cable.

One indication of the growth in messages is the number of computers in the federal government, which has grown from 22,000 in 1983 to over 100,000 in 1985. Another rough measure is a National Security Agency estimate that all the modems in 1972 could transmit about 600,000 characters per second, while in 1984 the number was 220 million characters per second.

4.6. U.S. NATIONAL SECURITY LAWS

In 1968, the U.S. Congress passed the Omnibus Crime Control and Safe Streets Act, which contained a provision protecting the privacy of verbal and wire communication. The law prohibits electronic eavesdropping of phone conversations except for law enforcement surveillance under a court order. Violations are punishable by prison terms and fines. The law allows telephone company monitoring of communication to ensure adequate services or to protect company property and permits surveillance done with the consent of a participant in the conversations.

Since the preceding act applied to wire communications and did not cover new technology such as microwave and satellite transmission, the Electronic Communications Act of 1985 that was passed by the U.S. Congress extends protection against interception from voice transmission to virtually all electronic communications. Any digitized portion of a telephone call, the transmission of data over telephone lines, and the transmission of images by microwave or by any other combination of transmission media were covered. Civil and criminal penalties for any violation were included in the bill.

Other federal laws governing national information security are the following:

- The Right to Financial Privacy Act (1978)—outlines conditions under which federal agencies are allowed access to customer records held by banks and lending institutions
- The Foreign Intelligence Surveillance Act (1978)—sets standards for the use of electronic spying devices to collect foreign intelligence and counterintelligence in the United States
- The Electronic Funds Transfer Act (1980)—requires banks and other institutions to notify customers about third party access to electronic funds transfers
- The Cable Communications Policy Act (1984)—restricts the type of information that cable companies can collect and disclose about a customer and creates a subscriber's right to privacy from government surveillance

4.7. EMPLOYMENT

Information processing and telecommunications affect employment in different ways. Some countries fear that the storage of data abroad instead of in their country may result in loss or transfer of jobs. The results are a loss of government revenues, support services, and value added taxes and impact their balance of payments. On the other hand, many governments recognize that national economies are interdependent and that international trade and finance are affected by foreign competitors.

4.8. INTERNATIONAL INFORMATION FLOW

International information flow (IIF) or transborder data flow (TDF) signifies units of information coded electronically by a digital computer that transfers or processes the information in one or more countries.

The interchange of information among countries is required to communicate freely within the enterprise and with its suppliers and customers. Information is a major resource in national and international investment, trade, commerce, and industry. In an independent community of nations, information can play that role if it is allowed to flow freely and can be accessed and exchanged on a worldwide basis.

Information technology is important to developing nations to develop skills and expertise and the transfer of technology. Restrictions on the collection, storage, and distribution of data will affect the national economy, information processing facilities, services, and products.

4.8.1. Examples of International Information Flow

The following are examples of international information flow:

- *Word processing networks.* International companies compose letters and messages on word processing equipment for transmission to computers in other countries. In turn, letters and messages from foreign countries are received directly on the same word processors.
- *Financial flow.* In many developing countries, many of the computers are in the banking and financial sector. Large U.S., European, and Japanese banks and international credit card companies can establish profiles on individuals based on age, education, income, spending for entertainment and travel, countries visited, and major retail stores used to purchase goods and services. As information flow between companies and interenterprise transfers increase, companies will be able to create files for higher-income-earning individuals and companies that are more accurate than those collected by the national governments.
- *International databases.* Several international service bureaus provide local access to their network and connect to remote applications and databases in other countries. In addition, critical databases can be stored in another country for backup. Database services offer information concerning science and technology, and bibliographic, agricultural, economic, and national resources.
- *Airline data traffic.* Most international airlines are connected to and dependent on the Societe Internationale de Telecommunications Aeronautiques (SITA) for reservations and bookings. Airlines also transmit special handling arrangements, information on deportees and refugees, baggage information, tracing procedures for missed checked baggage, advice on security matters, and procedures for persons traveling on a space-available basis.

4.9. INTERNAL INFORMATION SECURITY

The growth in national and international data networks and in the access and use of terminals, PCs, and small computers, combined with knowledgeable users, has increased the vulnerability of internal information systems. Unauthorized or criminal access to systems owned by businesses, universities, medical facilities, and governments has led to data theft, financial fraud, embezzlement, and sabotage. Concern for these matters has led to new procedures for implementing the safety of information assets.

Security is a total system commitment for the protection of data and programs from unauthorized use, modification, or destruction through intentional or unintentional action. The components are the hardware, software, organizational, and control features of the operating system. Data security is the combination of physical protection, administrative controls, and hardware and software controls. Physical protection includes site planning and construction, fire detection and extinguishing, air conditioning, water, drainage, and physical access control to restricted areas. Administrative controls include security strategies, education, and data processing monitoring and control. Hardware and software can provide logical access control, data encryption, and media protection.

Data security includes access control, auditability, and fallback or recovery. Access control features individual user access codes for persons who require access to personal data in order to perform their job. Auditability helps to establish the validity and accuracy of results and evaluates the adequacy of controls and compliance with standards. Fallback and recovery procedures provide continued operation after a major failure of information processing services.

A central authority or data controller is required to protect, establish, and maintain authorization lists; assure authorized access; and control problems. The degree of protection is determined by the amount of overhead that is allocated to system performance, the number of passwords that users will tolerate for access to data, and the limits of access (free to all or a portion). Access can be at the user program, system program, user data file, or system data file level. Security access facilities should provide different categories of read only authority, read and write authority, destroy and erase authority, and execute authority. Security hardware features to be considered are terminal identification numbers (IDs), badge readers, and cryptography, which involves the encryption and decryption of information based on a data encryption standard (DES).

4.10. CRYPTOGRAPHY

Encryption is the scrambling or transformation of data to obtain a secure means of protecting information transmitted over telephone lines, microwaves, or satellite links. It is used to protect stored and transmitted data by means of a cryptographic algorithm, such as the DES standard of the National Bureau of Standards, which went into effect in 1977.

The most secure cryptographic systems use a cipher system that requires two basic elements, an algorithm or procedure and a set of variable cryptographic keys. A key is a short user-selected sequence of alphanumeric characters. An effective security system involves secure procedures for initiating, installing, maintaining, and managing keys.

Note: A full description of cryptography may be found in Meyer and Matyas, Cryptography, A New Dimension in Computer Data Security (see the Bibliography, Appendix C).

Encryption requirements are increasing as the most effective way of securing transmitted or stored data for satellite data transmission. The ability to tap and analyze communications links will become less expensive with technology. The user requirement for encryption is a password or sign-on procedure. Encryption may be applied to an entire file, all sessions on a communications line, selected fields of a data set, or management functions for individual user applications. Procedures for

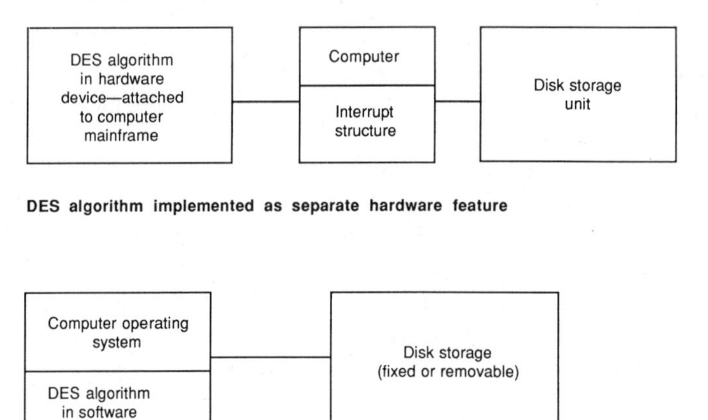

Figure 4.1. Encryption/decryption implementation for stored and transmitted data.

key management must be easy to understand, allowing such things as encrypted logging of two keys in a session, whereby recovery is possible only with the collaboration of two persons, each knowing a different master key.

There are several methods of implementing encryption for data files in a computer system or in transmission over carrier or PTT facilities. The DES algorithm may be implemented in a computer hardware attachment or in software routines within the computer or may be embedded into a terminal or a separate stand-alone unit that is installed between the computer communications interface and the data communications equipment. Figure 4.1 is an example of two types of implementation. The DES standard may be implemented in a separate hardware attachment to the computer's interrupt structure or as a separate software routine. Both units can be called by program to encrypt or decrypt information to or from disk storage, to or from an application, or for transmission over a telecommunications network.

In Figure 4.2, two separate units are used to protect information over Telco or PTT facilities. One unit is used at the sending end to encrypt data, and at the receiving end another comparable unit is used to decrypt the incoming data. The encryption and decryption devices will fit between the computer's line attachment unit and a modem. This arrangement is the same for both ends of the line.

Figure 4.3 represents an implementation of the DES algorithm in a terminal or as a hardware/software combination in the computer host system. Depending on information flow, data is encrypted on transfer out and decrypted at the receiving end. The modem on each side looks at a digital data stream.

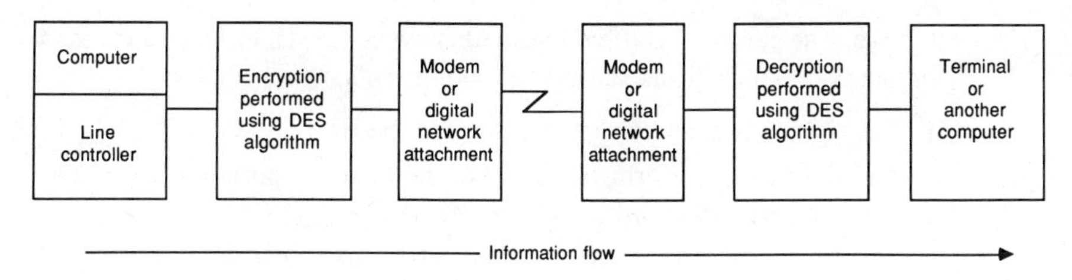

Figure 4.2. DES algorithm: stand-alone devices.

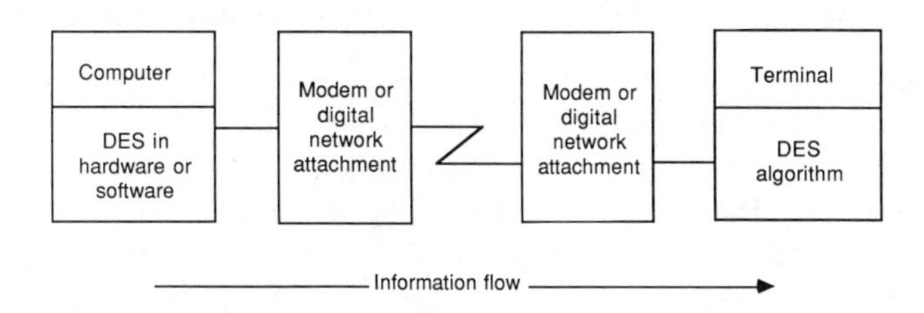

Figure 4.3. DES algorithm: integrated devices.

4.11. IN-HOUSE IMPLEMENTATION PLAN

Installing and maintaining a database on persons and nationally sensitive items in Europe and the remainder of the world are serious processes. Several data protection acts require a formal procedure for compliance with national laws. The following is a list of steps that may be used to implement an in-house procedure:

1. *Appoint a data controller.* A data controller is a person responsible for defining an appropriate data security system and the level of security for the company. That person will be required to select a combination of organization, administrative, physical, and technical measures. A data controller will be required to know national registration procedures for in-house requirements as well as international information flow regulations. In a large organization, a full-time person may be required.
2. *Educate your staff.* Classes should be held to provide guidance on the principles and implementation of national laws and the transferring of information to other countries. Information is a unique resource and is very valuable in supporting the production and marketing of goods and services. Most international information flows are with the same enterprise or intraen-

terprise. The company staff will need to know certain rights granted to them for handling personal and sensitive economic data:

a. The right to respect how information will be used
b. The right to use information for specific business purposes and not to maintain irrelevant information on any person
c. The right of the individual in the database to access that database on demand
d. The right to cease using, amend, or completely erase eroneous information
e. The right to stop using the information for unauthorized purposes

3. *Find out where personal data is collected in the organization.* The proliferation of PCs within an organization presents an enormous tracking problem. Many large high technology and insurance firms purchase PCs by the thousands. Databases or electronic files on personal data must be located within departments or groups. A data controller or coordinator should determine if individuals are knowledgeable of present laws and determine the relevance of the data to the business, the length of time the data will be retained, and if the person responsible for maintaining the database has a plan to have the data subjects review the information and correct any errors.
4. *Check compliance and developments.* The data controller should periodically confirm data security, the accuracy of data, and the currentness of material. The data controller must understand that he or she will be held accountable and be subject to either fine or prison for noncompliance if individuals seek compensation for unauthorized disclosure or improper use of data.
5. *Prepare for registration.* The data controller must determine whether one or several registrations are required and must consider internal control as well as future needs. Trade associations may be of some help. The data controller must establish procedures to monitor compliance with good data protection principles.
6. *Know the rules.* It is a criminal offense for any person to knowingly provide personal database services without having registered as a computer bureau. Data users are liable for paying compensation for damage or associated distress suffered because of inaccuracy of personal data.

4.12. TRADE BARRIERS TO INTERNATIONAL INFORMATION FLOW

The development of information technology has had an effect on the interchange of business information both within and across national boundaries. In today's environment, information is a key resource in national and international trade, commerce, and industry. Most businesses are dependent on many nations for the marketing and manufacturing of materials, goods, and services. Information can stimulate economies if allowed to flow freely and if it can be accessed and exchanged on a worldwide basis.

Companies of all sizes use international data flows to manage their worldwide operations with better efficiency. Specific application areas include worldwide inventory, sales, production, marketing, planning, and financial coordination. However, there are serious concerns within the international business community that the flow of information may be inhibited by new laws or restrictions. The debate on data privacy in national and international forums has had the unfortunate consequence of causing data flows, particularly the international exchange of data (also called transborder data flow), to be legislated and regulated.

Several trade barriers to telecommunications, data, and information services have been published by the Office of the U.S. Trade Representative. Discriminatory licensing practices and procedures have been reported in several European countries. Other restrictions include limitation of foreign equity in a firm and the extension of laws to legal persons. One of the most severe restrictions for service bureaus is the restriction on international leased lines. Germany (FRG) prohibits international leased lines from being connected to the German public switched or dial-up networks unless the connection is made by a computer in Germany that provides some processing of the data before connection to the switched network. Within Germany, international leased lines are available only if there are guarantees that the lines will not be used to transmit unprocessed data to foreign telecommunications networks. That is, the data must be processed within the country. Many banking authorities interpret this as a requirement to process financial information within the country. This procedure excludes any on-line processing from another country and forces the purchase of data processing equipment within Germany.

Personal restrictions regarding foreign personnel include commercial visas, discriminatory regulations, and administration practices that make it difficult for qualified personnel to enter the country to work, and to continue to work after they are allowed to enter.

There have been many cases reported of excessive tariffs for data communications equipment and electronics. Many countries also impose heavy import fees on computer systems needed to meet the U.S. airline reservations standards and required for integration into existing international networks. For example, note the following:

1. Egypt: 41% import duty on foreign equipment for airline applications
2. Greece: 31.5% import duty on airline reservations equipment
3. European community (now 12 nations): 17% input duty imposed on integrated circuits
4. Israel: 57% import duty on airline reservations equipment
5. Italy: 25% import duty on airline reservations equipment
6. Spain: 57% import duty on airline reservations equipment
7. United Kingdom: 4% duty and 8% VAT (value added tax) on all microfilm documents for resale to the public (the tax applies to microfilmed information and not to its paper equivalent)

Many data processing (DP), telecommunications, and electronics industries are subsidized by national governments. For example, in a 3-year period spanning 1980–1983, the French government subsidized the DP industry with $500 million, German (FRG) subsidies were $350 million, and Italian subsidies were $355 million. Swedish DP subsidies were $111 million, and the United Kindom spent $550 million on DP subsidies.

4.13. COPYRIGHT PROTECTION FOR DATABASES

The ability to store, update, transfer, read, and copy information in electronic form has raised the issue of whether database protection applies to patent law or copyright law. One of the main arguments against the application of patent law is that computer software is not considered an invention warranting a patent, and so many governments defer the question to their copyright laws. Most governments argue that protection of computer software should apply for a shorter duration than copyright laws. Table 4.2 shows the status of copyright laws with respect to the protection of databases.

4.14. INTERNATIONAL ORGANIZATIONS

There are several organizations concerned with the implementation of data privacy and international information flow. Two leading organizations in the cross-Atlantic community are the Office for Economic Cooperation and Development (OECD) and the Council of Europe (CofE).

The OECD is located in Paris and has a membership of 19 European countries plus Australia, Canada, New Zealand, Japan, and the United States. In 1980, the OECD council adopted the "Guidelines Governing the Protection of Privacy and Transborder Flows of Personal Data." This set of guidelines covers the protection of privacy and international information flow of personal data for both manual and automatic (or computer) files and stresses self-regulation. Each member government is asked to sign the agreement and refrain from restricting transborder flows of personal data between itself and another member country. Adherence to the guidelines may facilitate access by companies to data from abroad. It is a two-way street; if the private sector does not take action to implement the guidelines, there is risk that other OECD countries (24) will restrict or prohibit the flow of personal data.

In addition, the OECD is a forum for the study of and debate on information policy and makes recommendations and adopts policies as well as business practices for member countries. While representation is limited to national governments, business reactions have an impact, and businesses are represented through one of its committees, the Business & Industry Advisory Committee (BIAC). The OECD emphasizes free flow of information for all types of files and restricts information flow to natural persons.

TABLE 4.2 Government Protection for Databases

Country	Database Protection
Australia	Government sources indicate that databases may be covered.
Austria	There are no specific rules; the criteria for eligibility to copyright protection are an intellectual creation.
Belgium	Databases are not covered in copyright laws.
Brazil	Databases are not covered in copyright laws; there is a bill pending to give legal protection to computer software.
Canada	There are no specific provisons for databases.
Denmark	A catalog rule introduced in the 1960s prohibits unauthorized reproduction of catalogs, tables, etc., where a large number of particulars have been summarized.
Finland	A catalog rule applies that prohibits unauthorized reproduction of catalogs, tables, etc., where a large number of particular or discrete information has been summarized.
France	There are no specific provisons for databases.
Germany (FRG)	Databases are protected under copyright laws.
Hungary	Special legislation has been passed for copyright protection.
India	Database protection legislation is pending.
Japan	Special legislation has been passed for copyright protection.
Luxembourg	The government indicates that copyright laws may not cover databases.
Mexico	Copyright laws may apply to database protection.
Netherlands	Databases are not covered in copyright laws.
Norway	A catalog rule applies that prohibits unauthorized reproduction of catalogs, tables, etc., where a large number of particulars have been summarized.
Philippines	Special legislation has been passed for copyright protection.
South Africa	Copyright laws may apply to database protection.
South Korea	A Copyright law is to take effect in January 1987.
Spain	There are no specific provisons for databases.
Sweden	The Copyright Act, Section 49, protects databases for 10 years from the year the database was published.
United Kingdom	The Copyright (Computer Software) Amendment Act of 1985 includes enhancements to the 1956 Copyright Act covering databases subject to United Nations Educational, Scientific, and Cultural Organization (UNESCO) recommendations.
United States	Databases are covered by copyright laws.

Source: Transnational data and communications reports, Washington, D.C.

The Council of Europe (CofE) is a Western European intergovernmental organization consisting of 21 countries and headquartered in Strasbourg, France. CofE is for the protection of individuals with regard to automatic processing of personal data. Member nations must implement the principles for data protection in the convention that was adopted in 1980, "Convention for the Protection of Individuals with Regard to Automatic Processing of Personal Data." It does not cover manual files, which means that a person could fill a briefcase with data files and not be covered under the convention. The CofE states that international information flow may be restricted if data goes to a nonsignatory country and that country allows data to flow to a nonsignatory country. All members signed the convention in 1980, and Sweden, Austria, Denmark, France, Iceland, and Norway have ratified the convention.

There are several other committees and organizations that are concerned with telecommunications and international information flow. Some of the key organizations are the following:

- *International Telephone and Telegraph Consultative Committee* (CCITT). The committee is a permanent member of the International Telecommunications Union (ITU), located in Geneva, and is concerned with technical, operating, and tariff questions.
- *Conference of European Postal and Telecommunications Administrations* (CEPT). This organization of European PTTs studies and coordinates technical and policy issues including standards and tariffs. CEPT discusses volume charging on national and international leased lines.
- *Conference of Interamerican Telecommunications* (CITEL). An Interamerican conference, part of the Organization of American States, this conference serves as a forum on telecommunications and prepares regional views and positions for meetings between the states.
- *European Community* (EC). The EC is located in Brussels and promotes economic unification of 12 European countries. The EC has been involved in recommendations for harmonizing telecommunications tariffs and creating a communitywide market for telecommunications products. The 12 members are Belgium, Denmark, Germany (FRG), France, Greece, Ireland, Italy, Luxembourg, the Netherlands, Portugal, Spain, and the United Kingdom
- *General Agreement on Tariffs and Trade* (GATT). This international trade organization provides international tariff reduction negotiations based on the most favored nation principle. Its members are from developed and developing countries.
- *Intergovernmental Bureau for Informatics* (IBI). Headquartered in Rome and recognized by the UN, most of this bureau's 35 members are developing countries. Its charter is to assist countries to understand the impact of information flow and to derive some benefit from the increased technology. Its membership is from Central and South America, Europe, Africa, and the Middle East.

- *International Telecommunications Users Group* (INTUG). This organization is a voice for business users of telecommunications in 13 developed countries. It works with the EC, OECD, and CCITT committees to use its expertise from national users groups and companies.
- *United Nations Commission on International Trade Law* (UNCITRAL). A UN committee to harmonize international trade, UNCITRAL studies technology and the legal issues of computer records, the authentication of electronic transactions, and liabilities for errors or delays in data transmission.
- *United Nations Center on Transnational Corporations* (UNCTC). This organization provides staff work for the UN Commission on Transnational Corporations and aims to strengthen the position of developing countries with respect to international corporations and international information flow.

5

NETWORK ATTACHMENT PRODUCTS

In this chapter we will

- Discuss standards committees and their relevance to worldwide telecommunications.
- Present protocol converters and device emulation.
- Discuss analog and digital network attachments and different types of multiplexers.
- Study the impact of open systems interconnection (OSI) on future products.

Network attachment products are required whenever a voice, data, text, video, or image device is attached to a Telco, PTT, or specialized common carrier. Network attachment products are modems, acoustical couplers, multiplexers, concentrators, and data service units. Network attachment products are also referred to as data circuit terminating equipment (DCE). Depending on the device, attachment may be a stand-alone device or an integrated unit "under the covers" contained within the device it supports. To have compatibility between devices that must communicate throughout the world, a number of national, regional, and international standards committees representing public and private interests interact with each other to recommend and approve national and worldwide standards. Figure 5.1 has a layout of the types of network attachment products.

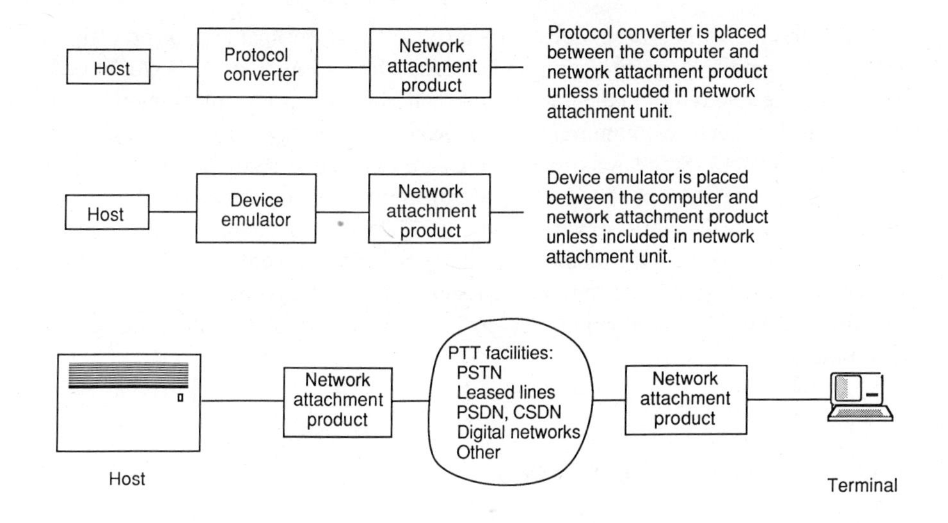

Network Attachment Products

[M] Modem is a term used throughout industry for attaching to analog networks (PSTN, leased line). Under strict CCITT definition, it would be called a DCE.

[DCE] Data circuit terminating equipment for digital networks. CCITT definition of DCE applies to both analog and digital equipment. In this context, it applies to digital networks. Called DSU (data service unit) in the United States.

[Concentrator or multiplexer]—[M] Concentrates or multiplexes multiple inputs into one analog line to PTT facilities.

[Concentrator or multiplexer]—[DCE] Concentrates or multiplexes multiple inputs into one digital line to PTT facilities.

Notes:

1. The network attachment product (modem or DCE) may be integrated within the protocol converter or device emulator or, as shown, may be a stand-alone device.
2. The modem or DCE may be integrated within the concentrator or multiplexer or, as shown, may be a stand-alone device.
3. Compatible network attachment units are required at both ends.

Figure 5.1. Network attachment products.

Rules and regulations are required to make electrical, mechanical, or procedural formats compatible to ensure communications. The various standards committees coordinate their activities in the form of international standards or recommendations that define architectures, communications protocols, and equipment interfaces for compatible devices. Standards organizations are founded in recognition of the need and benefits that can be obtained by standardization. The committees study various proposals, review the merits of various suppliers, arbitrate the relative merits of each proposal, and work with the members to agree on a final standard. In addition, they interact with other standards committees to harmonize their proposed standards on a worldwide basis. The following section is an overview of the more influential committees.

5.1. STANDARDS COMMITTEES

5.1.1. CCITT

The International Telecommunications Union (ITU) is based in Geneva and represents 161 nations in promoting the development of international standards to facilitate technical transfer of information throughout its member nations. The ITU consists of two main committees, the International Telegraph and Telephone Consultative Committee (CCITT) and the International Radio Consultative Committee (CCIR). The latter is responsible for setting radio, television, satellite, and other frequencies that are transmitted through the air.

The CCITT is made up of several categories of members. The top two levels are the only voting levels in the CCITT and the levels that make the final decision. Members of these levels are the governmental administrations (e.g., the PTTs and in the United States, the State Department) and recognized private operating agencies (RPOAs), e.g., AT&T, British Telecom, and NTT of Japan. Other categories include the scientific and industrial organizations (SIOs) (e.g., Bull, IBM, and Siemens) and international standards organizations (e.g., ISO). Members of these last two categories contribute or propose standards that may, after some period of debate, compromise, and modification, become formal CCITT recommendations, if the voting levels approve them. Compliance is controlled by the PTTs through tariffs or a homologation process.

For worldwide compatibility, network attachment products should comply with the CCITT recommendations.

Following is an outline of the ITU and CCITT relationships.

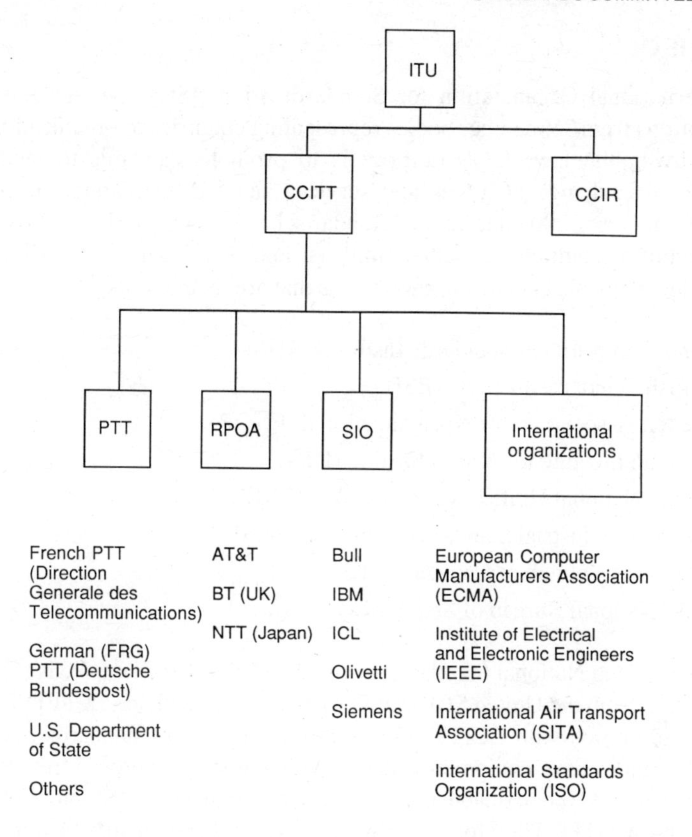

The CCITT is responsible for the worldwide standardization of international communication facilities:

- Transmission channels
- Limits for transmission performance on the transmission path, including data circuit terminating equipment (DCE)
- Interfaces between DCE and data terminal equipment (DTE) (in collaboration with ISO)
- Signaling on switched networks
- Error detection and correction procedures (in collaboration with ISO)
- Alphabets and code (in collaboration with ISO)
- Signal quality (in collaboration with ISO)

5.1.2. ISO

The International Organization for Standardization (ISO) consists of standards organizations from 75 member bodies representing countries or organizations incorporated by public law. ISO's purpose is to promote standards to facilitate the international exchange of goods and services and the international exchange of information. Over 5000 international standards have been published from over 2000 active technical committees, subcommittees, and working groups. Following are some of the national standards organizations that are member bodies:

- American National Standards Institute (ANSI)
- British Standards Institute (BSI)
- French Association for Normalization (AFNOR)
- German Institute for Normalization (DIN)
- Italian National Unification Group (UNIPREA)
- Japanese Industrial Standards Committee (JISC)
- Standards Council of Canada (SCC)
- U.S. National Bureau of Standards

The American National Standards Institute (ANSI), for example, is the member organization from the United States. ISO works closely with the CCITT. In addition to the collaboration with CCITT, ISO writes and/or adopts standards covering the physical attachment of equipment and compatible operating procedures. For example, ISO standards specify cable connector pin assignments on the interface between a DCE and a DTE. Therefore, compliance with ISO standards is important for equipment compatibility throughout the world.

5.1.3. IEC

The International Electrotechnical Commission (IEC), based in Geneva, Switzerland, is a voluntary standards organization with 44 member nations. Each member is a national committee representing its country's interest in electrical and electronic standards. The United States is represented by ANSI through it's U.S. National Committee (USNC) consisting of 22 different organizations including trade associations, government organizations, and test laboratories.

Over 1700 international standards and documents have been published by the IEC. The IEC has a coordinating group that works with other international bodies, mainly CCITT and ISO, on information technology.

5.1.4. ECMA

The European Computer Manufacturers Association (ECMA) acts as a coordinator for the cooperative development by computer manufacturers in Europe of standards applicable to the functional design and use of data processing equipment. The group consists of most of the information-associated manufacturers from all European countries. ECMA standards may be introduced to ISO directly or through the national standards organization of any European country. In some cases, the national standards groups of other countries may also introduce their own standards to ISO, but in general, these national standards are aligned with international standards.

5.1.5. ANSI

The American National Standards Institute (ANSI) is a coordinator for the development of standards and provides a mechanism for the development and U.S. approval of national standards. ANSI acts as the U.S. representative to the ISO committee.

ANSI covers many areas of standardization including mining, information systems, construction, electricity, electronics, and highway traffic safety. ANSI developed the American National Standard Code for Information Interchange (ASCII). U.S. organizations normally present their proposals to international organizations such as ISO through ANSI.

5.1.6. EIA

The Electronic Industries Association (EIA) consists of companies manufacturing electronic products in the United States and representatives of the U.S. government. It provides a vehicle for documenting ideas and agreement among the members and publishes them in the form of recommended standards. Although their recommended standards may or may not be adopted by ANSI as official national standards, they often become the industry-recognized standard through common practice. One example is the EIA RS-232-C standard, which defines an interface between DCE and DTE. Since there is no ANSI standard covering this interface, the DCE and DTE manufacturers (who originally wrote the specifications) chose to follow it in their equipment design. The term EIA RS-232-C is widely accepted and quoted in U.S. modem specifications.

EIA recommended standards and CCITT recommendations are often very similar or the same, for two reasons: First, EIA recommended standards are often processed through channels as CCITT contributions, which leads to a CCITT recommendation. Second, CCITT recommendations may be adopted by the EIA to meet specific industry requirements in the United States.

5.1.7. IEEE

The Institute of Electrical and Electronic Engineers (IEEE) is a professional group of engineers based in New York City. Members are mainly from the United States, but the membership includes participants from other countries. The IEEE sets standards in many fields of the electronics and electrical industries.

An example of their standards work is in the area of local area networks or LANs. The pioneering work was performed by IEEE and ECMA committees and other national bodies. Some of these standards are now becoming international standards

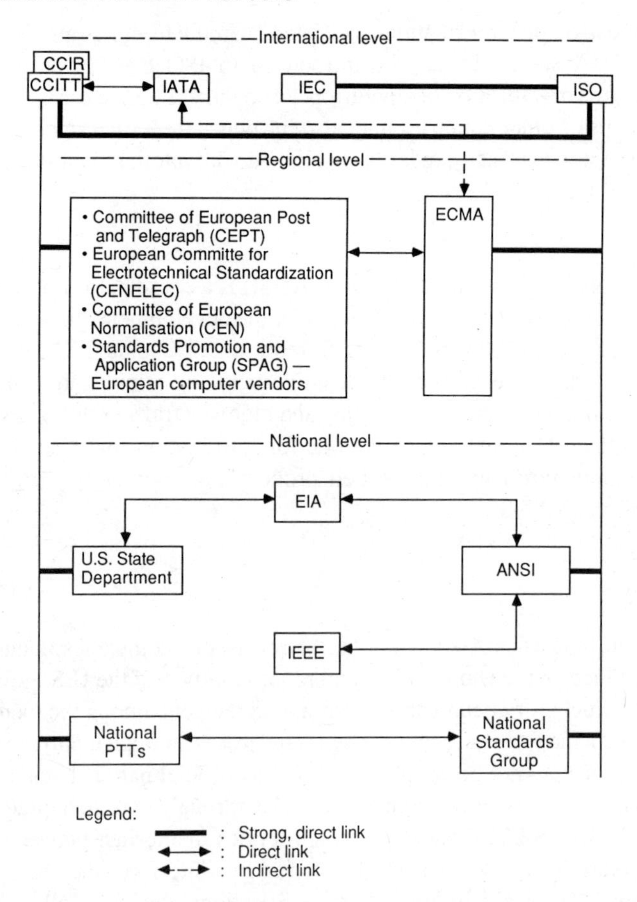

Figure 5.2. International, regional, and national standards relationships.

through the International Organization for Standardization (ISO). IEEE standards are usually considered to be U.S. standards and may be introduced to ISO through ANSI.

Figure 5.2 shows the relationship among the various worldwide standards groups. For overall worldwide acceptance of a product, CCITT recommendations are mandatory.

5.2. HOMOLOGATION PROCESS

Once the national PTT administrations accept the recommendations of the CCITT regional committees or their national standards, the device type to be attached to their networks must go through a homologation process. Homologation is a procedure for ensuring that the attaching device will not cause harm to the attached network, subscribers using the network, or personnel operating the network.

Harm to the network may include electrical hazards to PTT personnel, damage to PTT equipment, malfunction of PTT billing equipment, or degradation of service to persons other than the user of the subject terminal equipment, their calling party, or their called party.

The procedures vary by country and may be a simple written letter by the supplier or may require extensive documentation, written in the national language. In addition, many countries now require one unit for testing purposes or safety evaluation or both. After a successful test, a certificate of acceptance is given to the supplier, which may be required on each unit that allows attachment of the device to the tested network. A certificate is a formal statement made on the successful completion of a verification process. It provides official approval and grants permission for connection to a PTT or Telco network. Approval by the PTT may be an agreement to attach to a line or a more general type of approval for a specific network, e.g., a PSTN or leased line.

Recently, there has been some consideration of self-certification by the supplier. There is also a move to set up a separate agency, either governmental or independent, to make the certification independent of the PTT. In most countries, the supplier of the equipment is allowed to install and maintain the equipment. There are exceptions that require a PTT representative at the site when the unit is installed or that require PTT maintenance of the unit.

5.3. PROTOCOL CONVERSION

Protocol conversion may be required for terminals of different makes and variety so that they will work with existing host systems. These conversion programs protect an investment in equipment. Instead of purchasing new equipment, existing products may be attached to different protocols supported by the protocol converter. Attachment depends on the protocol or line procedure that will be used between the sending and receiving ends of the line. Programs may be required at both ends to ensure communications. A protocol is the line procedure between two systems and includes the procedural, electrical, functional, and mechanical interfaces that allow communications.

Protocol converters are usually microprocessor-based machines that interface a terminal's simple protocol to the host's more complex protocol. The converter communicates in two languages: the language of the terminal whose data it transforms and reformats before sending it to the host and the host's protocol, which performs similar functions when data is returned from the host. The terminal can be a plotter, display, PC, microcomputer, minicomputer, or another host.

A protocol converter actually changes one protocol to another by removing all bits except the raw data and repackaging it according to the new protocol. During the conversion sequence, the protocol converter accepts blocks of data in one protocol, adds or deletes the necessary control characters, reformats the block, and calculates the required check characters. The receiving device receives characters formatted according to its requirements. One example is an ASCII/synchronous data link control (SDLC) conversion where a string of characters is accepted, the start and stop bits are eliminated, the characters are assembled into a block, and the appropriate headers and trailers are added to create complete frames.

Another type of protocol converter is a gateway. Gateways provide an interface from one network to another network, for example, between a private leased line network to an X.25 packet switched data network or between national X.25 networks or from one architecture to another.

Gateway devices can be included in PTT networks or can be products that provide access between incompatible networks, for example, between IBM's System Network Architecture (SNA) and Digital Equipment Corporation's DECnet, or between SNA and Ethernet, or between a data communications device and X.25 public data network. Gateway products provide compatibility between network architectures, which have different protocols, codes, and interfaces.

The largest subset of gateway products are packet assembler/disassemblers (PADs). These devices permit host computers and peripheral equipment that use a communications protocol other than X.25 to be interconnected via a public packet switched network. On the terminal side, most PADs support the connection of several devices, which can be terminals, CPU ports, printers, or PCs. On the network side, a high-speed port usually provides a link to the X.25 network. PADs usually perform concentrating and multiplexing functions as well as protocol conversion. PADs are normally supplied by the packet switched data network provider. One exception is the AT&T ACCUNET® packet switched network, which is restricted by the FCC from offering a PAD function.

Most PAD products actually adapt a protocol rather than change it completely. The adaptation allows data in one protocol to pass through a network that uses another protocol. The transmitting PAD receives messages from the host or peripheral in the protocol of the sending device, converts and packetizes the information according to X.29 standards, and sends the packet through the network. At the receiving end of the X.25 link, another PAD performs error checking, disassembles the packets, and converts messages back to the native protocol.

Data link control protocols are either bit oriented or byte oriented. Byte-oriented protocols require that data be transmitted in 8-bit bytes (an octet), and an acknowledgment is required after each transmitted block before the next block may be sent. Bit-oriented protocols allow data to be transmitted in blocks of any length up to a specified maximum. Depending on the protocol, an acknowledgment takes place after one or several blocks have been transmitted.

There are four major line procedures and protocols that normally will be encountered when planning international communications. Of the four procedures and protocols used (see the following subsections), as a percentage of total line usage, asynchronous (ASCII) is increasing, mostly as a result of communicating PCs, BSC is decreasing, and SDLC and HDLC are increasing. Some of the most common protocols are the following.

5.3.1. Asynchronous Data Link Communications

Asynchronous transmission is a data link procedure. Data is transmitted and received at random based on a start-stop code. The characters or bytes are usually 5–8 bits in length plus start and stop bits.

An asynchronous procedure has very little error checking. Data may enter the communications line at any time. The end of the link detects the start bit and the transition within each transmitted character.

Most asynchronous data communications use an ASCII character set that consists of 128 characters. Of these, 95 are graphic characters and 33 are control characters. All characters are represented by 7 bits, binary ones and zeros, and were defined by ANSI. ASCII was adopted as an official U.S. government standard for data processing equipment in 1969 by the National Bureau of Standards and is now used worldwide for interactive communications. The ASCII code is often referred to as a Teletype (TTY) procedure since it was associated with the ASR-33 Teleprinter that was used extensively as the basic keyboard printer terminal for many minicomputers in the 1965–1975 period. Terminal data rates range from 10 characters (or bytes) to 1200 characters per second.

Asynchronous procedures accommodate communications links with a variety of configurations. Connections can be switched or leased line, point to point or multipoint. Data can be transmitted in either a half-duplex or full-duplex mode. In half-duplex, information flows in both directions but in one direction at a time, but in full-duplex, information may flow in both directions at the same time. See Figure 5.3.

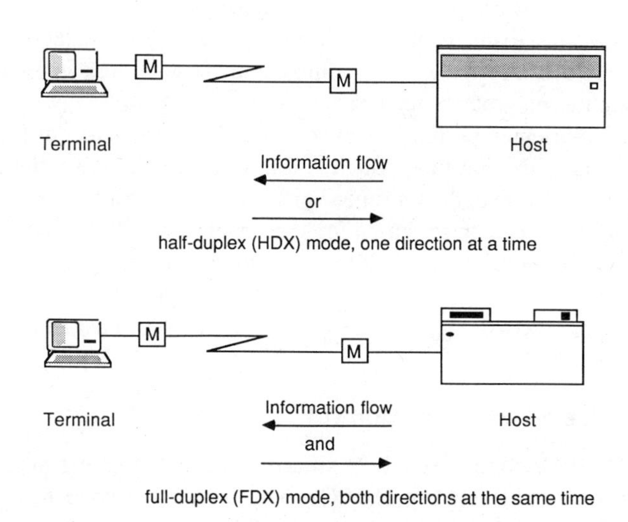

Figure 5.3. Comparison between half-duplex and full-duplex lines.

5.3.2. Binary Synchronous Communications

Binary synchronous communication (BSC) is a character-oriented protocol that uses an 8-bit byte for synchronous transmission of binary coded data. In BSC transmission, data, data link control characters, and other characters are transmitted together in a continuous series of bits. Synchronization is performed at the start of each message. BSC uses clock pulses to control the synchronization of data and control characters. A transmission in BSC consists of a number of synchronizing (SYN) characters that ensures synchronizing at both ends of the communications link. The SYN characters are followed by a start-of-text (STX) character, a block of text, an end-of-text (ETX) character, and a block error-checking character (BCC). Each block must be acknowledged before the next block may be sent.

Terminal data rates usually range from 300 characters (or bytes) to 1200 characters per second. Higher rates are possible, to 8000 characters per second.

5.3.3 Synchronous Data Link Control

Synchronous data link control (SDLC) is a synchronous method for transferring information over a data communications line. Transmission exchanges may be duplex (two-way transmission simultaneously) or half duplex (two-way transmission alternately). The line may be point-to-point switched (dial-up) or point-to-point non-switched (leased line) or multipoint. Synchronization is maintained by self-clocking modems or by internal clocking at each session. Normally, the primary station is the controlling station (host) and the secondary station is the responding station (terminal). In IBM SNA applications, the primary station controls the communications link by sending commands to the secondary station, and the secondary station responds to these commands. All information, including commands and responses, is transmitted between stations in groups called frames.

SDLC is a bit-oriented protocol that uses a synchronized series of frames. Each frame has a synchronization flag, which is followed by an address field, a control field that tells the purpose of the transmission, a data field, a frame-check field, and a trailing flag. The flag character is used for synchronization. SDLC extended allows up to 128 frames to be outstanding before an acknowledgment is received.

Terminal data rates using SDLC usually range from 300 characters (or bytes) to 1200 characters per second. Higher rates are possible, to 8000 characters per second.

5.3.4. High-Level Data Link Control

High-level data link control (HDLC) procedures are used for transferring data between a terminal and a network. HDLC is a bit-oriented protocol that is similar to the SDLC procedures used in IBM's SNA. HDLC is the main link protocol for packet switched data networks (X.25). The link protocol formats data and consists of procedures for setting up calls, transferring frames, and disconnecting calls and of procedures for the first level of recovery. HDLC offers a peer-to-peer transfer, or any equipped device may initiate the session.

5.3.5. Other Protocols

There are several other proprietary protocols available from equipment suppliers such as Burrough's BLDC and Multipoint Poll Select, CDC's 200 UT, Sperry's UDLC, and Univac's U200. Digital Equipment Corporation's DECnet networking product uses DDCMP (digital data communication message protocol), a byte-oriented protocol that can handle up to 255 unacknowledged transmissions, an advantage in satellite transmissions.

5.4. DEVICE EMULATION

A great many protocol converters provide both protocol conversion and emulation. All emulation devices provide protocol conversion, but not all protocol converters provide emulation. See Table 5.1 for the types of support packages for the four classes of computer systems.

While protocol converters handle incompatibility problems between sets of rules that particular devices use to communicate information, an emulator must handle incompatibilities between sending and receiving units, including differences in protocol, code, interface, device characteristics, and link characteristics. The difference between a protocol converter and an emulator is that the protocol converter actually strips down data and rewraps it according to a new set of rules, while the emulator reads the text in a whole message and emulates that text to the specifications of a different device.

There are different levels of emulation. As an example, the protocol converter can emulate a stand-alone display device. In this mode, it must handle all communications to and from the host. If the protocol converter emulates a control unit, then the system can send many messages to the converter, which will distribute them to the attached devices. Conversion in the controller is more efficient since each terminal does not go through the conversion process.

An emulation program allows a terminal to communicate with applications on a host supporting asynchronous procedure, BSC, or SNA/SDLC protocols with little or no change to the application programs. A terminal with an emulation program for a particular device can be connected to existing networks and appear similar to other control units' attached devices. Usually, it consists of an external device or a card that plugs into the processor and that may include software within the terminal and/or microcode (within the card) that makes the terminal look like a supported device to the larger system.

An example is the 3270 BSC emulation of a 3271 Model Control Unit. The 3271 Control Unit uses a BSC line protocol on a leased line in a multipoint tributary network. Multipoint means that the terminal attaches to a multipoint line starting from the host.

Note: Multipoint lines are continuous leased lines that are used to connect multiple locations within a country to reduce the overall line costs. Each device on the multipoint line shares the line and is in contention with other devices in other locations.

TABLE 5.1. Protocol Converters and Emulation Packages

Package	Personal Computers	Small Computers	Medium Computers	Large Computers
Asynchronous transmission	Emulate ASR-33 (TTY) Emulate DEC VT-100 display Emulate IBM 3101 display	Emulate multiple ASR-33s (TTY) Emulate DEC VT 100 Emulate multiple IBM 3101 displays	ASR-33 (TTY) host support Usually IBM 3101 support in front-end processor Support for VT 100 in DEC host computers	Usually ASR-33 (TTY) support in front-end processor (FEP) software Usually IBM 3101 support in FEP software
BSC, binary synchronous communications	Emulate 3270 BSC display and printer for single terminal or work station Emulate 2780/3780 BSC RJE work station	Emulate a 3270 BSC cluster controller, handle several terminals Emulate 2780/3780 BSC RJE work station Emulate 3741 BSC RJE work station Emulate 3770 BSC RJE work station	3270 BSC host support for single or cluster controllers 2780/3780 BSC host support 3741 BSC host support 3770 BSC host support	3270 BSC host support for single or cluster controllers 2780/3780 BSC host support 3741 BSC host support 3770 BSC host support
SDLC, synchronous data link control	Emulate 3270 SDLC display and printer for single terminal or work station	Emulate a 3270 SDLC cluster controllers, handle several terminals	3270 SDLC host support for single or cluster controllers 3770 SNA/SDLC RJE host support	3270 SDLC host support for single or cluster controllers 3770 SNA/SDLC RJE host support
HDLC, high-level synchronous data link control	Attached to a network interface via an asynchronous or BSC protocol	Packet network support from multiple asynchronous, BSC, or SDLC terminals	Host support for remote terminals and/or network node controller	Host support for remote terminals and/or network node controller

5.4.1. ASR-33/35 (TTY) Emulation

The ASR-33/35 is a terminal that looks like a Teletype Model ASR-33 or ASR-35 Teleprinter. ASR is automatic send and receive. The first minicomputer, DEC's PDP-8 in 1968, used the ASR-33 as a system console. As more minis were introduced, they continued to use the ASR-33 as system consoles, and the unit became an industry standard for asynchronous communications, both for keyboard/printer attachment and remote terminals to time sharing services.

5.4.2. VT100

Digital Equipment Corporation's VT100 is an asynchronous device that operates over a switched line and offers full-screen support. Other popular full-screen asynchronous terminals are Lear Siegler's ADM, the Televideo 900, and TI's Silent 700.

5.4.3. 3101 Emulation

The 3101 Emulator is a terminal that looks like an IBM 3101 asynchronous Display Terminal when the host interprets the incoming data stream. Specifically, it emulates the IBM Model 3101 Model 20.

5.4.4. 3270 BSC Emulation

The 3270 BSC Emulator is an external device or card that plugs into a terminal to look like a 3270 BSC Display Station when the host interprets the incoming data stream.

Caution when ordering: Note whether the host supports either the switched or leased line attachment.

5.4.5. 3270 SNA/SDLC Emulation

The 3270 SNA/SDLC Emulator is a terminal device that looks like a 3270 Display Station to a host IBM System/370 using SNA with SDLC protocol.

For example, a 3270 emulation program would make a keyboard display printer appear as a 3277 Model 2 Display Station and 3287 Model 1 or 2 Printer (SNA). An IBM 3270 Personal Computer emulates a 3270 Display Station and can also be used separately as a stand-alone PC.

5.4.6. 3741 BSC Emulation

The 3741 Emulator is a work station with support to make the terminal resemble an IBM 3741 BSC Data Station. This data station was the follow on product to the original card punches or "heads down" key entry punch cards. The IBM 3741 is used for data entry to magnetic tape or diskette in many office systems.

5.4.7. 3770 RJE Terminal

The 3770 RJE Terminal provides an emulation of either 3770 BSC or 3770 SNA/SDLC support. On receiving data, the host interprets the data stream to be the 3770 RJE terminal. As an RJE or remote job entry terminal, the terminal can enter batch processing jobs into host systems. Transmission will support the ASCII and the Extended Binary Coded Decimal Interchange Code (EBCDIC).

Note: The terminal looks like a 3770 to the host. However, the host must know a little more. Specifically, the incoming data stream looks like a Logical Unit Type 1 (LU1) to the host. Under IBM's SNA, LU1 resembles a keyboard/printer terminal.

5.4.8. 3780/2780 Batch BSC

The 3780/2780 Emulator is a terminal that appears to the host as a 3780 Batch or 2780 Batch BSC RJE Work Station.

5.4.9. Multileaving Remote Job Entry

Multileaving remote job entry (MRJE) allows a terminal to send and receive data at the same time. It requires host support of this feature [e.g., IBM Operating System/ Multiple Virtual System (OS/MVS, also called VS2) using Job Entry System 2 or 3 (JES2 or JES3).]

5.4.10. SNA Remote Job Entry

An SNA remote job entry (SRJE) device allows the device to submit data and instructions to an IBM System/370, 4300 Series, or other host supporting SDLC protocol and SNA. The output of these jobs can be returned to the device, to another work station, or to the host input and output devices.

5.5. ANALOG NETWORK ATTACHMENT

Analog networks are widely used for voice, data, and information transfer. Two of the more widely used networks are the public switched telephone network (PSTN) and the private leased line network. There are limitations to both networks, and cost trade-offs determine the applicability of each. Table 5.2 shows a comparison of a PSTN and a leased line network.

Many large users attempt to overcome the limitations of the PSTN (call setup, line quality, and high error rates) by leasing private circuits and building their own networks. The limitations of analog networks is due to the fact that they were designed for voice transmission. Leased lines are designed to operate with a wider performance range than the PSTN and have separate facilities from the PSTN.

Analog lines used in data communications are classified by voice-grade quality and operate over a frequency range of about 300–3400 hertz to handle normal

TABLE 5.2. Comparison of Analog Leased Circuits and the PSTN

Features	Public Switched Telephone Network (PSTN)	Private Leased Circuits
Call connect time	Could be as high as 20 seconds	Permanent connection; no call required to establish connection
Transmission speeds	Up to 4800 bps, full duplex (FDX) up to 9600 bps	Wide range of new FDX modems offered: 14.4, 16.8 and 19.2 kbps now available
Switched connections	Unlimited	Restricted; may not be attached to switched networks in some countries
Reliability	Reasonable for short transmission calls (minutes)	Good; very good with line conditioning
Error rates	May be quite high; not suitable for large amounts of data over international circuits	Good; improved with line conditioning on CCITT M1020 and M1040 lines
Geographical coverage	Available throughout the world	Good; could take up to 3 months for a line in some Telco locations; in some countries, the Telco or PTT may limit use, application, or availability

conversation. To transmit information at high speeds over these same facilities, DCE using advanced technology can efficiently move information over this very limited pipeline.

Analog circuits may be operated as either half duplex (HDX) or full duplex (FDX). Half-duplex operation means that information can pass in either direction but in only one direction at a time. Half-duplex connections are normally used for terminal-to-computer interactions. Most host systems support terminal applications in half duplex because persons normally react to information before taking any action. That is, the tendency is to read the information before responding. Half-duplex data links can be supported by either a two-wire or four-wire connection at line speeds of 9600 bps for BSC, SDLC, or HDLC synchronous operation. Usually, leased lines are four-wire connections. Normal asynchronous HDX communications are in the range of 300–2400 bps. Today, most PCs communicate using a 300-, 1200- or 2400-bps asynchronous modem.

Full duplex allows simultaneous information flow in both directions at the same time. Computer-to-computer applications use FDX because of the large volumes of information exchange and to avoid line turnaround time. A two-wire circuit allows FDX operation up to 9600 bps for switched line connections. Above 9600 bps, FDX modems use four-wire leased line connections to support 14,400-, and 16,800-, and

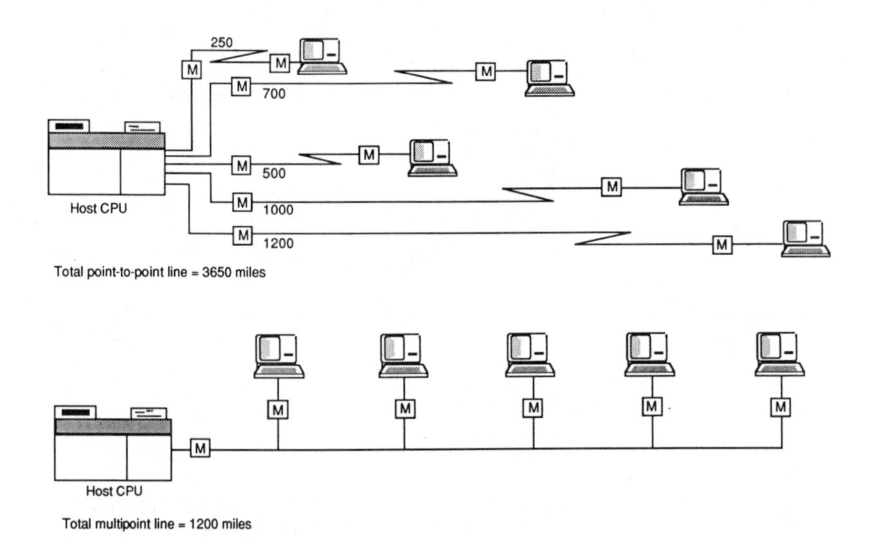

Figure 5.4. Difference between point-to-point and multipoint lines.

19,200-bps data rates in synchronous operation over the analog network. The sending and receiving circuits, each on two separate pairs of wires, are logically different sections in the modem.

A cost-effective leased line arrangement is a multipoint or multidrop line. Multipoint lines, from a Telco or PTT standpoint, are a continuous line with drops along the way. The advantage of a multipoint leased line is that the cost is determined by the total length of the line. In addition, the total length is usually less expensive than point-to-point service to all work stations. See Figure 5.4 for a layout of point-to-point and multipoint lines.

One drawback of a multipoint arrangement is that the line is shared between users rather than dedicated to a single user. A multipoint service requires host management of the line, and there are limitations on response time due to the polling and contention for use of the line.

Polling requires a master station (usually at the host) to send a request for data to each attached slave station (remote work stations) at frequent intervals (within seconds). If half-duplex lines are used, the master station or controlling modem is subject to line turnaround times due to the single directional flow of the control signals. The turnaround delay also affects the data capacity of a line. For large batch or RJE jobs or remote printing, FDX would eliminate the turnaround delay and reduce the time required to receive, return, or print a large job.

5.5.1. Modems

A modem is a MOdulation DEModulation device that accepts data from a computer or terminal device in the form of a digital signal and transforms the data into an analog signal for transmission over an analog switched telephone line or private analog leased line. Another compatible modem is required at the remote end of the communications line.

The two major types of modems are asynchronous and synchronous. Asynchronous modems are generally offered in the lower speed class with typical speeds of 110, 300, and 1200 bps. Synchronous modems have typical speeds of 2400, 4800, and 9600 bps.

Modems are offered in several categories rated by speed in bits per second. There are at least 150 models in the lower speed categories (up to 9600 bps) supplied by over 110 vendors in North America. The rest of the world may have restrictions for or limit the number of options available to a subscriber.

Low-speed modems (up to 600 bps) operate in North America (Canada and the United States) over an AT&T voice-grade line. In other countries, normal-quality lines for international telecommunications are covered by CCITT Recommendation M1040. Low-speed modems are used extensively on a switched line since they are an economical alternative to the more expensive higher-speed modems.

Medium-speed modems (1200–2400 bps) operate over analog or voice-grade lines at unlimited distances. The analog line is used in North America, and CCITT Recommendation M1020 or M1025 (four-wire only) covers international telecommunications to other countries.

High-speed modems [4.8–19.2 kilobits per second (kbps)] operate over analog voice-grade lines. Depending on the modem design, the line may require C2 line conditioning on data calls in North America. Line conditioning is used to reduce phase and amplitude distortions and random circuit noise. Line conditioning is a separate feature that is rented from the Telco (North America) and is applied at both ends of the line. In the rest of the world, line conditioning is included in the monthly cost for data-quality lines for PTT-supplied CCITT M1020 and M1025 Recommendations. An analog line with C2 conditioning in North America is comparable to an M1020 four-wire special-quality telephone line suitable for data transmission in other countries.

There are many modems on the market that do not require line conditioning at 14.4 kbps. A slight increase in the initial cost of these high-performance modems will reduce a subscriber's ongoing monthly line cost.

High-speed modems fall into two major categories, low-function modems for data transmission with two to four channels for multiplexing and high-performance modems with diagnostics and network management features. High-performance modems handle a full range of diagnostics, including local and remote analog and digital loopback tests and self testing features. Most offer at least two speeds, which can be switch selectable, although newer designs provide switching under program control for speeds of 2400, 4800, and 9600 bps. A significant number of devices now handle line speeds of 14.4, 16.8, and 19.2 kbps.

Wideband modems operate over a group or supergroup voice-multiplexed link at speeds over 56 kbps in North America and 48 kbps in countries using CCITT recommendations. Wideband modems also perform a reverse multiplexing function in that they can break down the high-speed bit stream into lower speeds. For example, a 56-kbps unit can handle six 9600-bps and one or two other low-speed devices depending on the modem's design characteristics. The high-speed serial interface is used for channel links between computers, communications controllers, or graphic applications.

Limited distance modems, short-haul modems, or line drivers operate over short distances of less than 50 miles. Transmission speeds on limited distance modems range from 2400 to 19.2 kbps, while a few units support speeds up to 64 kbps. New limited distance modems operate over two-wire lines.

Microcomputer modems can be stand-alone or integrated plug-in modem modules designed for PCs. Most microcomputer modems come with software programs for automatic dialing and automatic disconnect. The growth in PC users with on-line database applications, electronic mail, file transfer, and remote computing has led to volume-produced modems priced at a few hundred dollars for asynchronous 1200- or 2400-bps switched line modems.

Intelligent microcomputer modems are providing new applications integrated into the basic modem features. One example is Watson™, an intelligent asynchronous 300- or 1200-bps modem that combines the telephone with a PC to create a personal office automation system. This $698 modem plugs into a PC and lets the PC, combined with a touch tone or push-button telephone, become an electronic phone book and database that stores in a card file format hundreds of telephone numbers plus additional information about the caller. A user then can get the information whenever it's needed, either through the computer or over the phone. Watson™ includes the following:

- An automatic telephone dialer that dials the number on a card, including long distance access numbers
- An intelligent phone answering machine that can leave individualized messages for specific persons and that allows each caller to give an individual response
- An automatic time billing system that lets professionals track the amount of time they devote to individual clients
- An electronic calendar and date book that records appointments and that can automatically place a phone call to the user at the appropriate time and date to remind the person of an appointment

Watson is a trademark of the Natural MicroSystems Corp. in Natick, Massachusetts.

Acoustic couplers are devices that are coupled to the line through a standard telephone handset without any wiring. They have two rubber cups that hold a telephone receiver. Most couplers have manual dialing. There is no automatic retry as in a programmed microcomputer modem. However, acoustic couplers can be easily transported or integrated into terminals. A common advertising scenario is the marketing person who dials the home office from a hotel room to report the day's activities and obtain written messages.

Modem eliminators offer a replacement for costly modems when distances are short or within the same computer room. They replace two modems and are located physically about half the cable length between the two devices to be connected. Distances are usually limited to 1000 feet (305 meters).

Table 5.3 provides a list of AT&T compatibility modems that are quoted by several vendors as plug-compatible replacements. All AT&T and AT&T-compatible modems have an EIA RS-232-C interface.

TABLE 5.3. AT&T and AT&T-Compatible Modems

AT&T Device	Data Rate (bps)	Synchronization	AA/AD[a]	Mode	Line Interface
103	0–300	Async	Yes	HDX	2W
113B	0–300	Async	Yes	HDX	2W
202	0–1800	Async	Yes	HDX	2W/4W
212	0–300 1200	Async/ sync	Yes	HDX	2W
201	2400	Sync	Yes	HDX	2W/4W
208	4800	Sync	Yes	HDX	2W/4W
209	9600	Sync	Yes	HDX	2W/4W
1212	1200	Asynch/ synch	Yes	FDX	2W
2424	2400	Asynch/ synch	Yes	FDX	2W
4848	4800	Asynch/ synch	Yes	FDX	2W

[a]AA = autoanswer or automatic answering
AD = autodial or automatic dialing

Caution: Vendor modems may be compatible with AT&T modems but may have a lower probability of being compatible with each other.

AT&T modems and OEM (Original Equipment Manufacturers) modems in the U.S. are not following the CCITT Recommendations.

CCITT recommendations cover a range of 300 bps to 168 kbps and are normally compatible with another vendor's modems. However, it would be wise to have a demonstration prior to ordering equipment.

TABLE 5.4. Commonly Used CCITT Modem Standards

AT&T Device	Data Rate (bps)	Synchronization	AA/AD[a]	Mode	Line Interface
V.21	0–300	Async	Yes	FDX	2W
V.22	1.2k	Sync	Yes	FDX	2W
V.22 bis	2.4k	Sync	Yes	FDX	2W
V.23	0–1.2k	Async	Yes	HDX/FDX	2W
V.26	2.4k	Sync	N/A[b]	FDX	4W
V.26 bis	2.4k	Sync	Yes	HDX	2W
V.26 ter	2.4k	Sync	Yes	FDX	2W
V.27	4.8k	Sync	N/A	FDX	4W
V.27 bis	4.8k	Sync	N/A	FDX	4W
V.27 ter	4.8k	Sync	Yes	HDX	2W
V.29	9.6k	Sync	N/A	FDX	4W
V.32	9.6k	Sync	Yes	FDX	2W
V.33	14.4k	Sync	N/A	FDX	4W
V.35	48k	Sync	N/A	FDX	4W
V.36	48k/56k 64k/72k	Sync	N/A	FDX	4W
V.37	96k/112k 128k/144k 168k	Sync	N/A	FDX	4W

[a]AA = autoanswer or automatic answering
AD = autodial or automatic dialing
[b]N/A = not applicable

TABLE 5.5. Selection of CCITT Series V Recommendations: Data Transmission in the Telephone and Telex Networks

V.21	300-bps FDX modem for use in the public switched telephone network (PSTN)
V.22	1200-bps FDX modem for use in the PSTN and on point-to-point two-wire private leased line circuits
V.22 bis	2400-bps FDX modem for use in the PSTN and on point-to-point two-wire private leased line circuits
V.23	600/1200 baud modem standardized for general use in the public switched telephone network
V.24	List of definitions for interchange circuits between data terminal equipment (DTE) and data circuit terminating equipment (DCE)

Table 5.4 contains a list of commonly used and new emerging CCITT modems.

Table 5.5 provides definitions of the more popular CCITT V Series Recommendations.

TABLE 5.5. Selection of CCITT Series V Recommendations: Data Transmission in the Telephone and Telex Networks

V.25	Automatic calling and/or answering on the public switched telephone network including procedures for disabling of echo control devices for both manual and automatically established calls
V.25 bis	Sequence of events for establishing a connection between a serial automatic calling data station and an automatic answering data station using the 100 Series V.24 interface on the PSTN.
V.26	2400-bps modem for use on four-wire private leased line circuits
V.26 bis	2400-bps modem for use in the public switched telephone network
V.26 ter	2400-bps FDX modem based on echo canceling techniques for use in the PSTN and point-to-point two-wire private leased line circuits
V.27	4800-bps modem with manual equalizer for use on private leased line circuits
V.27 bis	4800-bps modem with automatic equalizer for use on private leased line circuits
V.27 ter	4800-bps modem for use in the public switched telephone network
V.28	Electrical characteristics for unbalanced double-current interchange circuits below 20,000 bps
V.29	9600-bps modem for use on point-to-point four-wire private leased line circuits
V.32	9600-bps FDX modem for use on the PSTN and point-to-point two-wire private leased line circuits
V.33	14,400-bps FDX modem standarized for use on four-wire private leased circuits
V.35	Data transmission at 56-kbps (North America) and 48 kbps (remainder of the world) using 60- to 108-kHz group band circuits
V.36	64-kbps modems for synchronous data transmission using 60- to 108-kHz group band circuits
V.37	Synchronous data transmission at a data signaling rate higher than 72 kbps using 60- to 108-kHz group band circuits

5.6. DIGITAL NETWORK ATTACHMENT

5.6.1. Data Service Units/Channel Service Units

Data service units (DSUs) and channel service units (CSUs) are used on digital data networks and replace the modem for digital transmission. The terminal side looking toward the network consists of a combination of status signals and data signals carrying address information plus information for selection of supplementary services. The DSU interprets these signals, provides timing, and converts or interfaces them to digital data network signals required to transmit in digital form. In the opposite receive direction, the DSU converts the data network signals to terminal signals. A DSU would be required for attachment to an AT&T Dataphone Digital Service (DDS) line. The CSU terminates the DDS line, amplifies the signal to be presented to the terminal, and provides remote loopback. DSUs and CSUs may be acquired separately or integrated into one package (DCE) in North America. In other countries, the DSU/CSU equivalent is included in the PTT service. Outside North America, there are a few exceptions where a digital terminating device can be purchased directly.

Table 5.6 shows a listing of the CCITT X Series Recommendations for digital data network connections.

5.7. MULTIPLEXERS AND CONCENTRATORS

Multiplexers are communications devices that allow a number of low-speed devices to share a higher-speed line on a private network. They are normally justified on an economical basis by the savings in lines, cabling, and labor costs. In telecommunications, one of the highest monthly costs is the line charges associated with renting public and private networks. Multiplexers or concentrators help to lower line costs by combining many relatively lightly used paths into one single, more heavily used path. The line cost reduction is due to fewer lines.

Multiplexers are applied in pairs, one at each end of the private leased line or private internal cable. They collect or concentrate low-speed lines into one single line at one end and distribute them out to the same number of devices at the other end. Multiplexers are transparent since they do not change protocols, formats, or the number of paths.

Several types of multiplexers that make use of line sharing for data will be discussed:

1. Frequency division multiplexers (FDMs)
2. Time division multiplexers (TDMs)
3. Statistical time division multiplexers (STDMs)
4. T1 multiplexers
5. Limited distance multiplexers
6. Fiber optic multiplexers
7. X.25 multiplexers

TABLE 5.6. Selection of Series X Recommendations Applicable to Public Digital Data Networks

X.3	Packet assembly/disassembly (PAD) facility in a public data network; allows interworking between an asynchronous DTE and a PSTN, a PSDN and a private leased line circuit, and a packet mode DTE and another asyncrhonous mode DTE
X.20 bis	V.24-compatible DTE for full-duplex asynchronous transmission services on public data networks
X.21	General-purpose interface between a DTE and DCE for synchronous operation on public data networks
X.21 bis	DTE used on public data network that is designed for interfacing to synchronous V Series modems; compatible with V.24
X.22	Interface between a DTE and a multiplex DCE, operating at 48,000 bps and multiplexing a number of X.21 subscriber channels using synchronous transmission.
X.24	List of definitions of interchange circuits between a DTE and DCE on public data networks
X.25	Interface between a DTE and DCE for terminals operating in the packet mode and connected to public data networks by a dedicated circuit
X.26	Electrical characteristics for unbalanced double-current interchange circuits for use with integrated circuit equipment for data communications
X.27	Electrical characteristics for balanced double-current interchange circuits equipment in the field of data communications
X.28	DTE/DCE interface for a start-stop data DTE accessing the packet assembly/disassembly facility in a public data network within the same country
X.29	Procedures for the exchange of control information and user data between a packet assembly/disassembly facility and a packet mode DTE or another PAD
X.30	Support of X.21 and X.21 bis-based DTE by an integrated services digital network (ISDN)
X.31	Support of packet mode terminal equipment by an ISDN
X.32	Interface between DTE and DCE for terminals operating in the packet mode and accessing a PSDN through a PSTN or a CSDN
X.71	Interworking between synchronous data networks using decentralized control signalling on international circuits
X.75	Call control procedures and data transfer system for international circuits between Packet Switched Data Networks
X.96	Call progress signals in public data networks
X.121	International numbering plan for public data networks

5.7.1. Frequency Division Multiplexers

Frequency division multiplexing (FDM) is one of the oldest multiplexing techniques and separates users by assigning them different frequencies within the analog bandwidth. Bandwidth is the range, from low to high frequencies, that is available for transmission. For example, a frequency division multiplexer (FDM) slices a communications channel into subchannels defined by frequencies. The terminal's bit pattern is converted into a corresponding transmission of one of two frequencies (0 and 1) within each subchannel. The FDM combines the subchannel's output onto one analog line for transmission. Data at the remote end of the analog line is subdivided into subchannels, each at a different frequency. Terminals are connected to a subchannel and receive data only when there is information on the subchannel's frequency. Subchannel selection is accomplished by tuning into preset frequencies. Operation is like a regular radio receiver where the terminal device selects one channel or frequency to receive information. FDMs can be effective for a multipoint application where there are a limited number of single drops distributed over a wide area. They are limited in speed and are not in wide use today.

5.7.2. Time Division Multiplexers

Time division multiplexers (TDMs) operate on a time slice for each user. The space is permanently assigned to a user and will travel empty if the user is not transmitting or receiving data. Inputs are received from several users and transmitted in sequence, one at a time. There is no overlapping of data as each pulse occupies its time space. Each device attached to the TDM is scanned on a fixed periodic basis; each line is scanned for a bit or character and interleaved into one high-speed line. The aggregate rate can be the rated speed of the TDM's modem. Concentrators perform functions similar to those of a TDM. In addition, concentrators may include code conversion, speed translation, and error-correcting procedures. In comparison, TDMs perform a straight bit or character passthrough. TDMs handle similar types of terminals, while concentrators are able to mix different types of remote terminals.

Table 5.7 shows a comparison of FDMs and TDMs under a set of given network situations.

5.7.3. Statistical Time Division Multiplexers

Statistical time division multiplexers, statistical multiplexers, intelligent multiplexers, or stat muxes have embedded microcomputers to control low-speed terminals. Stat muxes collect data from incoming lines and insert data and an identifier of the source line onto a common digital trunk. The received data is embedded into a high-level data link protocol (generally HDLC), and the original user data and terminal control characters remain unchanged (they are transmitted as user data). The opposite multiplexer at the remote end then extracts the data and source identification and gives the data to the equivalent line on the other side. If there is no data on the line, the space is allocated to another device. The main advantage of a stat mux over a TDM is the additional terminals that can be carried over the same line. The trunk line

TABLE 5.7. FDM and TDM Feature Comparison

Feature	FDM	TDM
Total bit speed	2400 bps or less	Can be over 2400 bps
Number of subchannels	Up to 16	Capable of more than 16 subchannels
Line operation	Analog high speed	Digital high speed
Different terminal speeds allowed	Yes; asynchronous terminals	Yes; asynchronous and synchronous
Terminal locations	Can be distributed to multiple locations on a multipoint line	Point-to point or between two locations
Multipoint operation	Yes; to several locations, depending on vendor equipment	No; would require special vendor product

is used in a full-duplex mode, while terminals work in half duplex. This means that a trunk line, with the same nominal speed as a downstream line, has roughly twice the throughput. This throughput then becomes four times that of the downstream line because it is generally true that terminals are normally active (receiving or transmitting) for only 50% of the time.

A TDM has a fixed number of preset devices and must allocate enough capacity to handle the maximum data rate. For example, a TDM could combine four 2400-bps terminals onto one 9600-bps line, but not more. This is by nature of the TDM design: It does not have enough intelligence to use the empty time slots from an additional user when one of the allocated terminals in its time sequence is not in use. A stat mux is not limited by the number of lines and can combine data traffic from active terminals. If 2400-bps lines are actually used at 20%, then many terminals can use the same 9600-bps high-speed line capacity.

The average data rate over the lines may be only 480 bps (2400 bps $\times$ 10 terminals $\times$ 20% $=$ 480 bps) at 20% utilization. However, data is not distributed evenly and usually arrives at random intervals. There are times when the number of characters waiting will exceed the capacity of the common line, a fault called *queuing delay*. The stat mux is, therefore, designed to reach a line utilization capacity of about 70% to effectively handle most of the incoming terminals and not cause queuing delays.

Most stat muxes will handle asynchronous and synchronous input devices with an input range up to 9600 bps and accept ASCII and EBCDIC codes. They are rated by the maximum combined input rate accepted from all channels, or the aggregate rate. Depending on the manufacturer, this rate varies between 38 kbps and 1.2 Mbps, which leads to an efficiency rating of 200–800%. At 800%, a stat mux would handle about four times the input of a TDM.

Figure 5.5 shows the differences between time division multiplexing and statistical multiplexing for equal amounts of data from four input devices. Note that the STDM requires less time to transmit the four blocks of data.

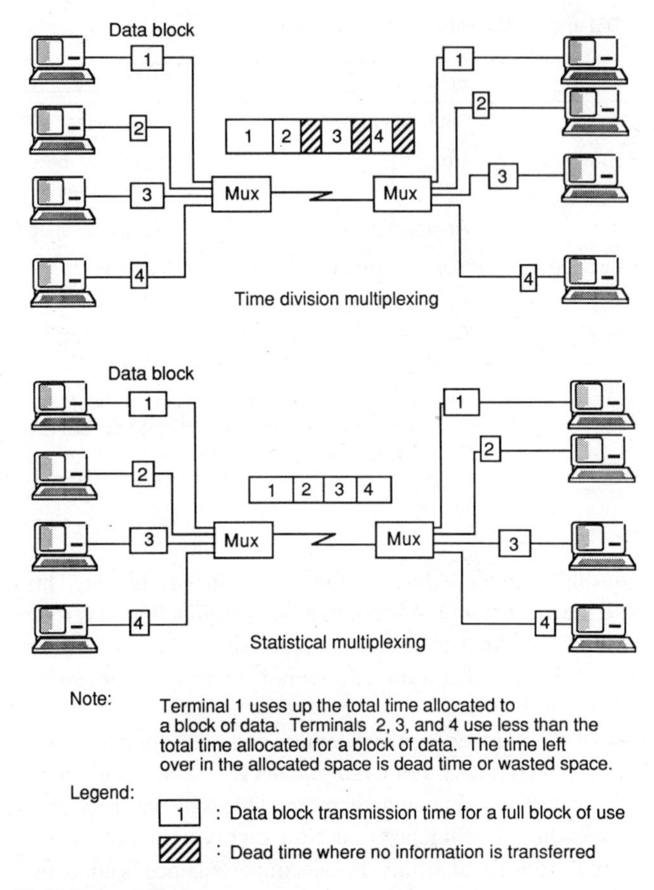

Figure 5.5. Multiplexing techniques.

5.7.4. T1 Multiplexers

Tl transmission, which has been used by AT&T and the Regional Bell Operating Companies (RBOCs) for over 20 years, is available for private line transmission. AT&T Communications (AT&T-C) offers their Accunet T1.5 Service to provide full-duplex, private line voice, data, and video transmission at 1.544 Mbps. The various Tl services provided by common carriers and satellite-based services will be discussed in later chapters.

Tl multiplexers interface terminals and voice equipment into the Tl facility provided by the common carriers. Tl multiplexers are bit-interleaved or character-interleaved TDMs handling primarily synchronous data and operate at 1.544 Mbps in North America. Internationally, the PTT backbone network transmission rate is 2.048 Mbps and corresponds to the CCITT G.703 Recommendation. The G.703 offering is similar but not compatible, and interconnection would require a gateway facility to handle differences in supervisory and signaling information and synchronization.

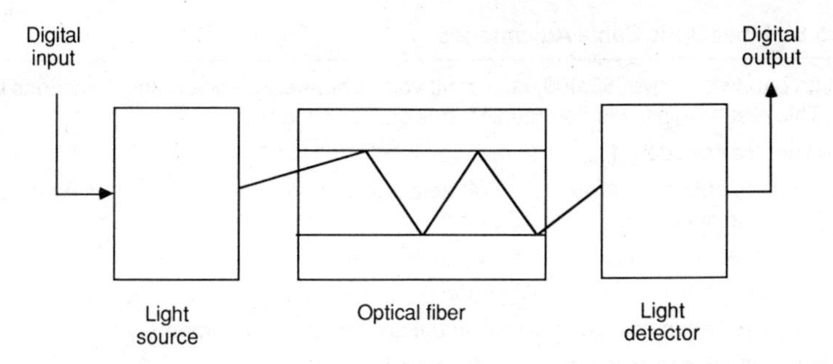

Figure 5.6. Fiber optic system.

A T1 multiplexer also handles voice by digitizing the voice signals and multiplexing them with data on the T1 line. Many T1 devices accommodate a wide range of units and a variety of media: coax, fiber optic cable, or twisted pair for point-to-point or multipoint operation. The T1 channel is subdivided into subslots of up to 64 kbps. For example, if a vendor offers a minimum rate of 4800 bps, terminals operating below this rate will take up the entire slot. If the input is 256 kbps, for example, it will require four subslots.

5.7.5. Limited Distance Multiplexers

Limited distance multiplexers or coax cable eliminators use a time division design to connect multiple local terminals to their controller. On the low-speed input side, they will accept up to 32 ports of 3270-type devices (over 2.2 million compatible devices are installed) or other devices in a transparent mode. Asynchronous or synchronous line speeds are supported up to 2 Mbps.

Limited distance muxes are very cost effective and result in large savings when compared with one cable between muxes versus several separate coax cables. For example, there are savings on coax cables, cable laying costs including labor, and potential new ductwork required for additional cable within a building.

5.7.6. Fiber Optic Multiplexers

Fiber optics is a communications technique that converts electrical energy to light energy, which is then propagated through an optical fiber, projected onto a light detector, and converted back to electrical energy. The propagation or transmission of light is through a plastic or glass fiber, and communication occurs through the transmission of light by the fiber, as shown in Figure 5.6.

For example, the digital signal from a computer interface is converted to light energy through a light source such as a light-emitting diode (LED) or injection laser. The light is emitted from the light source and propagates through the optical fiber by reflecting within the fiber as it is transmitted through the fiber. The light propagates down the core to the end of the fiber cable and is projected onto a light-detecting device, which produces an electrical current. The electric current is then converted to match the device it interfaces to.

TABLE 5.8. Fiber Optic Cable Advantages

- High bandwidth: over 50,000 equivalent voice channels on one communications link. This results in a lower communications cost per channel.
- Lower bit error rates (1 in 1 billion).
- Immune to electrical noise; does not radiate signal. No crosstalk between other cables in the same duct or cable carrying wells.
- Secure system: extremely difficult to tap into cable
- Electrically isolated; eliminates grounding and ground loops.
- Lower installation costs due to less space and weight requirements.
- High resistance to corrosion, fire, and chemicals.

Fiber optic cables have several advantages. The greatest advantage is high bandwidth or the frequency range that can be transmitted over one thin fiber. Fiber systems can transmit data at rates in excess of 1–3 gigahertz (GHz) or from 1 to 3 billion cycles per second. This bandwidth allows high-speed data transfer between processors or the multiplexing of several low-speed channels. The cost of fiber cable has dropped to a point where it is now feasible to install links between computers, buildings, or cities and between continents (underseas cable).

Another advantage of fiber optic cables is that they are immune to electrical noise, radio interference (electric motors, relays, power cables, or inductive fields), and radio or radar transmission sources. In addition, fiber cables do not radiate signals and are immune to crosstalk. This reduces the bit error rate (BER) for fiber optic cables to less than 1 in 1 billion compared to a bit error rate of 1 in 1 million in wire cables. The result is less error-checking overhead and a reduction in retransmission.

Fiber optics cables offer significant security benefits because the cable cannot be tapped without destroying the link. Due to the high capacity of fiber optic cables and the advantages listed in Table 5.8, fiber optic multiplexers are growing in use. They also act as a data concentrator for conventional multiplexers, computer interfaces, and terminals. Fiber optic multiplexers can accept up to 32 ports with speeds up to 1 million characters per port and any type of code since the channel is transparent.

5.7.7. X.25 Time Division Multiplexers

An X.25 or packet switched data multiplexer provides an interface for all types of start-stop asynchronous or ASCII terminals into X.25 synchronous lines. X.25 multiplexers implement a CCITT PAD (packet assembly/disassembly) function and are used mainly in private packet switched data networks.

An X.25 multiplexer would be used primarily for providing a PAD function or protocol converter or for other private packet switched networks built on private leased lines. A packet network built on private leased lines would require an X.25 multiplexer for start-stop or ASCII devices. X.25 multiplexers are also used to increase line concentration for remote sites, communications controllers, or front-end processors (FEPs).

In Europe, the PAD function is normally provided by some PTTs as an optional service on their packet switched data networks (PSDNs). Some private packet switched data networks exist in Europe, e.g. SWIFT.

PAD functions conform to three CCITT recommendations that define a PAD:

X.3: packet assembly/disassembly (PAD) facility in a public data network; CCITT V.24/V.28 interfaces (North America EIA RS-232-C) to the terminal

X.28: DTE/DCE interface for a start-stop mode data terminal equipment accessing the PAD in a public data network situated in the same country

X.29: Procedure for the exchange of control information and end user data between a PAD and a packet mode DTE or another PAD

5.8. COMMUNICATIONS CONTROLLERS

A communication system is a set of resources and operating procedures set up to allow access to data processing and database resources for all users and to furnish a gateway for other communications systems.

A main objective in planning a communication system is to unify teleprocessing and office automation applications and offer continuous service to multiple applications. The key is to separate the DP problems from the network problems and implement networks to support a large number of terminals and off-load the host system. A communication system, independent of the DP system, separates the application programming from communications. Any change in an application program should not require a change in the communications program. Also, communications programs should be independent of application programs. Network management and operations are isolated from the DP hosts. The separation of functions is normally supplied by off-loading the communications system into a communication controller.

Some of the communications requirements within an organization that are off-loaded to a communications controller are the following:

- **Off-load network control.** There are separate and dedicated networks and patterns for different communications flow. For example, papers, documents, data entry, internal and external mail, telex, and teleprocessing each have dedicated resources and separate communications procedures. Most organizations use more than four networks, e.g., analog switched line, analog leased line, digital leased line, and telex plus other internal networks.
- **Off-load terminal support and multiple protocols.** Many companies have large data networks with a high processing load on their DP centers. In addition, data communications networks are required to support thousands of terminals with different protocols supplied by multiple vendors. This leads to a greater load on DP systems and increases the difficulty of controlling and managing the networks. Resources must also be assigned to network problem determination and line recovery.

- **Offer fail-safe networks.** The more vital the communication processes, the more stringent the requirements of continuity of service. This increases, for example, the need for duplicate resources, alternate paths, as well as operating procedures and resources to prevent a failure or recover from a failure.
- **Offer flexible solutions.** There is a need to adapt solutions to the changing needs of the company or organization and to add new terminal applications such as text processing, image, Fax, and Videotex.

Communications controllers are separate units that connect remote terminals (or other hosts) to a host system. The communications controller may be locally attached to one or more host processors, or it may be attached by communication links to one or several other communications controllers. The same controller may be local to one host and remote to others. Communications controllers are often called front-end processors (FEPs).

A front-end processor can be justified when additional hardware and software development and maintenance costs are less than the cost of performing these functions on a mainframe.

Table 5.9 lists some of the more important functions provided by FEP hardware and software.

TABLE 5.9. Front End Processor Hardware/Software Resources

Functions	Hardware and Software Resources
Network Control	Command language and interpreter
	Terminal support routines
	Multitasking
	Networking programs
Network Services	Message switching, data collection
	Network control (start-up, close-down, sessions activation)
	Routing algorithms
	I/O access methods and utilities
	Performance analysis
	Accounting features (connect time, beginning and end times)
Line Handling	Line attachment
	Scanning multiple lines
	Maintain and update tables of attached devices
	Interrupt processing
	Support line disciplines (asynchronous, synchronous, BSC, SDLC, HDLC)
	Speed and protocol conversion
	Character assembly/disassembly
	Control character recognition
	Block checking
	Transmission status, posting

Tables 5.10 to 5.12 provide statistical data with which to project potential market demand for network attachment products.

TABLE 5.10. Network Termination Points Per 1000 Working Population

Country	Number of NTPS, 1991	NTPS per 1000 working population, 1991
United Kingdom	547,000	24.36
Germany (FRG)	515,200	23.96
France	400,300	19.85
Italy	331,300	22.53
Sweden	167,600	48.33
Spain	140,500	15.41
Netherlands	120,000	26.59
Finland	78,900	34.39
Switzerland	69,300	23.25
Denmark	64,200	21.22
Belgium	59,600	19.76
Norway	58,800	29.06
Austria	72,700	22.55
Ireland	14,800	19.68
Greece	14,000	8.1
Luxembourg	3,600	30.51
Portugal	5,800	2.32
Europe (17 countries)	2,663,600	22.82

TABLE 5.11. NTP Distribution of Switched Networks in 1991

	PTT Switched Networks		
Speed Class	PSTN	CSDN (X.21)	PSDN (X.25)
Voice-grade speed classes			
300 bps	64,300	6,700	17,800
1200 bps	167,100	10,600	17,300
2400 bps	208,550	211,300	102,100
4800 bps	17,600	43,900	85,600
9600 bps	4,650	49,800	123,500
Wideband speed classes			
Over 19,200–72,000 bps	—	3,000	2,200
Over 72,000 bps	—	800	—
Specialized service types			
Acoustic couplers	168,050	—	—
Push-button signaling units	26,750	—	—
Total all types	657,000	326,100	348,500

TABLE 5.12. NTP Distribution by Leased or Nonswitched Networks in 1991

	Leased Line or Nonswitched Networks			
Speed Class	Analog Leased Circuit	Telegraph Leased Circuit	Digital Leased Circuit	Public Network for Fixed Connections
Voice-grade speed classes				
300 bps	25,100	12,500	—	2,200
1200 bps	233,000	—	—	31,600
2400 bps	133,700	—	64,800	35,600
4800 bps	158,700	—	72,900	48,200
9600 bps	237,200	—	94,800	61,700
Wideband speed classes				
Over 19,200–72,000 bps	5,500	—	113,800	200
Over 72,000 bps	—	—	400	—
Total all types	793,200	12,500	346,700	179,500

5.9. NETWORK TERMINATION POINTS

Network termination points (NTPs) are the number of PTT lines that terminate on a customer's premises. For example, if there are 25 voice-grade telephone lines coming into a private branch exchange (PBX), two data lines and one telex line, the total is considered to be 28 NTPs. An indication of NTPs and the relative speeds of computer line outputs and terminal devices are projected by the Eurodata Foundation, a group of 18 European PTT administrations.

Tables 5.10 to 5.12 provide an excellent base to work with in predicting line speeds for remote applications since they cover 17 European countries: Austria, Belgium, Denmark, Finland, France, Germany (FRG), Greece, Ireland, Italy, Luxembourg, the Netherlands, Norway, Portugal, Spain, Sweden, Switzerland, and the United Kingdom. The material in Tables 5.10 to 5.12 was obtained from *Data Communications in Europe, 1983–91,* published by the Eurodata Foundation. Other publications are available from the Eurodata Foundation: 54 Fetter Lane, London EC4A 1AA, United Kingdom.

By 1991, France, Germany (FRG), Italy, and the United Kingdom will account for 67% of the NTPs in Western Europe. In terms of data communications development (measured by the number of NTPs per 1000 working population) the differences are less dramatic. Sweden continues to be the most advanced user of intelligent terminal devices with an increase from 20 NTPs in 1983 to 48 NTPs per 1000 working population in 1991. Western Europe will average 22.82 NTPs per 1000 working population in 1991.

5.10. OPEN SYSTEMS INTERCONNECTION

In many enterprises today, ther is a need for transferring information to multiple establishments within the enterprise and between enterprises. Usually, there are several different types of operating equipment and systems within an enterprise that cover a wide range of designs and implementation. If these different user systems are to communicate externally with other systems within the enterprise or to another enterprise, then they must become *open systems*.

OSI is concerned with the interconnection and cooperation of unlike systems for exchanging information between users. A user system can be one or more computers, associated software, peripherals, terminals, human operators, physical processes, information transfer means, etc., that form a separate entity capable of processing information and/or transferring information. For user systems with incompatible architectures to interconnect, each system must offer an interface that complies with a reference model that serves as the framework for the interaction of services, procedures, and protocols. Conformance with the standards defined within the framework of the model is required to enable an effective exchange of information between users.

A system that conforms to the reference model is referred to as an *open system,* and the interconnection is called the *open system interconnection* (OSI). The fact that a system is open does not imply any particular systems architecture, implementation, or technology, but refers to mutual recognition and support of the applicable standards. Therefore, the basic reference model defines interface standards and serves as a reference for positioning existing and emerging standards, and facilitates compatible interconnections for communicating applications processors. The reference model is defined in the ISO Standard IS 7498 and the identical CCITT Recommendation X.200.

There are several ongoing activities that have been identified for cooperation between open systems. They include interprocessor communications, data representation, data storage, process and resource management, information integrity, and security and program support for programs executed by OSI application processes.

Although the scope of the general architectural principles required for OSI is very broad, the standard is primarily concerned with systems comprising terminals, computers, and associated devices and the means for transferring information between these systems. The standards have been classified into a seven-layer model.

The advantage of layering is that it facilitates changes within a layer without affecting the entire model. Each layer is a separate entity and, starting with layer 1, a step-by-step enhancement of communications services. The boundary between two layers defines a stage in the enhancement process. General principles that were applied to the decision process in defining the layers were the following: Maintain at minimum difficulty—the system engineering task of describing and integrating the layers; create a boundary where the description of services can be small and

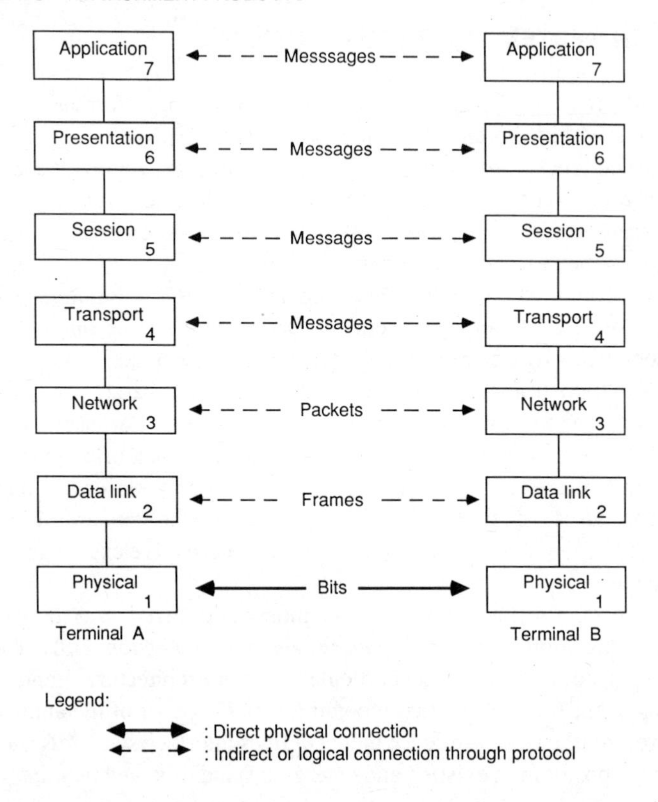

Figure 5.7. ISO OSI reference model.

interactions across the boundary are minimized; put similar functions into the same layer; and localize functions in a layer so it can be redesigned with new technology or architectural changes.

The seven layers, including the physical media provided by a public common carrier, private network supplier, or private in-house network, are shown in Figure 5.7. All data between two OSI peer-to-peer entities flows through the physical media connected to the physical layer at each end. A discussion of the seven layers follows.

5.10.1. Application Layer (7)

The applications layer directly serves the end user with application program support. This layer is composed of a user element (application process) that uses common application service elements to complete the telecommunications objectives. In addition, specific application service elements have been or are being defined to meet the needs of a particular applications, for example, virtual terminal protocol, job transfer (job transfer and management or JTM), file transfer (file transfer, access, and management or FTAM), and message handling (X.400 Message Handling System).

The application layer contains the actual program and supports the applications session between two systems, which includes the protocol exchanges for file transfer, virtual terminal, text communications, message handling, Videotex, and Teletex. International standards and industry standards (e.g., banking) are also covered.

5.10.2. Presentation Layer (6)

The presentation layer defines the method by which data is assembled and ensures that the information is delivered to the application process in a form that can be used and understood. Layer 6 includes facilities for the transfer of structured data. Other functions include requests to establish a session, requests to terminate a session, formatting, data transformation and transformation of syntax (a set of rules for the representation of user information), and special-purpose transformations.

Layer 6 defines functions for preparing data for applications use by performing data formatting and code conversion.

5.10.3. Session Layer (5)

A relationship or session must be established between communication applications in order for information to be exchanged. Layer 5 manages and controls the dialog, exchanges information, and terminates the session when all information has been accepted at the destination system.

The session service is divided into functional units for establishing a session, exchanging data in a synchronized manner, and releasing the connection when completed. The service may also establish synchronization points within the dialog that will allow an agreed point of reference to continue from in case of errors. Data exchange can be one way, two way alternate, or two way simultaneous.

5.10.4. Transport Layer (4)

The transport layer provides a consistent transport service for end-to-end control and information exchange in association with the lower three OSI layers. Layer 4 establishes transport connections for exchanging data between session entities, requesting and negotiating an agreed quality of service for the connection, transferring transport data units, and controlling the data rate. It provides the interface between the end users and the network.

5.10.5. Network Layer (3)

The network layer provides for the transparent transfer of data between transport entities. Layer 3 functions provide the means for establishing, maintaining, and releasing the network connections. Other services provided are expedited data transfer, error notification, flow control, sequencing data, and reset. It is also responsible for the switching and routing functions through switched telecommunications media, including the case where several subnetworks are used in tandem or in parallel.

5.10.6. Data Link Layer (2)

The data link layer is responsible for the transmission and error-free delivery of information over a physical link. The information content is transparent and not relevant since the layer's only concern is to deliver messages without errors. Functions within the data link layer are sequence control, error detection and recovery, control of data circuit interconnection, and data link layer management.

5.10.7. Physical Layer (1)

The physical layer is responsible for the transparent transmission of bit streams. It provides the electrical, mechanical, functional, and procedural means for establishing, maintaining, and terminating a connection for bit transmission between data link entities. This layer supports physical connectors conforming to specifications EIA RS-232-C (V.24/V.28) or CCITT X.21 or X.21 bis and other connections. All information flows between the physical layers through the transmission media in a point-to-point or multipoint configuration. Information can be one way, two way alternate, or two way simultaneous.

6

OFFICE SYSTEMS

In this chapter we will

- Discuss word processing, electronic mail, and management information systems
- Present 12 integrated office system products for office processing, electronic mail, and communications
- Discuss LANs, an attempt to tie incompatible devices together for the office
- Discuss the PABX and future voice/data integration

The evolution of office systems can be traced to DP and minicomputer equipment suppliers and their desire to create new markets for their equipment. Office systems are the result of enhancements to computer architectures and software application products since the early 1970s. The convergence of DP (with significant inputs from the scientific and engineering market), office systems, and communications is an evolutionary integration of additional functions into the basic product, offered at lower prices. Let's look at the origins.

6.1. OFFICE SYSTEMS AND COMPUTER ARCHITECTURE

Office systems are a result of developments in computer architecture and applications software development. There is no magic or mystery in the process. The basic goals are simple: process increasing amounts of information faster, move more information to its final destination quicker, and eliminate costly personnel steps in the process.

Initially, minicomputers consisted of limited basic instructions, a small-sized core memory, an external interface structure, limited peripherals and minimum software (editor, debugging routines, assembler), and a few application programs. Later, additional extensions in small computer architecture were added to meet competitive demand and market requirements. The initial components included the basic architecture, peripherals, and minimum applications. During this early period, office systems and office communicating products were noncomputerized; i.e., the products were fixed logic, wired for a specific function. The early components for small computer systems, office products, and office communications are shown in Table 6.1.

Mini suppliers expanded their product lines both up and down their price performance range resulting in small and medium systems, with additional enhancements to their basic architecture. Additional components and features were added. Basic processors or stripped down minis with a CPU and memory were offered at lower and lower prices and began replacing hardwired logic in office systems and office communications products. A major advantage of a stripped down or general-purpose computer instead of hardwired logic was that software allowed changes to programs or procedures without costly rewiring. Product development time and costs were reduced.

Enhancements were added to the basic architecture, including commercial instruction sets and decimal arithmetic. Peripherals and software enhancements were added to meet customer needs for letter-quality printing (matching an IBM Selectric typewriter) and high-speed printing, low-cost storage devices (cassettes, magnetic tape, diskettes), and new applications software (word processing, electronic mail). Additions to the key area of computer architecture, peripherals, and applications programs are shown in Table 6.2. A communications package includes a communications adapter to a modem or an integrated modem and a supporting communications software routine. Office systems communications consisted of asynchronous ASCII and BSC batch transfers.

Small and medium computers, basic CPU and memory components, and additional software tools all contributed to the upgrading of office systems. In all cases, information moved faster to one or more final destinations. Word processing dramatically improved productivity and can be measured by an increase in work output or a reduction in time to produce letters and correspondence. Table 6.3 lists the original equipment and the resulting products with applied technology.

The key to integration into the office structure is ease of use; e.g., functions are operated by pressing one or more keys or simple commands from a terminal keyboard.

The individual technologies applied to traditional work functions are useful, and greater utility is obtained because of their integration. Future systems will integrate voice, text, data, and image. Integration will produce capabilities such as appending voice messages to text messages, dictation that permits on-line editing for profes-

TABLE 6.1. Minicomputer Architecture and Office Systems

Small Computer Systems		
Minicomputer Architecture	Peripherals	Applications Programs
Basic instruction set	Card readers	Time sharing
Central processing unit	Terminals (ASCII, parallel, other)	Real-time applications
Memory	Disk storage	Event-driven programs
Hardware floating point processors (hexadecimal and octal output formats)	Magnetic tape storage	
Block (table) moves, search, compare	Printers to support engineering and scientific output	
Stack handling (push, pop)		
Internal bus structure		
Common bus arrangement		
Hierarchical (multilevel priority interrupt)		
Star design (central control point)		
Multitasking (shared logic)		

Office Systems	Office Communications
Typewriters	PBX or PABX (voice only)
Copiers	Telex
Dictation equipment	Facsimile
Keypunch	

TABLE 6.2. Small and Medium Computer Enhancements

Architecture	Peripherals	Applications Programs
Commercial instruction set Hardware decimal instruction set Cache memory Error-correcting memory	Keyboard displays with programmable function keys Quality line printer (to match the Selectric) High-speed laser printers Diskettes (floppy disks) High-volume disk drives	Office programs (accounting, inventory mangement, finance, word processing) Database design Word processing Management information systems

sionals, graphic output for electronic work stations, video conferencing for face-to-face meetings, and voice recognition. Access, independent of location, is proving to be an important feature in new systems.

6.2. WORD PROCESSING

Word processing is a system of people, procedures, and equipment for entering, revising, storing, and printing simple text documents, e.g., memoranda and reports. Word processing uses automated typewriters and storage devices for updating, merging, and retrieving files to be worked on. Word processors include typewriters or keyboard CRTs that work directly with temporary storage and have no command language or PCs and small computers with applications software that use a simple command language, disks, and diskettes. Minor corrections to existing documents can be done on line in less time than retyping a page. Word processing has gone beyond secretarial and WP operators and now includes principals and knowledge workers.

New approaches to word processing emphasize the value of text editors in enabling principals to construct ideas and translate them into text. Text databases are being formed in many organizations where previously entered text can be recalled, modified, merged with new material, or used as "boilerplate." In many offices of the future, administrators, principals, and managers will have a terminal on their desk to generate or correct text and send messages. Word processing has evolved into a machine that can handle both word and data processing functions. It has capabilities that not only handle clerical functions but also act as a tool for managers and principals for the distribution of ideas.

6.3. ELECTRONIC MAIL

Electronic mail is an extension of word processing where information is distributed over telephone lines or cables or by private networks. There are five major categories

TABLE 6.3. Office Systems Upgraded

Original Equipment	Applied Technology	Advantages
Typewriters	Typewriters with memory	Edit a line or a page
	Word processor (memory, peripherals applications software) Single station Shared logic systems (several users) Computer based (text editing programs)	Document stored for easy access Merge headings with standard letters; merge paragraphs or files Edit character, line, block; move data
Mail, telex	Communicating word processor (memory, applications software, communications package)	Files or text transmitted over telephone lines; reviewed, updated and printed at destination
	Teletex	Memory to memory transfer, send and receive simultaneously
Copiers	Communicating copier (memory, communications package)	Remote printing, document transfer Hard-copy distributed output High-speed (2400 bps), guaranteed error checked delivery, memory to memory
Dictating equipment	Voice store and forward (memory, communications package)	Call in to retrieve messages Send voice message to distribution list
Keypunch	Key to magnetic tape (memory, communications package)	Store information or magnetic tape Transmit in batch mode to host for processing
	Key to diskette (memory, communications package)	Store information on diskette Transmit in batch mode to host for processing
Facsimile	Facsimile to computer (intelligence in computer)	Send duplicate of originals to distribution list

of electronic mail: common carrier-based systems and public postal services, telex, facsimile, communicating word processors, and computer based message systems. Each of these categories is an important aspect of electronic mail. Telex, postal services (Bureaufax), and facsimile will be discussed in greater detail in later chapters. The major advantage for the users of electronic mail is that they can send and receive messages (voice and/or data), text, and files at their own pace.

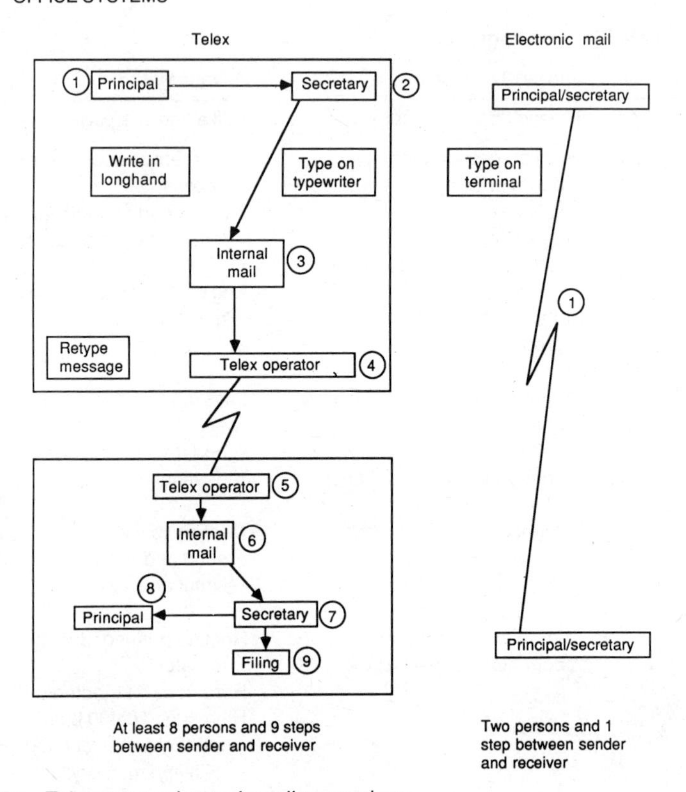

Figure 6.1. Telex versus electronic mail comparison.

Electronic mail has a number of benefits. Telephone tag is eliminated, fewer interruptions occur, and the employees can receive a message when they have a break in their work flow, enabling them to concentrate on an action item or compose a response. The quality of messages improves over less structured verbal communications. As a result, there can be greater flexibility in where and when people work. Electronic mail also provides a permanent, searchable stored record of all communications sent and received.

Figure 6.1 shows a comparison of the steps involved in sending a telex versus an intraenterprise memo. The principal or manager writes a telex, and it's typed by a secretary, typed and processed again by a telex operator, and sent to a remote operator who then distributes the telex through internal mail to the secretary or principal. In contrast, a principal or secretary creates the intraenterprise message on his or her terminal and presses a function key, the program searches for the final destination address, and away it goes—directly to the recipient.

Where is electronic mail heading? Most consultants' reports predict up to a $1 billion market in 1990. One international example is IBM's internal messaging network, called VNET, which had over 200,000 users in mid-1986. According to IBM's THINK publication, VNET results in an annual savings of $150 million.

Depending on traffic, most messages can be received at the remote user's terminal at his or her desk within an hour of sending, and often it takes only a few minutes. In a worldwide environment with time delays between continents, VNET offers daily access to persons and up-to-date information. For example, a person working in France can send a message to the eastern United States at 9:00 a.m. French time, and the message is waiting for the recipient when he or she logs onto the system the same morning, 6 hours later (9:00 a.m. eastern U.S. time). Most messages get a quick same-day response. An early morning message to Japan from France will arrive in Japan at the end of their working day. Messages can be returned from Japan the following morning and be in France before the person starts his or her working day. Another advantage is that the sender has a copy of the message sent and can compare the response. The system may also have a log of all outgoing and incoming messages.

6.3.1. Voice Mail

Voice mail, a new form of electronic mail, provides the same benefits as electronic mail systems in that it allows detailed messages to be stored and then forwarded to a particular recipient or group of people. It is digital and secure, works across time zones, and does not demand any keyboard skills. Voice-messaging products are available as private systems or as a service bureau offering or as a value added network service (VANS).

Another form of voice messaging is *voice recognition* that converts analog wave forms into ASCII streams of text. A practical example of the technology is a word processor that translates a limited number of the spoken words into typed text.

Voice synthesis converts electronically stored files, e.g., from a floppy disk, into spoken words via a work station speaker or telephone. Some office system vendors are now including this facility in their product lines. Examples are DECtalk from Digital Equipment and a similar product from Data General.

Voice annotation is a new product, originally developed by Information Technology of the United Kingdom, and includes a work station with a telephone handset that can be used to record alterations or *annotations* to prepared text. The design assumes the typist will prepare a rough draft and transmit the text to a manager or principal who will then correct it by recording additions and corrections. The system stops at the correct spot, and the secretary then listens to the recording and types in the corrections.

6.4. MANAGEMENT INFORMATION SYSTEMS

The integration of applications programs and communications packages has resulted in an automated office that provides a more effective system for storing, organizing, and managing information. These systems now have the performance and supporting software to effectively provide a central management information system (MIS). Information is a resource that is consumed by the knowledge worker, manager, or

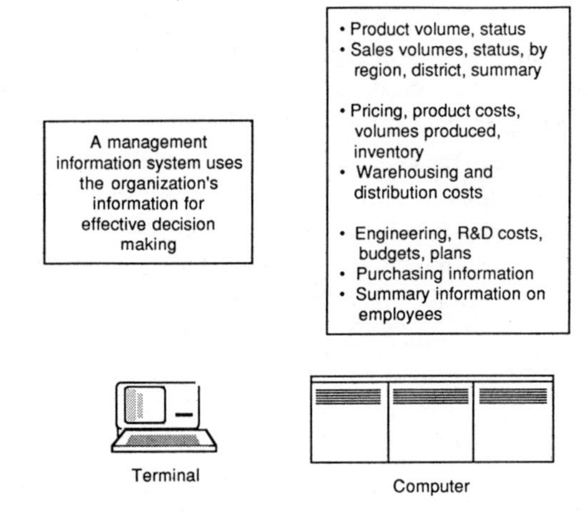

Figure 6.2. MIS for decision making.

principal and should be accessible and timely. Individuals are now able to interface more effectively with the knowledge base to access, reorganize, and use pertinent information.

Developments in ergonomics and the person/machine interface enable easier, more flexible, and friendly interaction with an organization's knowledge base. The central tool is the terminal. Once documents are created on a keyboard display, they can be stored electronically and then accessed on a terminal by simple commands using a query system of a textual database management system. It has now become possible for a person with no expertise in computers to quickly learn a system and create and access a database.

A MIS system connects a work station to a computer and may have managerial functions, such as a calendar, a tickler file, or a schedule of other managers, but the primary function of a MIS is to provide information to executives in an accurate, rapid, and readable form through a central computer system. Some of the usable information needs of management are shown in Figure 6.2. A manager should be able to obtain the latest information concerning the status of a broad range of activities such as sales, inventories, product costs, and other services. A MIS should inform managers of plans, schedules, comparisons, and operational and marketing trends.

6.5. INTEGRATED OFFICE SYSTEMS

Product integration of WP and electronic mail, DP, and management information systems are available from several computer suppliers. A comprehensive listing of integrated office systems support for 12 major suppliers will be presented.

Of the 12 suppliers, 4 originated as minicomputer suppliers (DEC, Data General, Hewlett Packard, and Wang), 4 as mainframe suppliers (Burroughs and Sperry

—now Unisys, Honeywell, and IBM), and 4 as word processing specialists (AES Data, Wordplex, WPS, and Xionics). Products were developed by architectural changes (mini suppliers and WP specialists) or as software-oriented applications products from mainframe vendors. Six suppliers are listed in Table 6.4 and 6 in Table 6.5. A description of the items in these tables follows:

1. *Specification*

Host machine(s)	Supplier and name of office systems product Computer systems product will operate on
Operating environment	Operating system required
Accessible from	Terminals that operate with the system
Protocols	Types of terminal and network support

2. *Decision support*

Spreadsheet	Capable of generating forms and tables and integrating math functions
Graphics	Software or separate attachments for generating graphics on terminal

3. *Data storage and provision*

Database query	Database inquiry and update
Document search method	Search for particular document by key word or key phrase
File transfer	File or document transfer
Access to other users' files	Private, public, or closed user group access, by owner's choice

4. *Communications*

Electronic mail	Messages and documents transfer
Registered mail	Degrees of confidentiality
Priority coding	None to multilevel
Document transfer	Electronic document distribution and document exchange
Integrated directory of users	List of users that use the system
ID and password security	Information protection from unauthorized users
Person or terminal based	Individual runs the system
Telex/Teletex	Public electronic mail attachment availability
Videotex	Access to public or private Videotex systems

Links to other systems	Use resources on another computer system
Voice messaging	Voice mail; leave, retrieve, and distribute voice messages
Voice output	Capable of generating voice response
5. Processing	
Executive WP	Simple, ease-of-use functions
Secretarial WP	Designed for production, highly interactive, full function
Spelling check	Incorrect word is highlighted and correct word can be inserted
Graphics, etc., incorporated	Line and curve drawing capability
Glossary	Key word assist
Records processing	Store, forward, and link to other systems
Personal computing	Calculator program available
Programming languages	Capable of programming in and accessing high-level languages
Access to other facilities	Access other networks via PC or gateway
6. Administration	
Personal diary	Keep personal records
"Tickler" file	Automatic reminders of meetings and "to-dos"
Action list	List of "to-dos" requiring immediate attention
Out sign	Not available
Meeting booking	Set up meetings with one or more persons; book electronically; reduces phoning
Resource booking	Book conference rooms, etc.
Delegation	Delegate work items
Indexes of messages	List by name, date, progress
Distribution lists	Capable of sending one document to multiple users automatically by defining a list
Phone directories	Electronic list from display terminal

Tables 6.4 and 6.5 were compiled by Roger Whitehead of Office Futures, 14 Amy Road, Oxted, Surrey RH8 OPX, United Kingdom, and published in *Fintech 2, Electronic Office*, a *Financial Times* publication, London, United Kingdom.

TABLE 6.4. Office Systems Comparison

Specification	AES Data, AES 7000 Series	Burroughs, Office Automation Package (OMS II)	Data General, Comprehensive Electronic Office	DEC, All-in-1 Electronic Desk	Hewlett Packard, Personal Productivity Center	Honeywell, Office Automation Software
Host machine(s)	AES 7300, 7600	B1900 upwards through 'A' and 'V' Series	Desk Top System 10 to ECLIPSE MV Systems	VAX range; office workstations	Series 3000 range and HP and IBM PCs	Micro System 6/10 & 6/22, DPS6 range
Operating environment	AES Operating System (AESOP) 7300 Unix V-7600	MCP	Advanced Operating System (AOS)	VMS	Multi-Programming Executive (MPE) V and MS-DOS (PC-DOS)	GCOS6 MOD400
Accessible from	AES Systems PC and compatibles	ET1100/ET2000 terminals, B21/B22/B25 micros, OW25 word processor TD830/MT983 terminals, OW400/OW300 word processors (Teletype devices)	DASHER terminals, DESKTOP GENERATION micros "The One" micro, NTI Display Phone, IBM PC, Wang WP	VT100 and VT200 series terminals, DECmate's DEC mini and micro-computers Rainbow & professional work stations	ALL HP terminals, all HP Series 100 micros, IBM PC	VIP 7300 & 7800 terminals, micro Systems executive and Office Work station, IBM PC

(continued)

TABLE 6.4. Office Systems Comparison

	AES Data, AES 7000 Series	Burroughs, Office Automation Package (OMS II)	Data General, Comprehensive Electronic Office	DEC, All-in-1 Electronic Desk	Hewlett Packard, Personal Productivity Center	Honeywell, Office Automation Software
Protocols	IBM 3270, 3780, 2780, 2770; DEC VT100, VT102; ICL CO3 asynchronous	Poll Select, Teletype 3270 SNA	XODIAC (X.25) network, SNA, 2780, 3780, 3270, HASP, CO3	X.25, Ethernet SNA (via bridge), DISOSS (DIA/DCA), DECnet	X.25; IEEE 802.3, 3270, and RJE in BSC and SNA; "Network Services" (NS); CO2, CO3	DSA, X25, Ethernet SNA, BSC, TTY, CO3
Decision support						
Spreadsheet	Yes—integrated with text or via third party packages	Multiplan, Lotus, Visicalc, Extended Multiplan	"Decision Base"	"DECalc" plus option of FCS-EPS	Visicalc, Lotus 1-2-3, PERTMaster, & FCS-EPS	"InfoCalc" and via CP/M, CCP/M, and PC DOS
Graphics	Line drawing-7000 Series, full graphics via linked PCs	If Graphics slice on B20	"Trendview" and "Present"	"Datatrieve" and DECgraph (and DECslide)	"HP Easy Chart," "HP Draw," DSG/3000, "Gallery Collection"	—
Data storage and provision						
Database query	AES and third party options	LINC, DMINQ, REPORTER	"Present" and "Decision Link" (to "Information Database")	"Datatrieve"	"HP Inform," "HP Query," and "HP Acess"	Yes

Document search method	By user-selected parameters and key words	Author, date, key words, subject heading, words, name, content addressable	Parameters (author, destination) or key words	Parameters, key words, or key phrases	By subject	By document name, author, comments, title, or key words
File transfer	By page, file, or work space (directory)	Files or documents	As documents	As documents	As files or documents, singly or as packages	Yes
Access to other users' files	Yes—with user-defined security levels	Owner's choice public + group and computer network	By owner's choice	Only to common libraries	By owner's choice	By owner's choice, also to "public" & "group" files
Communications						
Electronic mail	Messages and documents	Messages and documents	Messages and documents	"Electronic Mail System" (short- or long-form documents) and others	"HP DeskManager"	Messages, notes, documents, files, and spreadsheets
Registered mail	Yes	Yes, plus confidentiality	Yes, plus "confidential" if wanted	Received and "read"	Yes, plus "private" if wanted (and acknow-ledgment)	Yes

(continued)

TABLE 6.4. Office Systems Comparison

	AES Data, AES 7000 Series	Burroughs, Office Automation Package (OMS II)	Data General, Comprehensive Electronic Office	DEC, All-in-1 Electronic Desk	Hewlett Packard, Personal Productivity Center	Honeywell, Office Automation Software
Priority coding	—	Yes	—	—	—	—
Document transfer	Yes—also to IBM via AIF (AES interchange)	Yes (DIA/DCA)	Yes, and to Wang, IBM, DEC, Xerox (via "document exchange")	Yes, and to DEC, DISOSS, Wang OIS, and others	Yes, and to Wang	Yes
Integrated directory of users	Yes	Yes	—	—	—	Yes
ID & password security	Yes	Yes	Yes	Yes	Yes	Yes (2 levels of user—"entry" and "operator"
Person or terminal based	Both	Person	Person	Person	Person	Person
Telex/Teletex	Yes—using British Telecom approved methods	Yes	—	"VAXTEL" telex interface	"HP Telex" (in and out, person to person)	Yes
Videotex	Yes—Prestel or Private Viewdata	Yes	—	"VAX VTX"	Yes	"Themis"

Links to other systems	Yes—IBM, DEC, ICL, Honeywell, HP, etc.	(DIA/DCA)	IBM, ICL, Wang (incl. Mailway), ASCII/ANSI via "DG Gate" X	IBM (DISOSS) (via DDXF), Wang OIS	IBM	IBM (via "Gateway" ICL)
Voice messaging	No	No	—	Pending BABT approval	—	—
Voice output	No	No	—	"DECtalk" (also via remote phone)	—	—
Processing						
Executive WP	Yes—via Managerial Interface	Yes	—	Yes	"Executive Memo Maker"	—
Secretarial WP	Yes—fully interactive	Yes	Yes (file based) & "Word-View" (Eclipse only)	"WPS-Plus"	"HP Word"	Yes
Spelling check	Yes—also with multiple user dictionaries	Yes	Yes	70,000 words + 20,000 own	"HP Spell" (74,000 words + 3000 own)	80,000 words + own
Graphics etc., incorporated	Line drawing on 7000 Series, full graphics via linked PCs	Yes	Yes ("Trendview" & "Drawing Board")	Yes	Yes	Yes
Glossary	Yes	No	Yes	Yes	—	Yes

(continued)

TABLE 6.4. Office Systems Comparison

	AES Data, AES 7000 Series	Burroughs, Office Automation Package (OMS II)	Data General, Comprehensive Electronic Office	DEC, All-in-1 Electronic Desk	Hewlett Packard, Personal Productivity Center	Honeywell, Office Automation Software
Records processing	Yes	Yes	Yes	List processing on WPS	"HP List Keeper" & "Personal Cardfile"	Yes
Personal computing	On 7000 Series and linked PCs	With B25 as terminal	If PC used as terminal	"Desk calculator" plus application software	If PC used as terminal	If MSE used as terminal
Programming languages	DPL (AES Command File Language) plus other common languages	Many	Several	COBOL, BASIC	All commercial, plus Pascal, C, Prolog, & LISP	Several
Access to other facilities	Yes—PC and communications	Any on network	Any on network including DDP	Yes, via "Interrupt"	Any on network	Any on network

Administration						
Personal diary	Yes—networkwide	Yes	Yes	Yes (now networkwide)	Yes	Yes
“Tickler” file	Yes—networkwide	No	Yes	Yes	Yes	—
Action list	Yes—networkwide	Yes	Yes	Yes	Yes	—
“Out” sign	Yes—networkwide	Yes	Yes	Via user profile	Yes, and forwarding	—
Meeting booking	Yes—networkwide	Yes	Yes	Yes—networkwide	Yes	Yes
Resource booking	Yes—networkwide	Yes	Yes	Yes	Yes	—
Delegation	Yes	Yes	Yes	—	Yes	Yes
Indexes of messages	Yes	Yes	Yes	Several, by status	Yes, plus progress	Yes
Distribution lists	Yes	Yes	Yes	Yes	Yes	Yes
Phone directories	Yes	Yes	Yes	Yes	Yes	—

TABLE 6.5. Office Systems Comparison

	IBM, PROFS (Professional Office System)	Sperry, Sperry Office	Wang, Office	Wordplex, 8000	WPS, Office Master	Xionics
Specification						
Host machine(s)	4300 range; 30XX series (in S/370 mode)	1100 range, S80 model 6, Mapper 10 and System 11	VS range	8000	Full range of Alpha Micro systems	Xionics Masternode or Xionics Micronode
Operating environment	VM/CMS (Virtual Machine/ Conversational/ Monitoring System)	OS1100 and OS3	VS operating system	GEMINI	Alpha Micro Operating System (AMOS) with links to MS-DOS and CP/M	Xibos Workstation Operating System
Accessible from	3270 terminals, PC range, display writer	UTS 20, 30, 40 terminals, Sperry MS-DOS PC running DSP PC, IBM compatible PC running DSP-PC	Alliance and OIS terminals, 5300 series terminals (including audio work station), PC, PIC (Professional Image Computer)	Wordplex work stations and PCs	Any standard ASCII serial, Televideo 925, SOROC, Hazeltine, VT100, any MS/DOS-based PC via (Office Master Information Interchange System)	Xionics standard work station, PCs fitted with PC X 1 coprocessor card and remote VT-100 terminals

Protocols	BSC (binary synchronous communications) SNA (via gateways)	X25, UTS 20, Telex	WangNet, Fast LAN, 2780/3780, 3270 TTY, 3274 (SNA), WSN (Wang System Networking)	3270 Bisync & SNA, 2780/3780 Telex, TTY, ICL CO3, DEC VT100 NCR 769-301, Honeywell VIP 7800, Teletex, Viewdata Private/Public	X.25, 2780, 3780, 3270, ALpha-NET, Viewdata	Xionics network—HDLC external network—3270, 3780, ASCII TTY, DEC VT100, VIEWDATA. (ISO-LINK Package)
Decision support						
Spreadsheet	—	Any CP/M or MS-DOS based on PC integrated with WP and records processing	"Maths Planner" and via PC	Through PC or CP/M	AlphaCALC, QIKCALC, and MS-DOS spreadsheets if PC used	Supercalc, Lotus
Graphics	Via GDDM	See above	If graphics work station (6300) or PC	Through PC or CP/M	If PC used	No true graphics
Data storage and provision						
Database query	—	Link to PAL/UNIDAS, Mapper	"Office File Manager" or via "Easyquery" and "PACE"	Wordfile	MARGINS, STRIX (free text) ANDI, DataVUE, QIKFILE	Key word retrieval utility

(continued)

TABLE 6.5. Office Systems Comparison

	IBM, PROFS (Professional Office System)	Sperry, Sperry Office	Wang, Office	Wordplex, 8000	WPS, Office Master	Xionics
Document search method	By document author, date, subject, address, or key words, also for alien docs (via STAIRS)	Document retrieval by title, author, subject	By key words, free text search of abstracts, and document identifiers (also for images, DP files, and off-line docs)	Document search by drawer-subject, folder, topic, document —individual file/key words	By document name, author, category subject, or key words	Direct access by document name, or directory search by partial name, document search by word or string
File transfer	Central file	As documents	As files or documents, singly or as "packages"	As documents	As files, documents, singly or as packages	Files are transferred across Xionics networks in the form of documents
Access to other users' files	By owner's choice	By owner's choice	By administrator's arrangement and owner's choice	Owner's choice	By owner's choice, also to common library areas	By owner's permission or through default group access rights

Communications						
Electronic mail	Messages, notes, and documents	Documents and memos	Memos, WP documents, DP files, or "package" of documents	Yes, Wordmail	MARGINS—MEMO messages and notes only, TSASS—messages and documents	Messages and documents, urgent or nonurgent
Registered mail	Yes	Confidential, Auto Acknowledge	Yes, plus request for acknowledgment	Yes	Yes—with received and "read" on all above with full password protection	Received mail is registered, acknowledgment can be requested
Priority coding	—	Yes	Four levels	Yes	Nine levels on TSASS only	None
Document transfer	Yes	Yes, and to and from Wang and IBM via Document Interchange	Yes	Yes	Yes under TSASS—supports definable communications between different machines and allows EBCDIC to ASCII translation	Yes, and under supported protocols
Integrated directory of users	Yes	Yes	Yes	Yes	Yes—on all mail systems	Yes

(continued)

TABLE 6.5. Office Systems Comparison

	IBM, PROFS (Professional Office System)	Sperry, Sperry Office	Wang, Office	Wordplex, 8000	WPS, Office Master	Xionics
ID & Password Security	Yes	Yes	Yes	Yes	Yes	Yes
Person or terminal based	Person	Person	Person	Person	Person	Person
Telex/Teletex	—	Sperrylink Telex internal and external to all users via network processor	—	Supertex, telex package providing store and forward and group addressing, Teletex	AlphaTELEX and Office TELEX—depending on user requirements	Telex management software
Videotex	—	Via application package	Via PC	Yes, private or Prestel, UK standards supported	AlphaVISION—as private host system, various products including PCs for access to other Viewdata systems	Included in Xionics ISO-LINK package

Links to other systems	Via RCS to other VM systems	Via gateway to other than Sperry, hosts and public databases	Via WSN to OIS Alliance, other VS, & to IBM, DEC, ICL	Via SNA/SDLC	See protocols and TSASS document transfer	IBM, DEC, TTY, in ISOLINK; others via protocol convertors
Voice messaging	—	Yes via Voice Information processing System (VIPS), BABT approved	—	No	No	Implemented but not supported
Voice output	—	Yes via voice response from VIPS, PC, 11, or non-Sperry Last	—	No	No	Not available
Processing						
Executive WP	Basic (XEDIT)	Yes with NITS (see above) and DSP-PC	—	Yes	Margins—memo	Standard
Secretarial WP	Via DCF or if Displaywriter used as terminal	DSSP 30 and DSSP-PC	"WPS"	Yes	SuperVUE	Standard
Spelling check	134,000 words + own, plus style checker	90,000 words + own	86,000 words + own	Yes, 75,000 words + 750 word defined lexicon	Up to 80,000 words in any one dictionary	Only via PC work stations using package

(continued)

TABLE 6.5. Office Systems Comparison

	IBM, PROFS (Professional Office System)	Sperry, Sperry Office	Wang, Office	Wordplex, 8000	WPS, Office Master	Xionics
Graphics etc., incorporated	—	Graphics board on optional on UTS 30, software with PC	Data only	PC terminal only	Available through the glossary functions	Line drawing only
Glossary	—	—	Yes (WP & DP)	Wordaid	Two forms: definable glossary for single key stroke functions and glossary programming of all functions	No
Records processing	—	Yes	Yes	Yes, Supersort	Conditional list processing links to all other data on the system	Limited facility under CP/M
Personal computing	If Displaywriter or PC used as terminal	On both UTS 30 and PC	Yes, if PC used	Using PC as professional work station	Calendar, desk calculator, links to application software, PC software if PC used as terminal	Yes

Programming languages	—	Several	All VS languages (COBOL, BASIC, RPG)	BASIC/WPX ASAP/WPX WORAID 3 (near English programming languages)	BASIC, FORTRAN, PASCAL, C	PLZ for programmers
Access to other facilities	By entry to operating system	Via gateway support and link key	Any on network—	—	Any on network	Any on network
Administration						
Personal diary	Yes	Yes, including shared calendar and review of past and future dates	Yes	Yes	Yes	User designed, KPL procedures can be prepared for all routtine functions
"Tickler" file	Yes	Incorporates within calendar, includes document review and attachment	Yes	—	Yes	User designed, KPL procedures can be prepared for all routtine functions
Action list	Yes		Yes	—	Yes	
Out sign	—		—	—	On TSASS electronic mail	
Meeting booking	Yes	Yes, meeting scheduler	Yes	—	Yes	User designed, KPL procedures
Resource booking	Yes	Yes	Yes	Yes	Yes	User designed, KPL procedures

(continued)

TABLE 6.5. Office Systems Comparison

	IBM, PROFS (Professional Office System)	Sperry, Sperry Office	Wang, Office	Wordplex, 8000	WPS, Office Master	Xionics
Delegation	Yes	Yes	Yes	—	Yes	User designed, KPL procedures
Indexes of messages	Notes & documents only	Yes	Yes	Yes	Yes, plus progress	User designed, KPL procedures
Distribution lists	Yes	Yes, both mail and voice	Yes	Yes	Yes	User designed, KPL procedures
Phone directories	—	Personal and corporate	Yes, & "Personal Phone Book"	—	Yes	User designed, KPL procedures

6.6. OFFICE COMMUNICATIONS NETWORKS

Over 80% of information generated in most organizations moves within the same establishment. Information or data, text, and image are developed by word processors, PCs, intelligent work stations, graphics terminals, or shared devices by online attachment to a host or departmental computer. The current office trend is to arrange employees according to job functions to increase interaction among staff performing the same tasks.

Local area networks (LANs) are a means to provide internal connectivity between incompatible office products and are a cost-effective method of improving communications within a work group by allowing users to use equipment at their workplace and share information stored on this equipment with other members of the group. The intent of LANs is to provide a sharing of resources among several users within an establishment.

LANs may become the future foundation of distributed applications in offices and other commercial environments. Systems are being developed with LAN control and exchange of information among systems, terminals, and PCs. In the future, all products within an establishment—word processors, PCs, professional work stations, electronic copiers, facsimile devices, and optical character recognition (OCR) scanners—will be able to communicate with each other.

The main problem to overcome is incompatibility between office devices. Most PCs, facsimile, copiers, and OCR scanners that are tied to a LAN do not communicate. Each product must have a higher-level translator above the physical connection and data link and network layers, before they can communicate. The problem is diagrammed in Figure 6.3. You can dial China and establish a connection, but if you do not speak Chinese, there is no communication.

At this point, the LANs standards committees are concerned about connectivity. Meaningful communications between office systems (above the data link layer up through the applications layer) are being implemented individually by different office systems suppliers.

In an office, a LAN can physically link word processors, computers, printers, files, work stations, and the telephone PABX systems to provide document access, storage, and distribution; voice messaging; facsimile; and electronic mail services among like architectures. LANs must be extended above the data link level in order to achieve an automated office where dissimilar systems from different manufacturers can be used from a worker's display station.

6.6.1. Local Area Networks

The key reason for LANs introduction and future success is resource sharing or multiple users accessing the same expensive equipment and applications software. With the use of fiber optic cables, LANs will allow sharing of expensive resources such as high-speed printers, large-capacity disk drives, time sharing services, application programs, and database management systems.

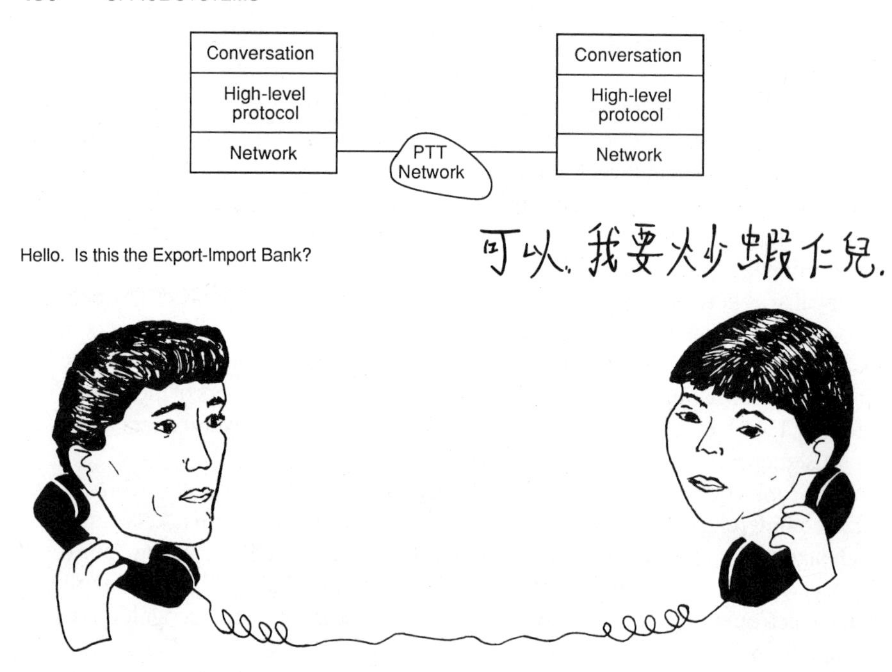

Figure 6.3. Connection is not communication.

Compared with conventional internal and external point-to-point networks, LANs offer higher speed (typically greater than 1 Mbps) and lower error rates over longer distances. LAN throughput or the amount of data that can move through the network can extend up to 500 Mbps, a rate much higher than traditional point-to-point external networks. To put throughput capacity in perspective, at 500 Mbps, the entire 43 million word *Encyclopaedia Brittanica* can be transmitted in about 4 seconds.

LANs solve the problem of moving large volumes of information at very high speeds around a relatively small area, a building or a building complex, at distances up to 10 kilometers (6 miles). Ethernet, developed by Xerox Corporation, is a LAN product designed for intraoffice communications designed to transmit information between work stations and support high-speed devices. Using Ethernet, a document can be created at a word processor and sent to another for review and approval. Next, the document can be sent to an electronic printer for hard-copy printout or to an on-line library for retention and future retrieval. Ethernet uses a coaxial cable system among work stations, printers, and storage devices and can move data at speeds up to 10 Mbps.

There are many LANs offering connectivity. Over 150 vendors have LANs products that can be differentiated by topology (star, ring, and bus), transmission medium (twisted pair, coaxial cable wire, and fiber optic cable), multiplexing method (time division or frequency division), and access method (contention methods like CSMA/CD or token passing). The key to selecting a network is to first

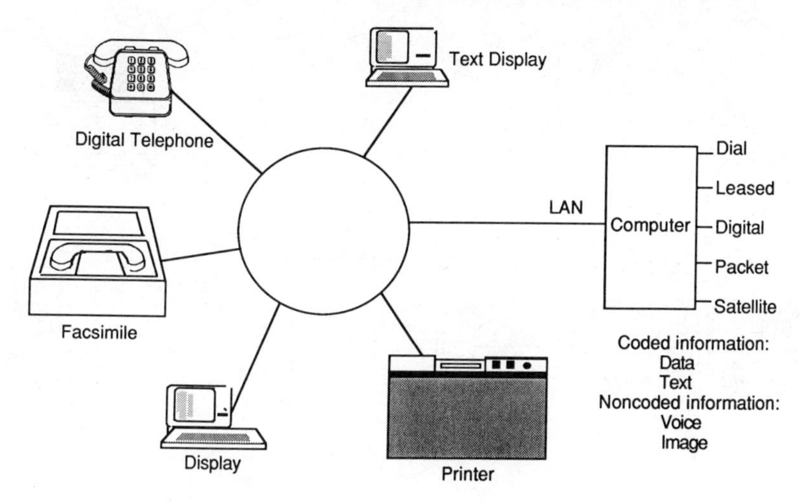

Figure 6.4. Local area network.

determine what the network should accomplish, what products are to be connected, and the type of information to be transmitted. Figure 6.4 is a conceptual view of the types of products that could be attached to a LAN.

6.6.2. LAN Configurations

LANs are defined by their network layout or topology, access methods, and cabling system transmission medium. The network layout is the physical and logical arrangement of stations in a network. Three basic LAN topologies are the star, ring, and bus, which are chosen based on the environment and control method—centralized or distributed. Centralized control provides access to the network and allocates the channel from a communication processor. Distributed control allows network components and work stations to establish connection and access the network independently.

Figures 6.5, 6.6, and 6.7 illustrate the three common LAN topologies.

Star

A star network is like a hub with spokes where the hub is the central controller connected to multiple links, one to each station on a point-to-point link. All communications must pass through the central controller. Two problems exist: If the controller goes down, the network goes down, plus separate cabling must be run from the controller to each device. A star network is used when control or resources are at a central location, e.g., a large disk, time sharing applications, application support, or special printer. Star networks are useful for small clustered controllers or if resources are in a central location or workers are clustered in groups.

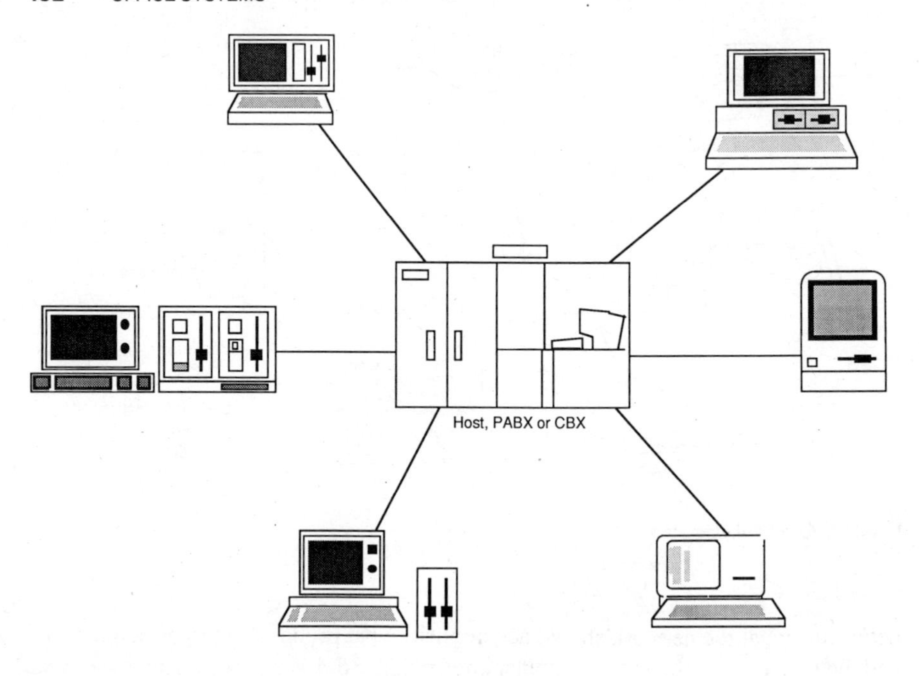

Figure 6.5. LAN star topology.

Ring

A ring is, in effect, a bus with the outer ends joined. Each node or connection receives each message but reads only those messages addressed to it. Messages from a source node flow in one direction from node to node around the ring until they reach their destination. Control in a ring network is usually assigned to one station and is accomplished by passing a token from node to node. A control node is designated to monitor the network for a lost token and to recreate the token when necessary. All nodes can contain the control logic, with the first one discovering the problem taking independent corrective action.

Ring and Loop

The ring and loop is a variation of the ring and normally uses a polled multidrop method for communications control. A time division multiplexer (TDM) is used to divide the available time slots so that each user on the network has equal access.

In token passing, one or more empty frames circulate around the ring. If a station has information to transmit, it looks for the first empty frame or token and replaces it with a frame containing its data and passes it on to the next station. Frames typically contain source, destination, and control information and make a full rotation on the ring being read and regenerated by each station on the ring. The frame is copied only by the destination station or stations.

The risk with a ring topology is that for certain limited failure modes if one termination device fails, the entire network is down. Failure can be overcome by

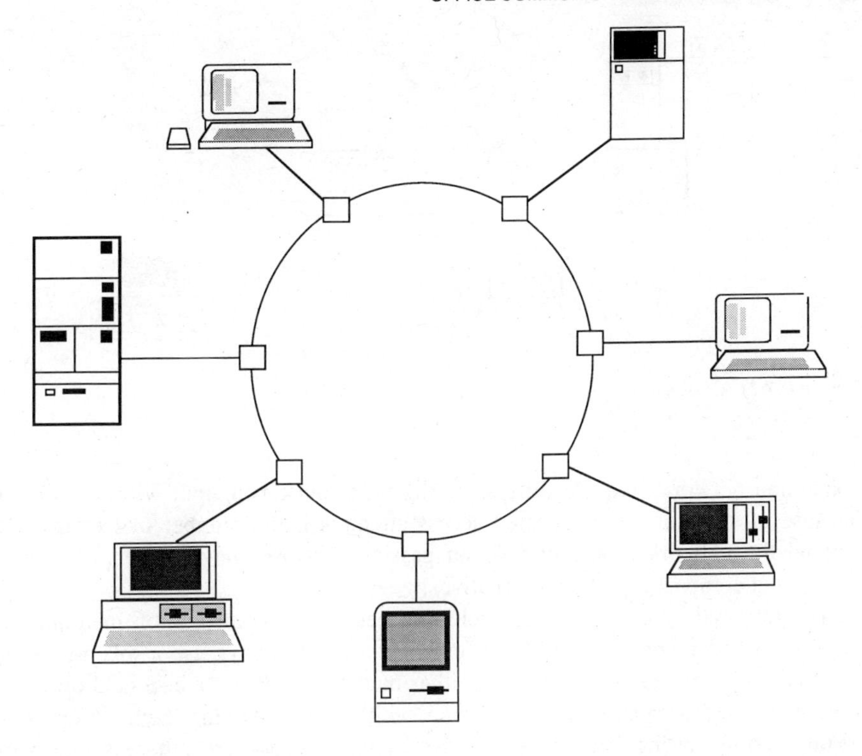

Figure 6.6. LAN ring topology.

redundant rings that allow failed nodes to be bypassed. The drawback with this arrangement is that every time a node is bypassed, the signal must travel twice as far without being repeated.

Bus

All broadband networks and many baseband networks use a bus topology. Devices connected to the bus share a common line and are in contention with each other for access.

Bus nodes are designed so that the transmitter is isolated from the transmission medium. If one node fails along the line, a node farther along the bus can continue to transmit and receive messages.

Messages or packets are broadcast simultaneously to all terminals on a bus. Access is allocated by a control node, or the nodes are treated as equals and contend for access.

Access Methods

The most important factor determining a LAN's throughput and reliability is the access method that controls traffic in the network. Dial-up networks are slow, with the dialing time being a significant portion of the connect time. Multiple access

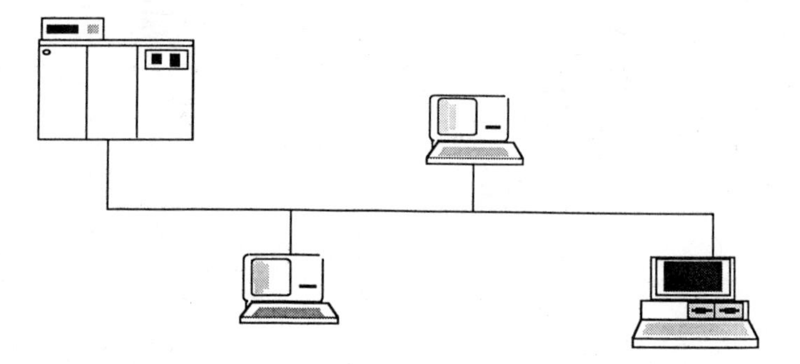

Figure 6.7. LAN bus topology.

contention networks require little signaling time, but throughput will be reduced because of collisions and repeated access attempts if the number of terminals is beyond the network's capacity. Token passing networks have more predictable throughput compared to other alternatives.

The two main methods for controlling access to the transmission medium are contention and token passing. Ring networks almost always use a token-passing method, while bus networks can use either method. Contention access allows any station to initiate a transmission at any time, while in token passing, each station takes its turn in transmitting. The problem with contention access is that there is no way to predict how many collisions will occur or how long it will take the network to recover before a transmission can be made. Token passing is a predictable method since only one token can be passed at one time.

Contention access networks perform best for a mix in traffic patterns where no one station dominates a line. As the number of station transmission requests increases and the distance between nodes increases, transmission quality is degraded due to propagation delays. Another problem with contention access is a lack of priority scheme. All messages are treated equally, independently of the urgency of request.

The most popular LAN access is a contention method called carrier-sense multiple access with collision detection (CSMA/CD). Message frames are sent, and others must wait for the current frames to travel around the network. All stations on the network must detect the presence of or sense (carrier-sense) another station's messages on the network. The frames are transmitted until complete, followed by a time gap, to allow another station to transmit.

A station that is ready to transmit must test or sense that no other carrier is present. It then sends a few bits or a preamble to ensure that the first bit has had time to travel to the farthest point on the network. If the line is free, the preamble is followed by a frame consisting of address, data, control, and check bits.

If two or more stations begin sending their preamble before they can detect the presence of the carrier, a "collision" occurs, and the message is garbled. Each station's controller delays its retry by a random number of slot times before attempting another transmission.

In the CSMA/CD scheme, a station waiting to transmit monitors the main data channel for a carrier signal to indicate that traffic is present. If the station picks up a carrier signal, it postpones its transmission for a random interval. Two stations sensing a clear channel may try to transmit at the same time, resulting in a collision—the signals from each transmitting station interfere with each other. CSMA/CD would enable stations to recognize collisions, immediately stop transmitting, and resume transmission after a random waiting period.

The advantage of CSMA/CD is that it permits network access control without the need for a central network controller. CSMA/CD is simple and easy to implement and has a simple, efficient control method.

In token-passing access, each station wishing to transmit circulates a special bit pattern called a token that assigns the right to transmit to the station that receives it. After a station receives the token, it transmits its data and then passes the token to the next station on the line. Only a station that has the token may transmit.

Token passing provides orderly access and allows transmission priority among stations. Token-passing networks are more difficult to configure than CSMA/CD networks because each station must have a logical address and a logical place in the token-passing sequence. Station failures present serious problems since the network's passing sequence must be bypassed.

As in other contention systems, a token passing or slotted network's performance is affected by the number of stations on the network. Depending on the priority configuration, a station may have to wait for all other stations to use their opportunities before it can transmit.

Transmission Media

LANs can use the same type of twisted-pair copper wire that connects telephones. Twisted-pair wire is inexpensive and easy to install and can carry data at rates up to 1 Mbps for several hundred feet without using repeaters.

A major disadvantage of twisted-pair wire is that it is susceptible to electrical interference or external noise. The interference limits transmission speeds and distances because the signal weakens and noise increases over distance. Shielding could reduce noise interference, but shielded wire costs are significantly higher.

Coaxial cable, used to carry video signals for cable television (CATV), can be used for both baseband and broadband LANs. Coaxial cable or coax offers more channels, is faster, permits longer distances, and is only slightly more expensive than twisted-pair wire.

Baseband and broadband networks have different electrical characteristics and applications. Baseband systems can have data rates of up to 50 Mbps and can connect hundreds of nodes spread over a distance up to 1.5 miles. In baseband systems, one channel's signal occupies the entire available bandwidth.

Baseband networks are good for small- to medium-sized data processing or office automation environments and work well within a building or building complex. Broadband systems apply CATV technology to carry several radio frequency signals simultaneously in full duplex at data rates up to 200 Mbps. Distances between nodes can be as long as 200 miles (320 kilometers).

TABLE 6.6. Cables Used in Local Networks

Transmission Medium	Maximum Data Rate	Typical Distance	Multipoint Capacity and Range of Network Users	Installed Cost Per Foot
Twisted pair	4 Mbps	1 km	Fair 25–200	0.40–2.25
Baseband coaxial cable	50 Mbps	3 km	Good 50–1000	0.35–1.50
Broadband coaxial cable	350 Mbps	10 km	Better 200–20,000	0.50–3.50
Fiber optics	800 Mpbs	10 km	Limited 2–8	2.00–7.00

Each device in a broadband LAN requires a modem, which adds significantly to the overall network cost. Since broadband networks carry many signals simultaneously, they must be specially configured to send and receive signals. One technique is to install a frequency translator. Another technique is to loop the cable to create a dual cable so that each station has a transmitter on one ring and a receiver on the other. A dual cable configuration offers twice the bandwidth at twice the price.

Fiber optics offers very high bandwidth to support high data transfer rates and is not susceptible to noise. The major advantages of fiber optic cable are durability, security, corrosion resistance, and immunity to electrical interference. The main disadvantage of optical fiber is the difficulty of aligning and connecting thousands of light filaments to ensure a continuous connection.

The cost of fiber optic cable is expected to decrease by the 1990s and become competitive with coaxial cable. Table 6.6 shows a comparison of cable media used in local networks; Figure 6.8 shows different cables used in LANs.

6.6.3. LANs and OSI

The key to LAN acceptance is the availability of low-cost interfaces to connect different types of equipment to a LAN. The complexity of LAN protocols requires wide acceptance among several computer suppliers in order for manufacturers to commit to a chip design whereby the interface can be reduced to a chip or single module. A widely accepted design enables independent devices from a variety of suppliers to communicate. Several key local area network standards were developed within the IEEE by a committee called Project 802. These design standards define a set of interfaces and protocols for the local network to ensure compatibility with equipment made by different manufacturers. The IEEE 802 Standards have gained acceptance by ANSI and are in the process of being accepted by the U.S. National Bureau of Standards and the International Organization for Standardization (ISO) in Geneva, Switzerland.

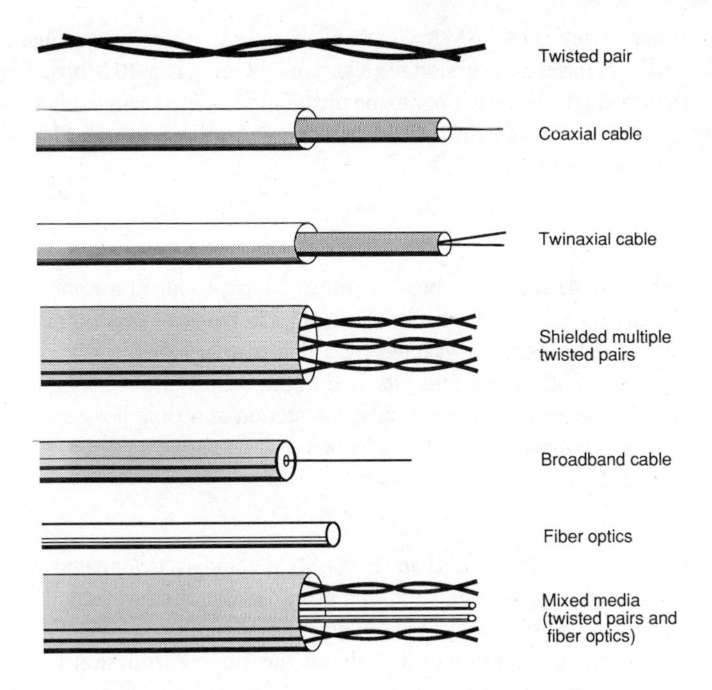

Figure 6.8. Local area network transmission media choices.

The scope of LANs is reflected within the 802 committee. There are six standards-writing committees organized by the lower three layers of the OSI reference model, i.e., the physical, data link, and network layers. The data link layer is handled by subcommittee 802.1 and further subdivided by media access.

The data link layer is divided into two sublayers for local area networks, called logical link control and medium access control. They perform frame exchange between peer data link layer entities. Frame exchange includes frame delimiting, frame transmission, address recognition, and error checking. Logical link control, defined by subcommittee 802.2, performs the remaining data link functions. The main task of media access control is to achieve successful transmission when multiple stations share the same transmission medium.

The following three IEEE subcommittees have produced IEEE standards for medium access control and the associated physical layers.

IEEE 802.3 CSMA/CD Bus

Subcommittee 802.3 has responsibility for local area networks (CSMA/CD) on a bus. The CSMA/CD protocol allows stations to transmit if they do not detect a carrier signal on the line. If two stations transmit simultaneously, their signals collide and are corrupted. Both stations stop transmitting when they detect the collision and attempt to transmit after a set period of time. The standards specify 10-Mbps operation on coaxial cable at a distance of 500 meters using baseband transmission.

Three other varieties of CSMA/CD standards are in preparation. One is lower-cost IEEE 802.3A baseband version for less than 200 meters at 10 Mbps. The second is another baseband proposal to operate on unshielded twisted-pair cable for less than 500 meters at 1 Mbps, and the third is a broadband transmission standard for distances over 3600 meters at 10 Mbps.

IEEE 802.4 Token Bus

Subcommittee 802.4, the token bus committee, has produced a joint IEEE/ANSI standard and an ISO standard for a token-passing scheme on coaxial cable. On a token bus, permission to transmit is granted by a data frame called a token that passes successively around all the stations attached to the bus. Stations may enter or leave the token-passing sequence because only one station at a time possesses the token. There is no loss of bandwidth due to collisions.

IEEE 802.5 Token Ring

Subcommittee 802.5 has produced an IEEE/ANSI standard designated ANSI/IEEE 802.5-1985. A token ring is comprised of a group of stations connected sequentially by a series of point-to-point links. Except when transmitting, each station acts as an active repeater and regenerator of signals on the ring. Permission to transmit is granted by a token that circulates on the ring until a station needs it. A transmitting station holds the token and then releases it after the transmission is completed. Unlike the token on the token bus, the token is not addressed to any particular station.

6.6.4. IEEE/ECMA/ISO

Figure 6.9 represents the relationships among the IEEE committee, ECMA, and the ISO organization. The three organizations are working together to meet multiple requirements for LANs in terms of speed, distances, media access, and baseband and broadband transmission.

Upper levels	Higher-layer interface standard: IEEE 802.1		
Data link layer 2	Logical link control standard: IEEE 802.2 ECMA 82		
	Media access: CSMA/CD IEEE 802.3 ECMA 81	Token bus IEEE 802.4 ECMA 90	Token ring IEEE 802.5 ECMA 89
Physical layer 1	CSMA/CD media	Token bus media	Token ring media

Figure 6.9. IEEE/ECMA/OSI relationships.

6.7. PABX—VOICE/DATA INTEGRATION

If an organization has an existing private automatic branch exchange (PABX) or circuit switching system for directing telephone calls, the addition of data lines to the switching system is an economical way to add a few terminals. Compared to the cost of adding a LAN, the incremental cost of adding terminals to an existing switching system is often negligible. In general, extra or spare wires exist in telephone ducts and cables throughout a building, and service technicians are trained to install equipment and maintain lines. Therefore, adding a few lines requires simple changes to existing operations. In addition, circuit switching is capable of operating over a wider range than many LANs. For example, LANs using contention access can operate at about 1500 feet (0.45 kilometer), while PABXs using limited distance modems can transmit data at 19.2 kbps over a mile (1.6 kilometers). Many PABXs are also designed to interface to X.25 packet switched data networks.

Although local area networks are capable of handling voice traffic, most LANs are inefficient handlers of voice compared to a PABX. In the future, LAN controllers may be integrated into a PABX or connected by a gateway for interconnection and transition between protocols. PABXs cannot handle the burst-type information traffic characteristic of the automated office. In this situation a PABX can take more time to establish a terminal-to-terminal connection than to send the message.

Some of the newer computerized PABX (sometimes called computer branch exchanges or CBXs) have the ability to collect usage information for billing circuit costs back to the user. By contrast, most LAN applications lack accounting and billing features and the ability to gather usage statistics. Despite their advantages, PABXs are not a universal solution to data communications problems in many offices. In many companies, data communications is under the DP department, and telephone is in a separate unrelated communications department. They tend to speak different languages (jargon) and do not normally integrate planning and budgeting for voice and data circuits.

A PABX consists of four elements, as shown in Figure 6.10. The switching network is a collection of electronic devices that connect circuits together. The more recently designed PABXs are computer based and use time division multiplexing. The incoming voice or data signal is sampled in the line port or in the station, and the

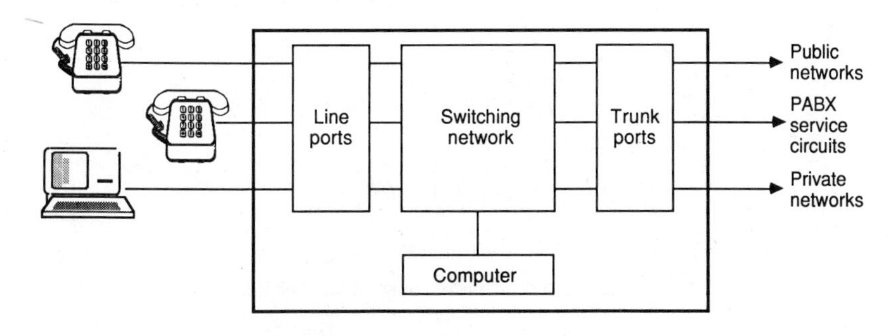

Figure 6.10. Major PABX components.

computer routes a signal from one line port to another or to a Telco trunk. Data terminals can be connected to dedicated line ports or can share ports with a telephone in an alternating voice-data arrangement. In addition, data and voice signals can be multiplexed over the same line and applied to a combination voice/data port from the PABX.

Since human speech is inherently analog, a conversion from analog to digital must take place either in the telephone or in each connection through the digital switch. In new digital PABXs, analog conversion occurs in the terminal, adding to the cost of the station. For stations that need to transmit data in addition to voice, the cost is marginal since both voice and data are digitized in the user's terminal. Figure 6.11 is a diagram of an existing PABX.

Today, many PABXs can integrate voice and data. A work station user can use a telephone and data terminal simulatneously. Many PABXs use separate pairs of wire to connect the telephone to the voice port and the data terminal to the data port of the PABX. The trend in current PABX design is to multiplex voice and data over a single high-speed circuit that is wired with one or two twisted pairs to s single port in the

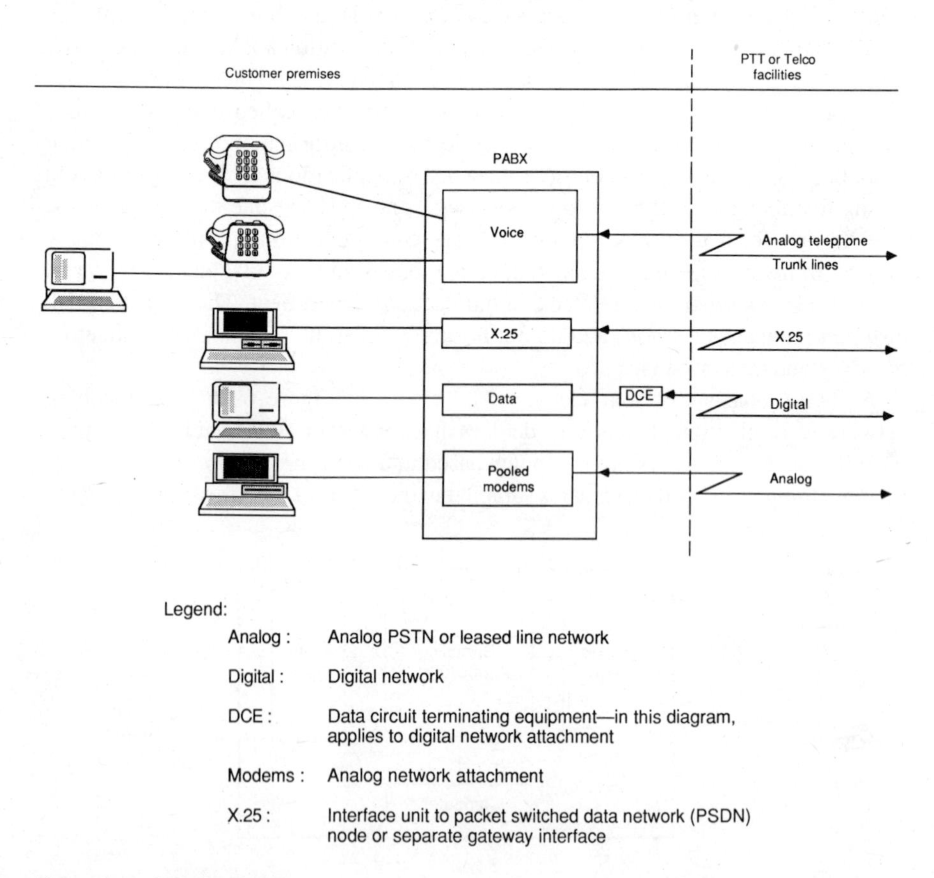

Figure 6.11. Existing PABX.

PABX. By using this method, voice is digitized to a 64-kbps signal at the terminal and integrated with a data signal at data rates up to 64 kbps. The two signals are transmitted at 128 kbps to the PABX line port where the data and voice signals are then separated and switched toward their final destinations.

Unless a PABX has been designed to accept direct digital inputs, the line ports are equipped only for analog, requiring the use of modems to convert the signal to analog before it is switched. New switching systems provide digital line ports as well as analog ports (required for existing equipment).

Figure 6.12 represents tomorrow's digital PABXs. Voice and data, as well as the digital network attachment units, will be integrated within the same unit. X.25 high-speed packet network attachments will be a feature offered by most PABX suppliers. A gateway or interface to high-speed LANs will be common as well as direct attachment to front-end processors.

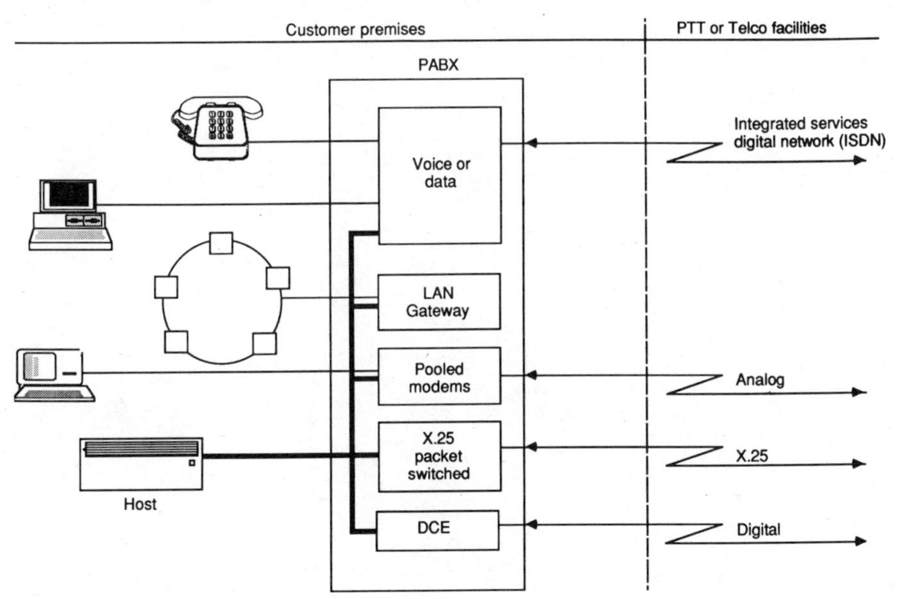

Note: ISDN, analog, X.25, and digital networks will coexist through the 1990s in most countries.

Legend:

Analog:	Analog PSTN or leased line network
Digital:	Digital network
DCE:	Data circuit terminating equipment—in this diagram, applies to digital network attachment
Modems:	Analog network attachment
LAN Gateway:	Interface LAN to a private network
X.25:	Interface unit to packet switched data network (PSDN) node or separate gateway interface

Figure 6.12. Future digital PABX.

ISDN attachments will be standard on new digital PABXs. Depending on the number of multiplexing channels or network termination points required for an establishment, attachments may be integrated within the PABX or as a separate external device.

In future office systems product offerings, voice and data will become standard features. One indication of the merging of voice and data communications into office systems is the acquisitions of and joint agreements with PABX suppliers by major office systems suppliers. Data General purchased Integrated Digital Networks (PABX); DEC signed a cooperative agreement with MITEL, who has 40% of the world market for small PABXs; IBM purchased ROLM (PABX/CBX); and Wang purchased 30% of Intecom, a Dallas-based PABX supplier.

7

DIGITAL VOICE AND DATA NETWORKS

In this chapter we will

- Describe circuit switched data networks (CSDNs)
- Discuss packet switched data networks (PSDNs)
- Examine the evolution of digital networks toward ISDNs
- Discuss ISDNs, the future voice and data networks that are being implemented by many countries

7.1. CIRCUIT SWITCHED DATA NETWORK

CSDN or X.21 networks are digital switching networks designed for data transmission. They are based on new digital switching techniques and CCITT recommendations and are available in Austria, France, Germany (FRG), Italy, Japan, the Nordic countries, and the United Kingdom. Some European countries have announced a service at the international level. That is, the country will provide an integrated network to other CSDNs or through a gateway (conversion between networks) connection to another country's digital network. For example, the French PTT will interface its digital networks to another country's X.21 network through an international gateway if that country subscribes to their Telecom 1 satellite program. The receiving country would require a compatible Telecom 1 earth station in order to receive the information.

In a CSDN network, a connection can be set up between any two subscribers belonging to the same class of service (i.e., using the same speed or, if asynchronous attachments are used, the same speed and code). Typically, the connection or session is held only for the time required to exchange information between the two end points, and then the call is cleared. The overall connection is similar to a dial-up or public switched telephone call. A call is made, messages are exchanged, and when completed, both parties hang up and the circuit is free for another call.

A major advantage to a CSDN is that user data is transparent. That is, once the call is established, any data line control can be used.

Data terminal equipment (DTE) connections are basically synchronous with clock signals to user equipment provided by the network at both ends. In addition, some networks allow asynchronous terminals up to 1200 bps.

The facilities of the CSDN network can be divided into two groups: basic facilities or the fundamental functions of the network parameters that users always have access to as subscribers and additional facilities or special features that are available to subscribers as a fixed service or on a per call basis. These facilities are usually requested by the subscriber at the time of initial subscription to the X.21 service. Additional features are provided in Table 7.1.

7.1.1. X.21/CSDN Network Connections

The CCITT X.21 Recommendation describes a synchronous interface between DTE and DCE for used with public data networks. It was designed with only four intelligent circuits: transmit (data to the network), receive (data from the network), control (signals to the network), and indication (signals from the network). Timing circuits and grounding are also required.

TABLE 7.1. Additional X.21 Network Features

Feature	Example
SIMPLIFIED CALL SET-UP	
Direct call Simplifies call for subscriber who always calls one given subscriber.	This facility is used by a terminal that always communicates with the same data processing bureau.
Selective direct call Simplifies call procedure for a subscriber with many calls to a limited number of other subscribers.	A chain of shops uses the data network for the collection of bookkeeping information, or the main branch in each area collects data from the other branches in that area at frequent intervals.
Abbreviated address calling Simplified call procedure for a limited number of other subscribers.	Simplifies dialing (e.g., two digits instead of six) and makes it simpler to remember numbers.
IMPROVED SECURITY	
Closed user group Protects a group of subscribers against calls from subscribers outside the group.	A bank has a computer center at its head office and terminals located in a number of branches. The bank wants to prevent unauthorized calls to the terminals in the branches.

TABLE 7.1. Additional X.21 Network Features

Feature	Example
Outgoing calls barred Terminals are used only for incoming calls.	A bank has a terminal that must be prepared to receive calls from the head office, but a local operator cannot initiate a data call from the terminal.
Outgoing international calls barred Prevents unauthorized use of terminal equipment.	Used if there is a risk of the terminal being used to transfer data on individuals to databases abroad.
Incoming calls barred Prevents terminals from accepting incoming calls.	A simple unmanned terminal is installed in a metering station to record weather information. A central computer is called periodically, and the terminal transmits the measured data.
Incoming international calls barred Prevents calls from terminals in other countries.	The service provides extra protection against access to database information from abroad.
INCREASE COMMUNICATIONS, CONVENIENCE FACILITIES	
Multiple lines at the same address (group number) Several connections use the same number.	Data traffic to a home office is so dense that several ports to the computer are allocated to a group number. The network searches for a free port, and the terminals avoid having to call alternative numbers.
Connect when free A call is placed in a queue, and the call is established when the connection is free.	Eliminates redialing active subscribers.
Redirection of calls Redirects calls to a terminal other than the one called when the called terminal cannot accept data.	A data processing bureau can serve its customers by redirecting the call to another bureau or backup center.
IMPROVED SECURITY	
Calling line identification Provides the called subscriber with information on who is calling.	If a data collection system has several types of terminals, the computer recognizes the number of the calling terminal and can direct the information to the correct database.
Called line identification Informs the calling subscriber of the number of the called subscriber.	A terminal operator can check that the network has recorded the correct number in conjunction with the call.
CHARGING FACILITIES	
Reverse charging Transfers charging of calls to the called subscriber.	A data collection system has terminals that initiate a call to its central computer. The charges are billed to the calling or called terminal.
Charge advice Advice on charge for each call.	A customer obtains a printout of the charges when calls are completed.

The subscriber may connect a computer, PC, or terminal, called DTE, to the network by a standard data network interface (X.21). Connection by X.21 will allow the user access to all the basic functions of the network and, if subscribed to, the ability to make use of all the additional facilities. To migrate to X.21, a subscriber can use existing equipment with a converting interface (X.21 bis) but with some restrictions on the full set of X.21 features. The X.21 bis transition interface has been defined as either the V.24 Recommendation or V.35 Recommendation depending on the signaling rate. The X.21 bis connection is a temporary solution in order that subscribers can use existing DTE with the network.

All channels in the data network are full duplex, which allows the simultaneous transmission of data in both directions at the full nominal speed. When the communication channel is established, all bit combintions can be transmitted, allowing the user freedom in choice of code, format, and transmission protocol. That is, the network is transparent to the transmitted information. All synchronous data terminal equipment connected to the network is synchronized from the network, which supplies clocking for each bit. A user can subscribe for operation at 600, 2400, 4800, or 9600 bps and 48 or 64 kbps. In all countries offering an X.21 CSDN, each individual subscriber is given a unique six-digit number within the network that is used for connection to the data network.

Depending on the type of interface, a call can be initiated either manually from the DTE or data circuit terminating equipment (DCE) or automatically from the DTE. The calling procedure is similar to a switched telephone or dial-up call. The subscriber can call any terminal within the same data rate class, and the call begins after the network has indicated that the circuit has been established.

Automatic answering, which is part of the recommendation, is provided from all connections when they are called. The call setup is the time elapsed from the instant when the number of the required subscriber is transferred to the data network until the network indicates that data transmission can begin. Normally, the total time is the time it takes to press a function key, remove a finger, and begin entering data. The call setup times provided by the Nordic data networks are as follows:

Below 100 milliseconds for 90% of the calls
Below 500 milliseconds for 99% of the calls
Below 2 seconds for 100% of the calls

The call clearing time is the time elapsed from the instant the subscriber gives a clear signal until the network has broken the circuit. Call clearing is quite fast, as noted by the following times quoted by the Nordic PTTs:

Below 50 milliseconds for 90% of the calls
Below 200 milliseconds for 100% of the calls

Some advantages of the X.21 circuit switched data network are the following:

- There is minor reconfiguration of the network when additional terminals are required.
- The subscriber pays only for connect time, plus basic and additional features.
- Most networks offer higher speeds compared to the public switched telephone network, which is limited to 14.4 kbps. Line speeds of 64 kbps and higher are possible in many countries.
- For the casual or intermittent user, the CSDN is less expensive than a fixed cost private leased line.

Figure 7.1 provides a comparison between a private leased line and the CSDN.

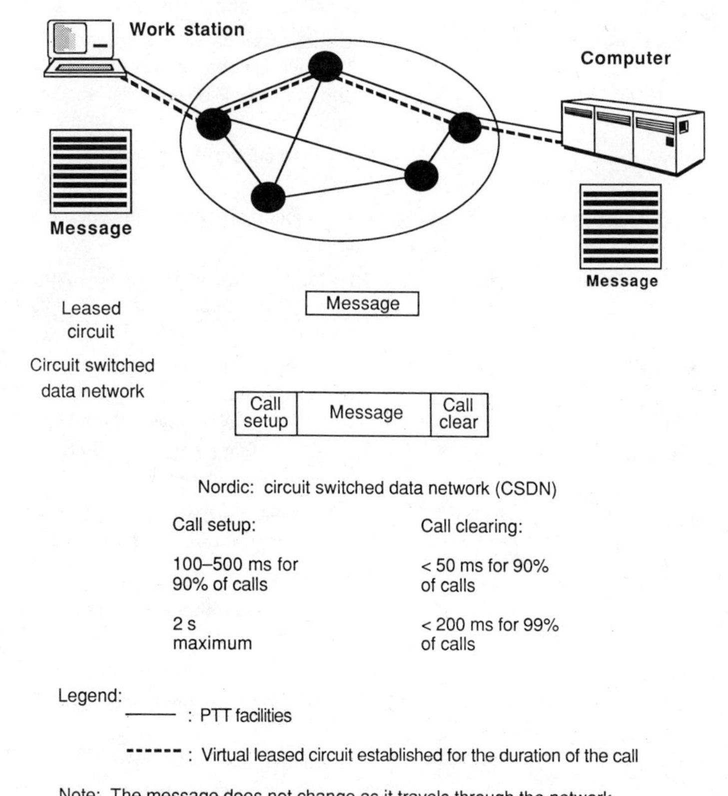

Figure 7.1. CSDN versus leased circuit comparison.

Call Progress Signals

All calls from DTE are answered automatically by the network. If the communication circuit cannot be established immediately, the network indicates the status of the call by a two-digit code number, as presented in Table 7.2.

7.1.2. X.21 Data Networks

Many countries have implemented high-speed X.21 circuit switched digital networks or data networks using the X.21 switching techniques. A list of these countries, including the network name and line speeds available, is given in Table 7.3.

TABLE 7.2. Call Progress Signals in the Nordic CSDN

No.	Application
02 Redirected call	The call has been redirected.
03 Connect when free	The subscriber called is engaged (all ports). The calling subscriber is placed in a queue, and the call will be established when a port is free.
21 Number busy	The called subscriber is engaged in another call.
22 Selection signal procedure error	The selection procedure has not been followed.
23 Selection signal transmission error	An error has occurred in the selection signal during transmission to the data swtiching exchange.
41 Access barred	The called and the calling subscribers cannot communicate, e.g., because the subscribers are not in the same closed user group.
42 Number changed	The number to the called subscriber has recently been changed.
43 Not obtainable	For example, the called subscriber belongs to a different data signaling rate class.
44 Out of order	Not working
45 Controlled not ready	The called subscriber is switched to the local mode.
46 Uncontrolled not ready	The called subscriber is out of service, e.g., due to power failure to data terminal equipment.
47 DCE power off	The called DCE has a power failure.
48 Invalid facility request	Erroneous use of the selection signal (in conjunction with a subscribed additional facility).
49 Network fault in local loop	Fault on the B terminal's subscriber circuit.
52 Incompatible user class of service	B terminal uses another data signaling rate.
61 Network congestion	Call is not established due to a temporary fault or congestion in the network.

TABLE 7.3. Country CSDN Offerings

Country	Service Name	Nominal Line Speeds	Asynch/ Synch	Interface
Austria	Datex-L	300 bps	Asynch	X.20 bis, X.20
		2.4, 4.8, 9.6 kbps	Synch	X.21
Denmark	Datex	300 bps	Asynch	X.20 bis
		2.4, 4.8, 9.6 kbps	Synch	X.21 bis, X.21
Finland	Datex	300 bps	Asynch	X.20 bis
		600 bps, 2.4, 4.8, 9.6 kbps	Synch	X.21 bis, X.21
France	Transcom	48(1), 64 kbps	Synch	X.21 or V.35
	Transdyn	Telecom 1 Network		
		2.4, 4.8, 9.6 kbps	Synch	V.24 plus V.28 or X.21
		48, 56, 64 kbps	Synch	X.21 or V.35
		128, 256, 512 kbps, 1, 2 Mbps	Synch	X.21
Germany	Datex-L	300 bps	Start-stop	X.20 bis, X.20
	Datex-L	2.4, 4.8, 9.6, 64 kbps	Synch	X.21 bis, X.21
	Inteldat	56 kbps	Synch	—
Italy	RFD	64 kbps	Synch	X.21
	Telex Dati	2.4, 4.8, 9.6 kpbs	Synch	X.21
Japan	DDX-C	200, 300, 1200 bps	Start-stop	X.20
		2.4, 4.8, 9.6, 48 kpbs	Synch	X.21 bis, X.21
Norway	Datex	300 bps	Asynch	X.20 bis, X.20
		600 bps, 2.4, 4.8, 9.6 kpbs	Synch	X.21 bis, X.21
Sweden	Datex	300 bps	Asynch	X.20 bis, X.20
		600 bps, 2.4, 4.8, 9.6 kbps	Synch	X.21 bis, X.21
United Kingdom	IDA	2.4, 4.8, 9.6, 48, 64 kbps	Synch	X.21 bis, X.21

Notes:
1. The French Transcom 48-kbps service will be available in September 1987.
2. The network attachments in the table that use the X.21 protocol for layer 1 may not implement the full call progress signals and additional subscriber features for each network.
3. X.20 is an interface between DTE and DCE for start-stop transmission services on public data networks.
4. X.20 bis is used on public data networks for DTE to interface to asynchronous FDX V-Series modems.
5. X.21 bis is an interim offering allowing existing terminals with V.24 types of connectors to attach to the CSDN.

Source: PTT published reports

7.1.3. Short Hold Mode

The improved performance enhancements in CSDN services compared to the public switched telephone network include a new feature called short hold mode (SHM). SHM allows the remote equipment to automatically be released from the line, which stops the charging equipment since there is no transmission. The feature is due to the main characteristics of the CSDN, that is, very rapid call setup and call clearing times and small billing increments. This service is ideal for remote point of sale (POS) terminals that require access to a host database for transaction updates or credit checking. The transaction time or the time the terminal is required access to the host is generally less than a minute. The SHM feature, which would be implemented in both the terminal and host communications program or the host's front-end processor, releases the line when no further transmission is imminent. This procedure is accomplished by timing out if there is no transfer of data in either direction for a predetermined period. However, the communications program or front-end processor is required to retain the control blocks, pointers, and counters to ensure rapid reconnection when another transaction is required.

From a host standpoint, the remote point of sale terminal is always connected. This means that the physical connection is broken, but not the logical connection as far as the host is concerned. When a new transaction from either the host or the same remote POS terminal is initiated, there is minimal handshaking or initialization to synchronize communications and identify the terminal for reconnection. The calling device uses the X.21 call establishing procedure and performs handshaking to acknowledge the same device. In many cases, the line is held for a predetermined time by the system software to sufficiently complete the request. After that time period the link is released again using the X.21 protocol.

7.1.4. Nordic CSDN Tariff Differences

Although the Nordic countries have the same interconnected CSDN or Datex Network, their billing increments, minimum time charges, and tariffs vary considerably. In each country, there is a separate charge for each connection, and, depending on the country, it varies by the time of transaction, as noted in Denmark and Finland, or is a flat charge, as in Norway and Sweden. The call charges apply within the country for 1 minute. If calls of a shorter duration are made for Denmark or Norway, multiply by 0.833 per 5-second increment for Denmark and 0.033 per 2-second increment for Finland. In any event, a great many transaction calls of short duration can be placed compared to the monthly charge for a leased line connection.

If an incremental charge for reestablishment of the DTE using the SHM feature is added, there is still a savings with the short hold feature when compared to the total connect time. Table 7.4 includes the basic charges for four Nordic countries.

7.1.5. Multiple Port Sharing

Many transactions from remote POS terminals require minimal communication time—about a minute's duration—and, therefore, the communications link is disconnected most of the time. To minimize the number of ports or adapters on the host or a

TABLE 7.4. Nordic Countries—CSDN Tariff Charges

Country	Minimum Time Period Charge	Per Call Charges (1 minute)		
		2.4 kbps	4.8 kbps	9.6 kbps
Denmark	5 seconds	$.05	$.09	$.19
Finland	2 seconds	.05	.10	.19
Norway	—	.09	.18	.35
Sweden	—	.03	.07	.13

Source: PTT published tariffs.

front-end processor, a new feature called multiple port sharing (MPS) is available from some vendors. MPS offers the ability to share ports among many devices by removing the tie between the physical port and the host's communications pointers. Each device is logically connected to the host. When a session is requested, the remote link is switched to the first available port, and the host's communications program or front-end processor dynamically reconnects the remote device and reestablishes the connection. By eliminating the need for the remote device to always be connected to the same port, which may be busy and would result in an uncompleted call, MPS completes the call and reduces the number of ports required to satisfy remote terminals in a company network.

7.2. PACKET SWITCHED DATA NETWORKS

Packet switching is a store and forward technique for transmitting packets between network nodes. A packet is a data block with header information that contains the destination address. Depending on the network design, each block has a maximum length (e.g., 64, 128, 256, 512) of octets, segments, or characters. Longer messages must be broken into two or more packets until the message to be transmitted is completely packetized. The packets are usually sent to the nearest node in the network and then transmitted over trunk lines or a channel on their way to their final destination. For the packet network providers to maximize the channel capacity between nodes, packets from many end users are intermixed over the same channel. The channel between nodes is busy only for the time necessary to send the packet.

A packet switching network is designed to transport data in the packet format. To have communications between end users or terminal to host, there must be additional end-to-end protocols above the packet network to interact with the network nodes. The protocol used by packet networks is the CCITT X.25 Recommendation, which defines the interface between the end user data terminal equipment (DTE) and the public data networks using the packet switching store and forward technique.

The CCITT X.25 Recommendation specifies the DTE to DCE interface for packet mode switching and is concerned with the delivery of packets to their final destination and in their proper sequence to duplicate the original message. Ultimately, the overall CCITT or worldwide objective is to assure effective communications

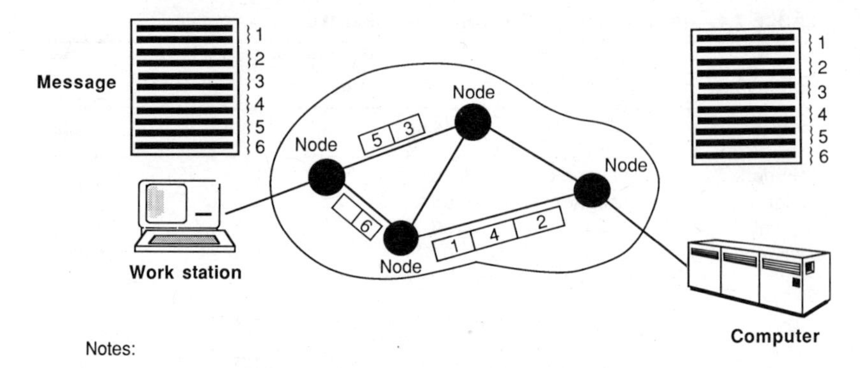

Notes:

1. The terminal's message is broken down into packets by the node processor.

2. As each packet travels through the network via different nodes on its way to the final destination, the packets are not kept in sequence. In addition, they are intermixed with other packets from other enterprises.

3. The packets are reassembled into the original message form before being presented to the remote DTE (computer).

Figure 7.2. X.25 packet switching data network.

between products of different manufacturers. X.25 will work with the OSI (open system interconnect) model. The OSI model's lower three layers are similar to the X.25 interface: the physical connection or layer 1, the data link control or layer 2, and the network or layer 3.

Figure 7.2 is an example of a packet switched data network. On the left, a message is created on a work station. The message is then sent to the nearest node in the network where it is disassembled into packets. The packets are sent through the network, and, finally, the packets are reassembled in their original order and sent out of the network to the computer.

The advantage of packet networks is the sharing of physical resources. For example, up to 4096 virtual connections can share the same physical port. It can provide multiple simultaneous "conversations" to a single terminal. The internal design of the network is the manufacturer's choice.

7.2.1. Virtual Circuits

There are two types of packet circuits: a permanent virtual circuit (PVC) and a switched virtual circuit (SVC). A PVC is established between the data terminal equipment (DTE), e.g., computer and terminal when the service is first subscribed to. A calling procedure is not required, since the terminal's messages are always sent to the same remote point. A PVC may be compared with a leased line. One difference is that each packet requires header (H) information, while normal leased line operation accepts larger block sizes or files with less repetitious header information. A switched virtual circuit (SVC) is like a switched or dial-up call. The connection

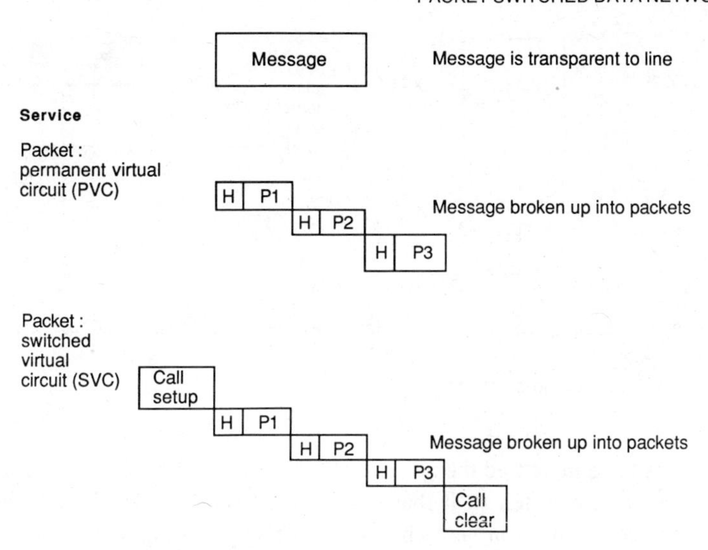

Figure 7.3. Permanent and switched virtual circuit comparison.

lasts while the two devices are in session or are connected. Compared to a PVC, once established, the SVC adds a call clearing procedure on the end of a message. The difference between sending a message by a PVC and SVC connection is shown in Figure 7.3. In this example, we consider three packets, although the number of packets would vary by message length. H refers to the header information, and Pl, P2, and P3 refer to the message broken up into three packets.

The key to packet networks is the node design. Every node can send any portion of a message on every channel and selects the most convenient channel on its way to its final destination. The channels between the specialized computer nodes of the network keep track of the incoming messages in storage, recognize the destination of every message, and select, according to the final destination ID, an open channel for forwarding the message toward its final destination. This procedure is accomplished by the use of routing tables and dynamic routing. That is, the node for any portion of the same message can select different channels according to the actual network situation. The path that the message, or portion of the message, follows to the next available node is an algorithm based on congestion, service availability, and input and output queues. Nodes provide end-to-end integrity and message acknowledgment. The node closest to the end user's destination must resequence the message into its original form. The resequencing is required because the message could have been transmitted over different channels.

7.2.2. X.25 PAD Facility

In most public packet switched data networks, the network provider includes a PAD function. A PAD performs a packet assembly and disassembly and allows DTEs such as start-stop or character-oriented terminals to send and receive messages over the PSDN.

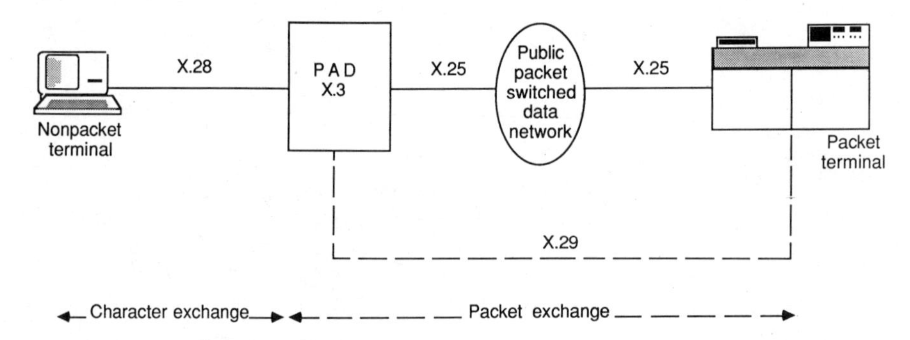

Figure 7.4. CCITT PAD recommendation.

PAD functions have increased the utility of PSDNs. Normally, each node in the network has a PAD connected to it that provides the interface to the telephone switched line or leased line from asynchronous start-stop terminals (DTEs). Normally, PADs are powerful medium-sized computers or specialized computers for implementing the basic PAD functions for speed, code, and protocol conversion. The PAD functions for start-stop terminals are designated by the CCITT X.3, X.28, and X.29 Recommendations and are sometimes called the *Triple X* functions.

Some of the basic functions that must be provided by the X.3 Recommendation are assembling characters from the terminal into packets; disassembling packets into characters for start-stop-type DTE, virtual call setup, and call clearing; forwarding packets; and recognizing and interpreting interrupts and service signals. In terms of DTE connections, RS-232-C applies to North America, and CCITT X.21 bis applies to other countries.

The CCITT X.28 Recommendation defines the DTE/DCE interface for start-stop mode data terminal equipment accessing the PAD in a public data network situated in the same country. It includes establishing the connection, initializating the sequence, and exchanging control information and user data. X.28 describes the connection between the DTE and leased line or the PSTN with full-duplex modems.

The CCITT X.29 Recommendation establishes the procedure for the exchange of control information and end user data between a PAD and a packet mode DTE or another PAD. The relationship between the X.3, X.28, and X.29 Recommendations is shown in Figure 7.4.

Figure 7.5 presents a typical node arrangement using a PAD facility. The uses of computer-based PAD facilities include DTE start-stop interfacing by either dial-up (PSTN) or private leased line. In addition, many network providers have added 3270 BSC interactive and 3780 BSC batch interface support to appeal to a broader market. BSC support does not conform to CCITT recommendations and is available only in Canada, the United States, and the Federal Republic of Germany. A typical node supports DTE direct attachment via leased lines and would have one or more medium- to high-speed connections to one or more nodes. PADs normally attach to the node processors through high-speed direct computer-to-computer connections. Figure 7.5 is a typical network node arrangement.

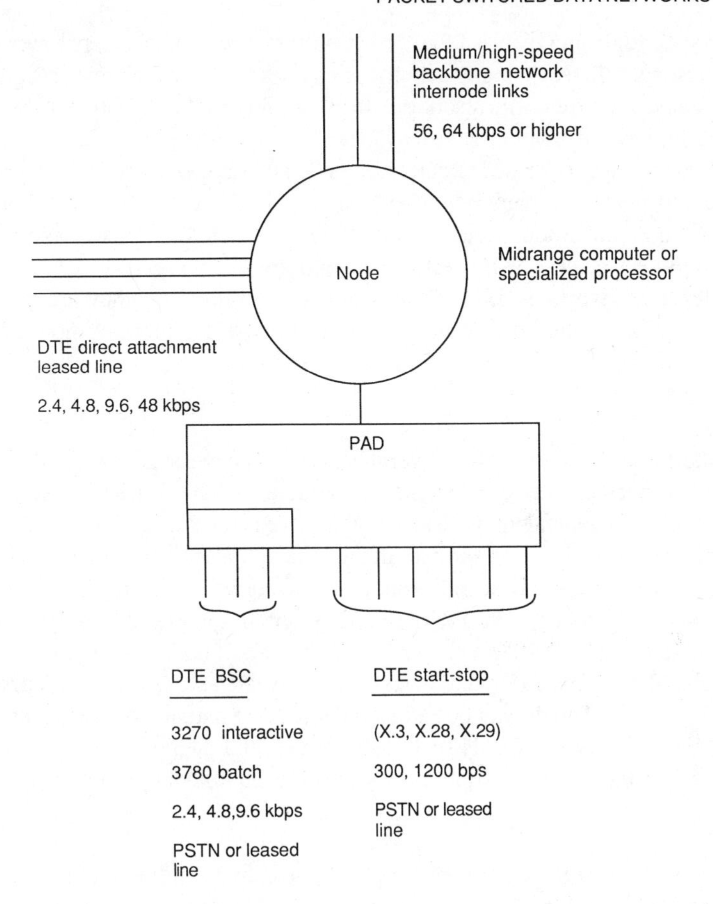

Figure 7.5. Typical node arrangement.

7.2.3. Benefits of PSDNs

PSDNs offer many potential benefits, both for the network provider and the user. Reduced message charges result from distance-independent charging. Normally, most PTTs and network providers charge for the number of segments, which are based on 64 octets or characters. Data connections are either by local dial-up calls or private leased lines to the nearest node. In many countries, nodes are widely dispersed based on equal service to all regions.

For many applications, existing equipment (DTE) can be used, resulting in little or no terminal investment. The use of PADs allows different speeds; for example, an asynchronous terminal operating at 300 bps into the network can "talk to" a host that has a direct private line at 9600 bps at the remote end of the network. Whatever the contents of the message, it is reproduced at the receiving end. In addition, different protocols and procedures, e.g., start-stop into the PAD and HDLC output to the host, are allowed in most PSDNs.

PSDNs are replacing many of the older Datel services. Datel offers lower speeds in comparison to PSDNs in many countries and does not have the wide acceptance of PSDNs. Leased circuits offer users greater flexibility and international leased circuits offer high quality, but PSDNs may offer a more economical solution where the particular advantages of leased circuits cannot be justified. Geographical distribution is difficult to obtain in many countries due to availability of facilities. For inter-enterprise communications, leased circuits are restricted for many applications, while PSDNs have little or no restrictions and are encouraged by PTTs. Several PTTs offer attractive rates for PSDNs, while at the same time they are increasing tariffs on leased circuits to drive customers to PSDNs to increase initialization of these facilities.

End Users

For an enterprise, X.25 may offer several benefits. The message cost is based on the number of segments, usually 64 octets or characters, that are transmitted over the network. Cost is independent of distance, and there is no charge based on the length that the message travels. For example, in the United States, Telenet and Tymnet are available from over 230 cities. In Europe, the node density is greater. For example, France has over 80 nodes in the TRANSPAC system in a land area equal to 5.88% of the United States, and Germany (FRG) has more than 21 nodes on Datex-P in a land area equal to 2.65% of the United States. These private and PTT network providers offer PAD facilities and direct node attachment in X.25 native mode. They also offer PAD for access over the PSTN or leased circuits and support a number of other protocols. For the end user, this allows existing equipment to be used with the network:

- Calls from the PSTN to the nearest node: terminals with asynchronous (start-stop), BSC, and HDLC procedures or protocols
- Leased line connection to the nearest node: terminal with direct X.25, asynchronous (start-stop), BSC, and HDLC procedures or protocols

X.25 from a PTT Standpoint

The basic advantage for the PTT is that the trunk circuits can be time-division-multiplexed between the two nodes. That is, a line is not dedicated to any given pair of remote users but is shared by many users. The data over the trunk lines can be multiplexed dynamically for traffic from different subscribers. The X.25 network is different from switched telephone or leased line networks. In packet switching, there is never a set of physical resources fully dedicated to a unique message. For any given period of time, the same network resources serve multiple concurrent users from different enterprises. Packet networks provide fast call setup procedures and high-quality transmission at a reasonable price.

Private Packet Networks

In a number of countries, including the United Kingdom and the United States, customers may install their own private packet switching networks by purchasing network nodes, complete with operational software, from a network node supplier, or they may build their own nodes and use high-speed leased lines from the PTT. Another alternative is a managed data network, which could be supplied by contracting with independent network providers for their own network or by sharing the provider's network with other users. There are advantages to either a private or public X.25 network. It allows an independent architecture and separation of network management from DP application management. Packet networks offer an enterprise with a significant number of terminals and host systems, the ability to have an independent network. As an alternative to a fixed private leased line arrangement, an X.25 network built on leased lines could offer a variety of traffic paths with dynamic routing instead of single (or multiple) predefined routes. Most public and private packet switching data networks offer the following advantages:

1. High throughput, dynamic routing, high reliability, and guaranteed receipt of messages.
2. Host systems and remote terminal equipment from different vendors are integrated into the network. However, the host complex must have the appropriate protocol conversion for application-dependent programs.
3. Packet networks are not hierarchical and therefore are not dependent on host software for the transmission of messages.
4. Packet networks reduce some of the network management responsibilities of application management.

7.2.4. X.25 and X.21 Relationships

The difference between an X.25 or packet switched data network and an X.21 or circuit switched data network can be shown as in Figure 7.6. A fully implemented direct X.25 network connection will use the X.21 physical level for direct connection to the nearest X.25 network's node. The CCITT X.25 Recommendation is the basis of the first three layers of the OSI Recommendation. X.21 is used in layer 1 of the CCITT X.25 Recommendation.

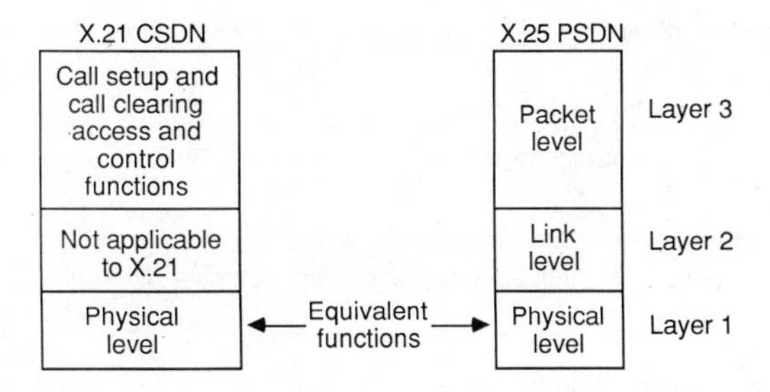

Figure 7.6. X.25 and X.21 comparison.

7.2.5. X.25 and OSI

The problem in dealing with terminals and computer systems is the wide variety of products that are incompatible with each other. OSI is an attempt to have standard properties and uniform interfaces between products developed by independent suppliers. If each product implements the various layers in the model, each terminal or remote device will represent an ideal or virtual terminal that can be handled by the applications program in the host computer. The main advantage is that applications programs can be designed independently of any particular terminal. X.25 is the basis of the first three layers of the OSI model. Figure 7.7 shows the OSI reference model and two intermediate X.25 network nodes.

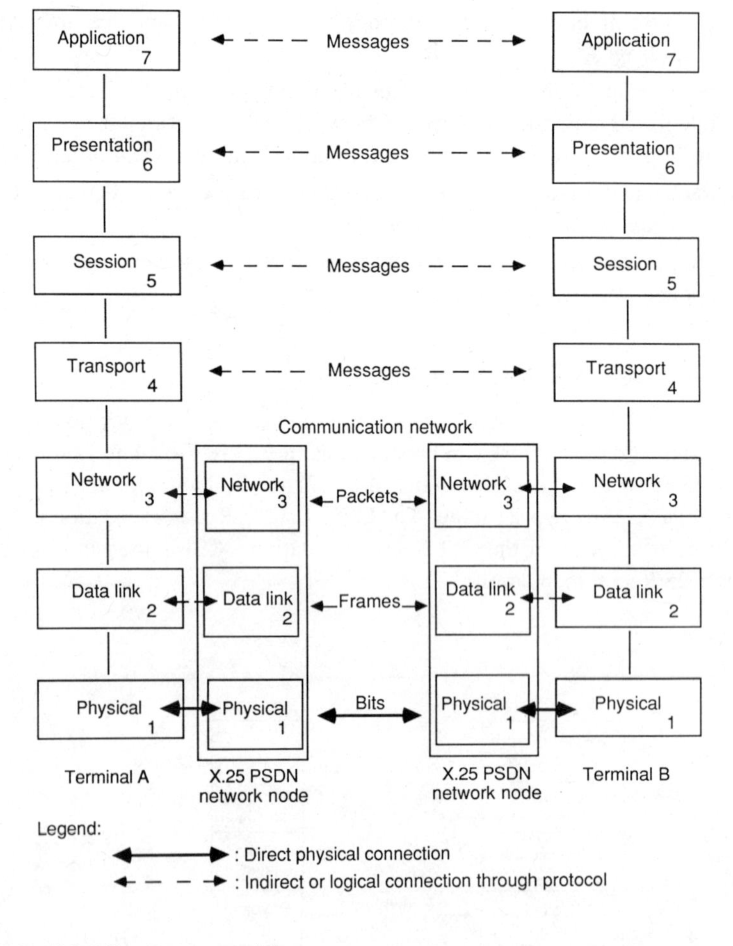

Figure 7.7. ISO OSI reference model.

7.2.6. X.21 and OSI

At the three lower levels of the OSI model, the X.21 interface is represented by a physical layer and a call setup, call clearing, access, and control function layer. Note that there is no link level in X.21 because each DTE has a dedicated link to the first switching node.

As in X.25, higher-level compatibility between open systems will not be a network function but an end-to-end function.

7.2.7. Country Status

X.25 networks are installed in over 62 countries or interwork through gateway countries (see Chapter 10). Interworking, or the ability to transfer information across national boundaries, is based on the CCITT X.75 Recommendation, which PTTs implement for international traffic.

Table 7.5 lists the 61 nations that have PSDN or privately supplied packet switched networks. In some countries, there are two X.25 PSDNs, one for national information transfer and another for international traffic. Depending on the country, the national network may or may not link to the international PSDN. In this case the international provider provides the complete connection within the country.

TABLE 7.5. Country Status

Country	Network Name	Country	Network Name
Argentina	ARPAC	Dubai	IDAS
Australia	AUSTPAC (national) MIDAS (international)	Finland	DATAPAK HTC-PSDN
Austria	DATEX-P	France	TRANSPAC
Bahrain	IDAS	Gabon	GABONPAC
Belgium	DCS	Germany (FRG)	Datex-P
Brazil	RENPAC	Greece	HELLASPAC
Cameroon	CAMPAC	Greenland	KANUPAK
Canada	DATAPAC (Telecom Canada) INFOSWITCH (CNPN)	Hong Kong	DATAPAK (national) INTELPAK (international)
Chile	ECOM	Hungary	NEDIX
China	CHINAPAC	Iceland	ERIPAX
Columbia	COLDAPAQ	Indonesia	INDOSAT PACKSATNET
Costa Rica	—	Ireland	EIRPAC
Denmark	DATAPAK, PAXNET		

(continued)

TABLE 7.5. Country Status

Country	Network Name	Country	Network Name
Israel	ISRANET	Saudi Arabia	—
Italy	ITAPAC	Senegal	—
Ivory Coast	SYTRANPAC	Singapore	TELEPAC
Japan	DDX-P (national) VENUS-P (international)	South Africa	SAPONET
		South Korea	DACOMNET
Kenya	—	Spain	IBERPAC
Kuwait	—	Sweden	DATAPAK
Luxembourg	LUXPAC	Switzerland	TELEPAC TYMNET
Malaysia	MAYPAC	Taiwan (ROC)	PACNET
Mexico	TELEPAC	Thailand	Packet Switched Network
Morocco	—		
Netherlands	DATANET 1	Tunisia	—
New Zealand	PACNET	Turkey	—
Norway	DATAPAK	United Arab Emirates (UAE)	—
Oman	—		
Panama	—	United Kingdom	PSS—Packet SwitchStream (national) IPSS (international)
Peru	ENTEL		
Philippines	—		
Portugal	TELEPAC	United States	Several
Puerto Rico	UDTS	Venezuela	VENEXPAQ (future)
Qatar	—	Yugoslavia	JUPAK

Source: Telenet (U.S. Sprint), Tymnet McDonnell Douglas, PTT published reports.

7.3. INTRODUCTION TO ISDN

A low growth rate in the basic telephone service market is leading PTTs to develop new services to increase revenue. Most of these new services have been implemented on dedicated networks because of the existing structure of the analog telephone network. However, it is becoming increasingly difficult for the PTTs to cope with the diverse communication requirements by continuing to build dedicated networks (e.g., X.21, X.25). A digital backbone network is now being implemented in most developed countries to upgrade the public telephone service. As a result, the technical difference between voice telephone and data transmission is disappearing. Most PTTs are exploiting the convergence and promoting the concept of an integrated services digital network (ISDN).

The ISDN is a concept that will evolve from the integrated digital telephone network (IDN) and provide end-to-end digital connectivity to support a wide range of services, including voice and nonvoice. Voice, as well as information services, can be handled over the existing local two-wire subscriber line. The main objective of ISDN is the combination of economy and flexibility with one common interface to a PTT or Telco telecommunications network. In the past, as a new data or nonvoice service was added, a separate network was set up by the PTT, resulting in a higher overhead cost to the PTT and the customer. ISDNs should reduce the number of voice and data networks coming into a customer's premises to one two-wire cable. The starting point for the integrated digital network is the digitization of the backbone network. Although data services are expanding more rapidly than telephone service in most countries, data subscribers represent 2 or 3% of the total number of telephone subscribers. Since telephone will remain the prime PTT service beyond the year 2000, a requirement for the successful introduction of ISDNs is that the charge for monthly service should be about that of the existing basic telephone service.

An ISDN is an all-digital network that can support simultaneous voice and data transmissions over a single high-speed access line coming into a customer's premises. An ISDN can mix voice and data simultaneously. Depending on the amount of services a customer subscribes to, annual savings could result by reducing or eliminating multiple incoming lines that now carry voice and data traffic. Some of the basic premises are as follows:

- ISDN is based on a digital telephone network
- ISDN permits digital end-to-end connections between subscriber installations (the base channel is 64 kbps).
- Subscribers can access multiple communication services via one digital subscriber line.
- Interfaces and protocols will comply with relevant CCITT recommendations, allowing consistent international interfaces and services.

An ISDN network consists of local loops, intercity channels, digital switching facilities between cities. Local channels can carry voice, data, or both for signaling and message information.

The basic ISDN service will be a 64-kbps connection for the subscriber. The ISDN will be a circuit switched system with circuit setup times on the order of seconds. All connections will be point to point.

The subscriber will access the network through the ISDN's new physical connection. As a migration path, the PTTs could provide for older equipment, e.g., X.21 and RS-232-C physical connections.

The ISDN is an expensive undertaking for the PTTs; new equipment is required, and the changes are labor intensive. Multiple low-speed analog circuits entering a customer's premises will be replaced with one high-speed digital line. The user will also have to make large investments in new equipment. To ease the users investment, it is expected that the PTTs and common carriers will offer analog adapters that allow the connection of existing analog devices.

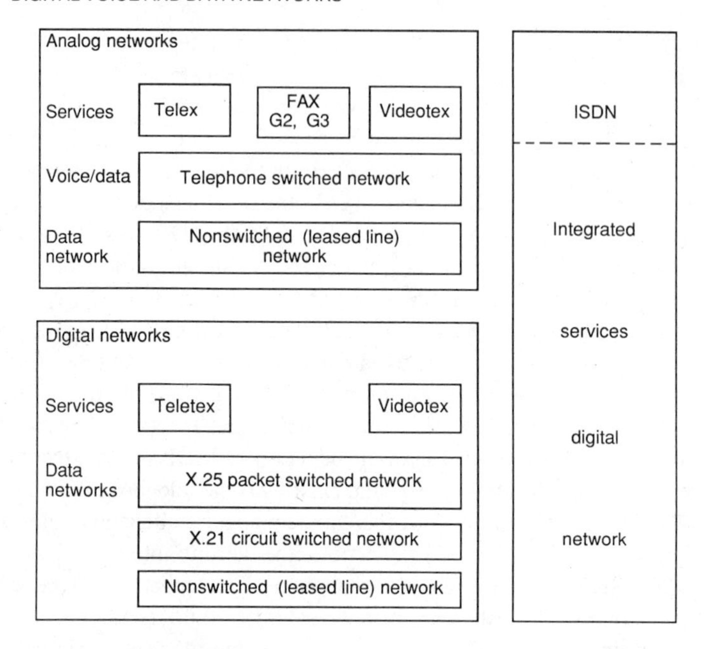

Figure 7.8. Evolution of PTT services.

Figure 7.8 is an example of existing services offered by the PTTs on analog and digital networks. Although all services are not available in every country, each service, telephone switched network, or data network is a separate network to be maintained and updated. The PTTs intent is to reduce the number of networks entering into a customer's premises by one two-wire cable. When fully implemented, this reduction of networks will dramatically reduce the PTTs' operating costs.

The ISDN provides two types of channels for the exchange of information:

1. A user information transfer channel or B channel carries user information up to the transmission rate of the channel. Several rates have already been defined using three different channels:

   - B channel: A 64-kbps channel
   - H0 channel: A 384-kbps channel
   - H1 channel: A 1536-kbps channel in Canada, Japan, and the United States or a 1920-kbps channel in Europe

   Additional channels with transmission rates ranging from 6 to 140 Mbps are being defined and are sometimes referred to as broadband channels.

2. A user-to-network control channel, also called the D channel, carries control or signaling information between the end user and the network and a packet switching service. The D channel protocol provides message handling, error detection and recovery, multiplexing, and addressing capabilities as part of data link layer 2. Call setup, call clearing, and additional call related capabilities are part of the network layer 3 function set.

7.3.1. ISDN User-Network Interfaces

There are two user-network interfaces presently defined for ISDN applications: the basic interface and the primary access interface.

The basic interface provides for network access by either small business users or residential users. It allows for one of three access arrangements: 2B + D, B + D, or D channels. In each case, the B channel operates at 64 kbps, and the D channel operates at 16 kbps. For example, a terminal or stand-alone work station will typically be connected to the local PTT's central office by one access channel, consisting of two 64-kbps B subchannels and one 16-kbps D subchannel, which is referred to as 2B + D (64 kbps + 64 kbps + 16 kbps = 144 kbps).

The primary access interface provides for network access by private automatic branch exchanges (PABXs), local area network (LAN) gateways, and other user nodes such as communication controllers and cluster controllers. In the United States and Canada, the primary channel consists of 23 B subchannels, each operating at 64 kbps, one 64-kbps D channel, and 8 kilobits of synchronization for a total channel rate of 1.544 Mbps, which is based on the T1 service. A B channel carries only message information, while the 64-kbps D channel is used primarily to carry signaling information. The European or CCITT equivalent is a 30B + D, which is based on a total channel rate of 2.048 Mbps, the CCITT G.703 Recommendation.

The primary access interface may consist of any combination of D, B, H0, and H1 channels where the total transmission rate does not exceed the primary rate. In this case the D channel operates at 64 kbps.

Another important feature of the ISDN is that it allows a user to reconfigure the network. For example, by using an on-site terminal or PC, a user can communicate with the ISDN control center via the D channel. By inputting specified codes into the PC keyboard, the user can initiate changes in the network. The network software and related digital switching equipment will then implement these instructions.

7.3.2. Subscriber Access Architecture

User network interfaces are presently proposed at reference points R, S, and T. R is the reference point closest to the user, while T is the farthest point from the user. A reference point "U" is being defined as a convenient theoretical line of demarcation between user premises equipment and the supporting network. See Figure 7.9.

CCITT subscriber basic access is defined in Figure 7.9 for two full-duplex B channels and one full-duplex D channel (signaling channel). Definitions for each block and for reference points are as follows:

NT1 = Network termination type 1: In Europe PTT-supplied interface on the customer's premises that performs functions equivalent to the physical layer or layer 1 of the OSI reference model (e.g., line interfacing).

NT2 = Network termination type 2: NT2 includes customer premises equipment such as PABXs, communications controllers, LANs, and terminal controllers. NT2 multiplexes more than one input line into one output line. Under the new ISDN passive bus, an NT2 is not required for eight or fewer terminals. In that case, the S reference and T reference would be identical.

TA = terminal adapter: Always located between R and S, the TA is a protocol converter that translates existing terminal protocols (e.g., X.21, X.25) or other non-CCITT protocols to an ISDN connection.

TE1 = ISDN terminal that includes a B (data) plus D (signaling) connection.

TE2 = Existing non-ISDN terminals or interfaces not included in CCITT recommendations.

R = Reference point between an existing terminal (V.24, X.21 bis, X.21, or X.25) interface and a TA.

S = Reference point between TE1 or a combination of TE2 and TA to NT2. The S interface always has a 2B + D connection.

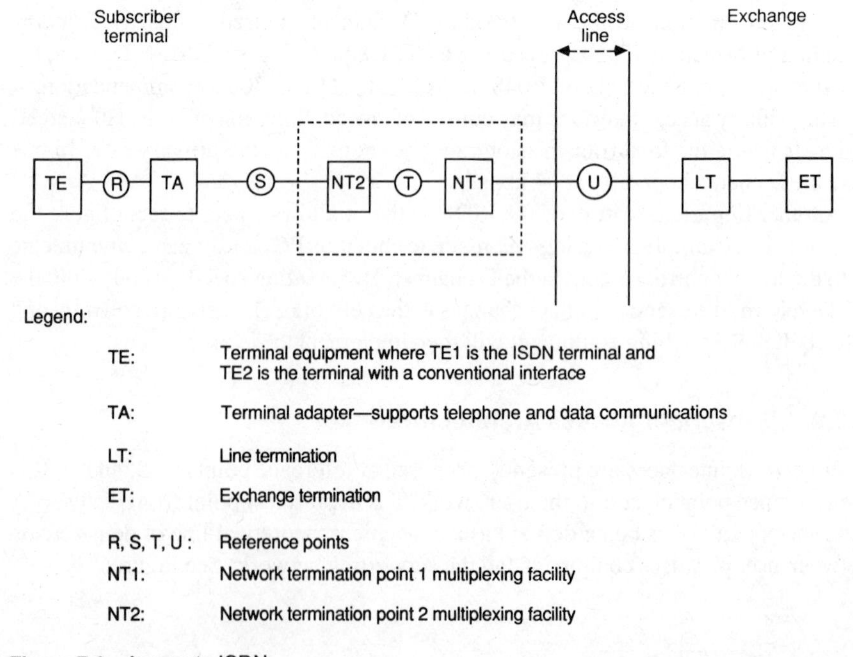

Figure 7.9. Access to ISDN.

B channels: There are two channels in the basic access configuration. Each channel permits 64 kbps of full-duplex transmission. The channels carry digital coded voice information, text, data, or image.

D channel: The D channel transmits the subscriber signaling information required for the two B channels. In the basic access configuration, the information carried in the D channel is packet oriented and transmitted at a rate of 16 kbps. The D channel controls external connection requests, charge-rate information, and the transfer of information regarding the requested service in the B channel. In addition to transmitting signaling information, the D channel can be used to transmit packet-oriented subscriber or telemetry data.

Subscriber interfaces: Terminals used with the integrated services network will provide access to voice and text services. Therefore, the subscriber interface at reference point T must be able to provide access to the two B channels and the D channel both for a single terminal device or for several terminal devices. Text and data terminal communications interfaces are linked to the terminal adapter's TA, whose function is to convert the transmission rate and signaling information between the channel of the interface at reference point R, used for transmitting both signals and useful information, and the dedicated signaling channel at reference point T.

Network access: Figure 7.9 shows the functional groupings for network terminations NT2 and NT1 and line termination LT. NT1 and NT2 terminate the network, and LT terminates the line at the exchange. LT and NT1 are responsible for performing the transmission tasks of transmitting and receiving signals at 144 kbps (64 kbps + 64 kbps + 16 kbps) over the local loop or subscriber access line. NT2 handles distribution and concentration tasks involved with different terminal device configurations.

Primary rate access: Primary rate access called *primary* because the network access speed matches the multiplexing structure that the PTTs and Telcos have in their backbone network. To combine a large number of B channels, CCITT has a structure made of 30 B channels and 1 D channel in Europe and 23 B channels and 1 D channel in Canada, Japan, and the United States. To support higher throughput, signaling information on the D channel is transmitted at 64 kbps.

Network interworking: Each ISDN subscriber must be able to access other ISDN subscribers and other networks. The conversion of signal information and transmission rates is the principal task of the network interworking facilities. ISDN is basically a circuit switching network but will provide packet-oriented terminals with access to packet switching networks.

7.3.3. ISDN Services

ISDN as defined by CCITT includes standards for transmission services called bearer services and value added services or teleservices. Value added services or teleservices are services in which the layers and terminal communications functions have been defined. Bearer services are used for the transparent transmission of voice or data.

Future services that may be realized are:

- Telephone will have enhanced speech quality with a better signal-to-noise ratio independent of distance, improved handsfree speaking, and speakers. Future developments may include stereo sound and audio conferencing.
- Teletex will offer enhancements over the existing Teletex service. Due to the higher transmission speed of 64 kbps, the transmission time for the text of one A4 page will be reduced to 2 seconds.
- Telefax will operate at 64 kbps, which will result in a transmission time of about 10 seconds for an A4 page or 8 1/2- by 11-inch page. Coding and resolution will be based on the Group 4 Fax and will conform to OSI layers 4–7.
- Texfax will be used for combined text and facsimile communication to transfer office documents containing text, diagrams, handwritten notes, letterheads, or signatures. Facsimile service characteristics such as character sets, resolution, communication protocols, and transmission times will correspond to ISDN Teletex and ISDN Telefax.
- Transmission services include the packet circuit and switched data communication services operating at transmission rates of 64 and 128 kbps. Packet switched services can be implemented in ISDN by means of central packet switching facilities.

Figure 7.10 includes introductory services and future services as well as new transmission services.

ISDN Videotex with a higher transmission rate of 64 kbps will have a faster image buildup to about 5 seconds depending on the image content and coding method during information retrieval. Faster input of Videotex frames into the Videotex centers by the information providers will increase information retrieval services and electronic mailbox applications. Voice mail and text mail are storage services for voice, text,

Services	Introductory services	Transmission services	Future services
Existing services moved over from the telephone network	Telephone Telefax (Group 2/3)	Data transmission with V series interfaces	Videotex
New ISDN services operating at 64 kbps	ISDN telephone ISDN video conference ISDN Teletex ISDN Telefax (Group 4) ISDN Texfax	Data transmission (circuit switched, packet switched)	ISDN Videotex Voice mail Text mail

Figure 7.10. ISDN services.

and data with central storage and mailbox functions to permit information retrieval by the recipient. The electronic mailboxes are implemented in the information storage centers of the information providers.

Alarm and emergency services can be provided for fire, theft, and accident reporting. Telemetry can provide remote meter reading and control of company and household systems such as heating, electricity, gas, and water. The ISDN will simplify the addition of alarm services and telemetry services because at 64 kbps, there is greater availability of lines at low error rates with digital techniques. The shorter transmission times for alarm and telemetry services could lead to more applications at a lower subscriber cost.

7.3.4. Pre-ISDNs

The ISDN is generally viewed by the PTTs as an evolutionary development resulting from the increased use of digital technology within existing telephone networks. As networks evolve to the stage where digital transmission and switching are employed internally, they are referred to as integrated digital networks. These networks or pre-ISDN networks will then evolve into ISDN's, providing transmission service for all types of information, including digitized voice, data, text, and image, which will result in the development of a wide range of new services. Thus, the ISDN is not a new and separate network but the result of the increasing use of digital technology in the public telephone networks.

Major U.S. network vendors predict that an AT&T nationwide public ISDN could be available to users as early as 1990. About half the trunk circuits linking central telephone offices are now digital, a prerequisite for the integrated voice and data service.

British Telecom (BT) in the United Kingdom has a pilot service called integrated digital access (IDA) that provides the subscriber with one 64-kbps channel, one 8-kbps channel (for data only), and one 8-kbps signaling channel. Access is gained by a digital telephone and the usual subscriber interfaces conforming to the CCITT X series recommendations. IDA will provide access to many BT services over a single physical link. It combines voice and data transmission over this digital link and replaces the need for separate access to the existing BT network. One single-line IDA is the customer's link to the local BT System X ISDN exchange and provides an 80-kbps full-duplex digital stream, composed of three channels: See Table 7.6 for the application of the band D channels.

TABLE 7.6. Application of the Band D Channels

Channel	Application
B1	Voice, or Synchronous data up to 64 kbps, or Asynchronous data up to 9.6 kbps
B2	Synchronous data up to 8 kbps, or Asynchronous data up to 1200 bps
D	8 kbps signalling channel

Two simultaneous calls to different destinations can be made over one single-line IDA link. BT developed two types of network terminating equipment for the initial service and offers a different mix of interfaces to a digital telephone: CCITT X.21 and/or X.21 bis standards.

In addition, a multiline IDA is designed to support connection of suitable digital PABXs. Each PABX is directly connected to the System X exchange via a 2-Mbps digital path for up to 30 simultaneous 64-kbps channels over one link. The PABX connects directly to the IDA line and switches between this line and its extensions. A variety of voice or non-voice terminals can be connected as extensions supporting a number of services. In effect, the PABX becomes an integrated services PABX.

The Japanese national PTT, Nippon Telegraph and Telephone Corporation (NTT), has implemented the Japanese version of ISDN called INS (information network system). Pilot services on the INS model system have been offered to selected businesses and residential users from September 1984. In Italy, an ISDN pilot project is planned with a subscriber access consisting of one 64-kbps channel and one 16-kbps signaling channel, both of which will have access to the packet switching network.

The pre-ISDN stage corresponds to the existing PTT situation where separate networks are used, each dedicated to a family of services. Some of these networks

TABLE 7.7. ISDN Strategy – Pilot or Public Service by 1990

Country	Service Type
Austria	Pilot Project
Belgium	Public Service
Canada	Pilot Project
Denmark	Stated Intention
Finland	Public Service
France	Public Service
Germany (FRG)	Public Service
Ireland	Public Service
Italy	Public Service
Japan	Public Service
Netherlands	Pilot Project
Norway	Pilot Project
Spain	Public Service
Sweden	Public Service
Switzerland	Pilot Project
United Kingdom	Public Service
United States	Public Service

Source: PTT published reports.

contain the prerequisites for becoming an ISDN network. Countries with plans for implementing ISDN networks or the preliminary pilot stages prior to year end 1990 are listed in Table 7.7. Responses are a stated intention, pilot project, or public full service. In cases where a public service is committed, partial coverage within the country will be in effect. Full implementation for all subscribers will not occur in most countries until the late 1990s or into the 21st century.

7.3.5. ISDNs in Germany (FRG)

The conversion of the Deutsche Bundespost's telephone network from analog to digital will make it possible to integrate all low-speed communication services into one single network. In 1986, several of the first digital switching centers were in operation. This development is an expansion of the existing digital network for text and data transmission operating at transmission speeds of 64 kbps.

The ISDN will exist in parallel with existing networks. One reason for this is the existence of special applications, particularly in data communications, that the ISDN will not be able to handle initially. The PTT is planning to introduce phase 1 or the integrated digital network (IDN), which will have local subscriber loops at 64 kbps. The basic access will be 2B + D (16 kbps). The S network interface termination will belong to the PTT, although it will be located on the subscriber's premises. By 1988, the PTT plans to offer users a digital telephone, a 64-kbps data service, Teletex Group 4 Fax, and access to the packet and circuit switched data networks. Adapters will be available so that current terminals can be used. The initial ISDN offerings will be modest: from eight local exchanges, each with 1000 ISDN subscriber lines in 1988, to a total of 600 local exchanges, each supporting 1500–2000 subscribers by 1993. It is projected that by the 21st century about 20% of telephone subscribers will subscribe to ISDN. Table 7.8 represents the progression of network integration by the Deutsche Bundespost.

7.3.6. ISDNs and OSI

The ISDN was designed according to the layered protocol of the open systems interconnection (OSI) reference model of the ISO. ISDN provides the lower three layers (physical, data link, and network) of the OSI.

Access to the ISDN provides a standardized connectivity tool common to several types of information: voice, text, data, and image. This group of services includes the circuit and packet switched transmission services operating at a 64-kbps rate.

In the case of circuit switched data network services, the standards cover the user-to-network signaling (e.g., the D channel protocols) of layer 1 and 3 functions for the user-to-user information transfer. For packet switched services, the standards cover, in addition to the X.21 CSDN, the layer 2 and 3 protocols of the user information packet transfer.

Text-oriented services (e.g., Teletex and Telefax) with a channel rate of 64 kbps will reduce the transmission time of an A4 page by a factor of 10, while the availability of two B channels at the user-to-network interface will provide a simul-

TABLE 7.8. Progression of Network Integration

<table>
<tr><th>Services</th><th>Current Network</th><th>From 1988</th><th>From 1990</th><th>After 1992</th></tr>
<tr><td>Telephone Data transmission in the telephone network
Telefax (Group 2/3)
Interactive Videotex</td><td>Telephone network</td><td rowspan="2">Narrow-band ISDN, 64 kbps</td><td rowspan="3">Narrow-band and broadband ISDN</td><td rowspan="4">Universal network</td></tr>
<tr><td>Telex
Datex-L (X.21)
Teletex
Datex-P (X.25)
Telefax (Group 4)</td><td>Integrated text and data network (IDN)</td></tr>
<tr><td>Video Telephony
Video conferencing</td><td>BIGFON</td><td>Video conference trial network</td></tr>
<tr><td>Broadcast radio and TV</td><td>Community antenna television systems</td><td>Broadband communications networks</td><td>Broadband communications networks</td></tr>
</table>

Source: Deutsche Bundespost, the Federal Republic of Germany.

taneous voice and text capability. For image-oriented services (e.g., Videotex), the channel rate of 64 kbps will guarantee a faster image buildup and quicker input of Videotex frames into the Videotex centers by information providers (IPs). From a user's standpoint, customer premises equipment (CPE) on the ISDN should include the applicable OSI layers in Table 7.9.

7.3.7. Standardization

The CCITT Study Group XVIII is responsible for the ISDN and has prepared a substantial number of draft recommendations covering general network structure and subscriber interfaces, called the I Series. The CCITT Study Group XI is involved in D channel signaling, and the CCITT Study Group VII has prepared recommendations covering the terminal connections to the ISDN with conventional X.21, X.21 bis, and X.25 interfaces.

The ISDN architecture is described in CCITT I Series recommendations. North American standards for ISDNs are being developed by the Exchange Carriers Standards Association (ECSA) Telephony-1 (T-1) Committee. European standards for ISDNs are being developed by each PTT with coordination by CEPT for compatibility (CEPT ISDN Specification TCS46XX). Complementary standards are being developed by the European Computer Manufacturing Association (ECMA) Technical Committee 32 Technical Group 1 (TC-32/TG-1).

TABLE 7.9. OSI Layers Applicable to Specific Customers' Premises Equipment

Customer Premises Equipment	Layers
Modems (Analog to Digital) Leased Line	1,2
Modems (Analog to Digital) Switched)	1,2,3
Telefax	4 to 7
Teletex	4 to 7
Videotex	6
X.400 Message Handling System	7

8

U.S. NETWORK SERVICES

In this chapter we will

- List private switched and leased line networks in the United States
- Discuss high-speed digital and packet switched networks
- Compare packet versus leased lines, and leased lines versus a multipoint line

In the past, whenever a communications line crossed a right-of-way or public conveyance (i.e., extended beyond a person's or firm's private property line), a common carrier was required to provide the communications line. This rule was changed with bypass, which allows microwave links and satellite links to operate across a right-of-way. This right to access is the ability, within the United States, to use local telephone, microwave, or satellite networks to send and receive intrastate (same state), interstate (another state), or international (another country) transmissions.

To provide communications between two or more states, a common carrier (provider of public transmission facilities) must have its services approved by the Federal Communications Commission (FCC). All U.S. common carriers are licensed and regulated by the FCC, Washington, D.C. The FCC provides federal regulations between states (interstate). Communications regulations within the state (intrastate) are regulated by a government body called a public utility (or service) commission.

Common carriers transmit voice, data, and messages to the general public for a fee. Within the United States, carriers are structured as telephone carriers and nontelephone carriers. Telephone carriers are certified by the FCC to provide communications services to the public such as telephone and message services. Nontelephone carriers are specialized common carriers that provide telex and electronic mail services, domestic satellite services, and value added network services (VANS).

8.1. PRIVATE NETWORK OFFERINGS

Public networks are available to everyone, and each customer shares the same network with others. Private leased line networks can be obtained that offer a dedicated line to connect different establishments within the same enterprise or between enterprises for the exclusive use of the connected parties. Telephone and nontelephone carriers that offer public and private networks for voice, data, video, and other applications are listed in Table 8.1. A description of each private network and the major carriers will be given following the table.

8.1.1. Private Nonswitched Networks

There are two categories of nonswitched networks available for private use: analog and digital. Most customers require both types of service to satisfy their data communications needs.

Analog Services

A voice-grade service is a line suitable for analog transmission signals in a frequency range of 300–3400 hertz—also called a 4000-hertz analog channel. Analog voice-grade circuits for voice and data are the most predominant nonswitched networks for business use in the United States, Nonswitched lines are available in a point-to-point or multipoint configuration. These channels can transmit voice, data, or other applications within the frequency range of 4000 hertz. From a usage standpoint, AT&T-C can supply either a voice channel or a data channel. Voice channels allow unlimited voice for 24 hours per day, 7 days per week.

Data calls require an acoustic coupler or modem to convert a computer's digital output signal to analog for operation on an analog line. Newer modems use various compression techniques and include microcomputers in their design in order to convert digital data at rates up to 19.2 kbps over an analog channel.

Private switched line services are provided by the Regional Bell Operating Companies (RBOCs), General Telephone and Electronics (GTE), and other independent telephone companies within the same local access and transport area or intra-LATA. The long-haul portion of a private line service between different local access and transport areas or inter-LATAs for both intrastate or interstate traffic are generally supplied by AT&T Communications (AT&T-C), U.S. Sprint Communications (U.S. Sprint), ITT/United States Transmission Systems (USTS), MCI Communications (MCI), Western Union Telegraph (WU), and others.

TABLE 8.1. Network Services for Voice, Data, and Other Applications

Private Network Offerings	
Service	Service Provider
Private non-switched services	
Analog Private Line	AT&T Communications ITT/United States Transmission Systems MCI Communications U.S. Sprint Communications Western Union Telegraph Company
Digital Private Line	AT&T Communications MCI Communications U.S. Sprint Communications
ACCUNET® T1.5 Service	AT&T Communications
Terrestrial digital services	MCI Communications
Switched Digital Services	
High-Speed Services	Local Exchange Carriers
ACCUNET®[a] Reserved 1.5	AT&T Communications
ACCUNET® Switched 56 Service	AT&T Communications
Satellite Services	
Domestic service	American Satellite Argo Communications AT&T Communications RCA American Communications (a unit of GE) GTE Spacenet Corp. Satellite Business Systems U.S. Sprint Communications Western Union Telegraph Company

[a]Registered service mark of American Telephone and Telegraph Company.

Digital Services

Digital data networks offer all digital channels for synchronous data at speeds from 2.4 to 56 kbps. At present, two companies provide point-to-point digital service, AT&T-C and U.S. Sprint.

AT&T-C DATAPHONE Digital Service (DDS) provides an interstate digital line at speeds ranging from 2.4 to 56 kbps. DDS provides interstate private line digital communications between major cities. The major advantage of a digital line is that noise and distortion are not amplified as in analog transmission. Digital signals are regenerated at regular intervals, or the pulse is detected and reconstructed into a new equivalent pulse. This process leads to higher-quality data transmission, greater availability, and less downtime. DDS is guaranteed by AT&T-C to provide an average performance exceeding 99.5% for operation at all speeds over a continuous 24-hour period. Credit is allowed if performance falls below this limit.

U.S. Sprint offers a digital data service between seven cities for voice, data, Fax, and other applications. Line speeds are offered from 2.4 to 9.6 kbps. The U.S. Sprint channel may handle up to 240 voice-grade channels.

1.544-Mbps Terrestrial Digital Services

T1 is a 1.544-Mbps facility designed to make more efficient use of voice and data traffic over terrestrial facilities. T1 circuits have digital repeaters about every 6000 feet, which regenerates the digital signal. At a 1.544-Mbps rate, time division multiplexers can be used to transmit 24 voice channels over a T1 facility.

AT&T-C offers ACCUNET® T1.5 Service, a 1.544 Mbps terrestrial digital service for high-volume transmission capabilities between two AT&T-C central offices. In addition, AT&T-C provides two multiplexers to subdivide the channel. They are the M-24 multiplexer, which permits up to 24 voice-grade channels, and the M-44 multiplexer, which permits signals from two T1.5 services, each containing 22 voice-grade signals to be compressed into one T1.5 service. Compatible multiplexers are required at each end of the line. A T.1.5 service may be an economical alternative for short distances or if there is sufficient voice and data traffic between two points. At each end of the line additional terminal equipment, such as a PABX or multiplexer, is required to subdivide the channel into lower-signal voice and data rates.

MCI also provides a point-to-point terrestrial digital service at transmission speeds to 1.544 Mbps. Their data service allows local access lines in increments of 2.4, 4.8, 9.6, and 56 kbps up to the maximum 1.544-Mbps rate.

8.1.2. Private Switched Digital Networks

AT&T-C offers two private switched digital network services for customer voice and data and other applications: ACCUNET® Reserved 1.5 Service and ACCUNET® Switched 56 Service, which will operate at 56 Kbps. The Reserved 1.5 Service is used to transmit voice and data channel circuits at a 1.544-Mbps digital signal rate over a terrestrial channel on a point-to-point line.

If there is high volume of voice and data traffic between two points, Reserved 1.5 can be a lower-cost networking solution. Each end of the line will require terminal equipment such as a PABX or multiplexer to subdivide the channel into lower signal rates.

Reserved 1.5 tariffs are based on a port charge, usage charge, call setup charge, and extension line charges. Twelve major cities offer the service: Atlanta, Boston, Chicago, Dallas, Detroit, Houston, Los Angeles, New York, Philadelphia, Pittsburgh, San Francisco, and Washington D.C. The service is used in conjunction with the ACCUNET® 1.5 Service to complete the local loop to the customer's premises. ACCUNET® Switched 56 Service is a two-way simultaneous 56-kbps digital link between certain major cities. A digital access line is required to connect customer premises equipment to the central office. Both services use a local loop or a twisted-pair cable from the customer's premises to the rate center. The monthly interoffice charge for the Switched 56 service has two components, a separate local private loop plus a long haul charge. The long haul is a usage charge based on an initial 30-second period plus an incremental charge for each additional 6 seconds.

8.1.3. Private Satellite Networks

Private satellite networks for domestic use are provided by the following companies. Domestic and international satellites will be covered in Chapter 13.

- *American Satellite Corp. (ASC):* This company provides point-to-point communications channels for voice, data, Fax, and other applications by domestic satellite in combination with local access facilities.
- *Argo Communications Corp:* This company provides single digital voice-grade channels between two specific cities where satellite earth stations have been constructed. A customer may obtain an exclusive service for a fixed monthly charge at speeds up to 1.544 Mbps.
- *AT&T-C:* This company offers dedicated high-capacity circuits for transmission via satellite at 1.544 Mbps or a whole transponder between two or more earth stations. Earth stations may be located on the customer's premises or at an AT&T-C central office.
- *GTE Spacenet:* A wholly owned subsidiary of GTE Corporation, this company provides a full range of digital satellite networks for business, education, and satellite news gathering. Applications include interactive data networks for retail, finance, and health care; electronic mail; and video conferencing. Customers can contract for occasional or full-time use.
- *RCA American Communications* (a unit of GE): This company offers point-to-point communications channels for voice, data, and other applications with local Telco connection. RCA also provides a 56-kbps digital data service.
- *Satellite Business Systems:* SBS provides domestic two-way or multipoint network service at the customer's premises or at network access centers. The satellite service includes networking features to ensure reliable transmission.
- *U.S. Sprint:* This company provides a point-to-point offering between 18 major cities. Service is available for analog channels.
- *Western Union Telegraph Company (WU):* Long known for its telex network, a world message standard, WU offers a whole transponder on a fixed term or month-to-month basis for transmission within the United States between two or more satellite earth stations.

8.2. PUBLIC SWITCHED NETWORK OFFERINGS

Public switched networks are available to everyone and are shared use facilities. That is, circuits are not dedicated to a company or an end user. One of the world's largest shared networks is the U.S. public switched telephone network (PSTN). Other public networks are Teletype and packet switched data networks. Table 8.2 gives a list of service providers for public switched offerings.

TABLE 8.2. Public Switched Offerings

Public Switched Network—Interchange and Local Exchange Carriers	
Service	Service Providers
Public switched telephone network (PSTN)	AT&T Communications Central Telephone & Utilities Cincinnati Bell Continental Telephone General Telephone and Electronics (GTE) Regional Bell Operating Companies (RBOC) Southern New England Telephone Company United Telephone Company Other independent Telcos
Circuit switched networks	
ACCUNET® Switched 56 Service	AT&T Communications
Teletype switched networks	
Telex and TWX	Western Union Telegraph plus international record carriers (ITT, RCA, TRT, and WUI may provide domestic service)

8.2.1. Public Switched Telephone Network

The PSTN or dial-up network allows direct dialing and operator-assisted calls to over 600 million telephones in more than 100 countries. Telephone tariffs or rates vary by the amount of assistance a customer requires. For example, on direct dial calls, the customers pay for the time on the line. Long distance direct dialing calls are based on distance, duration, time of day, day of the week, and type of call.

Most Telcos offer night and weekend discount rates, because their central switching facilities are designed to handle peak long distance business calls that normally occur from 8 a.m. to 5 p.m. local time, Monday through Friday. During the night and weekends, calls are at a minimum; there is plenty of spare capacity, and the same facilities sit idle. The incremental cost for long distance is minimal, and therefore, there is an incentive for discounting off-peak hours to build up traffic.

8.2.2. ACCUNET® Switched 56 Service

AT&T-C is offering a 56-kbps switched digital service between its central offices and customer premises. The local loop for a central office to a customer's premises is available via the Digital Data Service (DDS) local channel. Therefore, the complete connection uses a combination of both services. For large volumes of intermittent data traffic, Switched 56 can be cost effective. That is, it may be less expensive to make several calls via the Switched 56 Service compared to a permanent DDS connection. This of course depends on the distance between the two end points and the transmission time. For example, compared to DDS, it is less expensive to use the Switched 56 Service for distances of 500 miles if the total communications time is less than 1 hour per day for 20 working days per month.

8.2.3. Switched Teletype Networks

Telex is a worldwide standard for message exchange that allows any user to directly dial other subscribers. Communication is 66 words per minute based on an average of 6 characters per word plus interword spacing. Telex is an indispensable tool for companies doing business on an international basis. There are over 1,700,000 telex terminals installed throughout the world.

8.2.4. Packet Switched Data Networks

Most packet networks allow messages, data, files or documents to be transmitted over national and international networks. Basically, a packet network receives a message or data file and breaks it up into packet segments. Depending on the network provider, the size of a packet varies from 64 to 512 bytes or characters per packet. The breaking up of a message or data file into packets is called disassembly. After passing the packets through the network, the remote node (network concentration point) reassembles the packets into their original form before sending the message or data file to the destination computer or terminal. Figure 8.1 is an example of a national packet network. A message is created on the ASCII work station and sent to a packet assembly/disassembly (PAD) device. The PAD is usually integrated into the network node, but it may be placed in close proximity to the network node or connected to it by a high-speed link. The node then routes the message toward its final destination node for presentation to a host or another terminal.

Asynchronous terminals are normally dialed into a PAD device. To establish a session with a remote host, the user must start a local dialog with the PAD and request a virtual call to the host. PAD allows many types of terminals to access the network node. At the host site, a PAD or direct network attachment is required to complete the virtual call.

The node is the network backbone and is used to route hundreds of packets each second around the network. The node's software allows hosts to interface to the network and acts as a trunk switch so that in-transit traffic is properly directed to its final destination. The node interface function enables a host to establish a virtual circuit to a PAD terminal or another host, send and receive data, and control packets. The trunk interface allows the exchange of packets between nodes and routes packets to their final destination over the path with the shortest route or least congestion.

Depending on the amount of traffic, packet networks could offer reduced message cost over long distances since the user is charged for the number of packets transmitted and not the distance the message travels. Most packet network providers offer a PAD facility that will accept switched or dial-up inputs from asynchronous terminals, 3270 BSC terminals, or RJE work stations. In addition, network providers offer

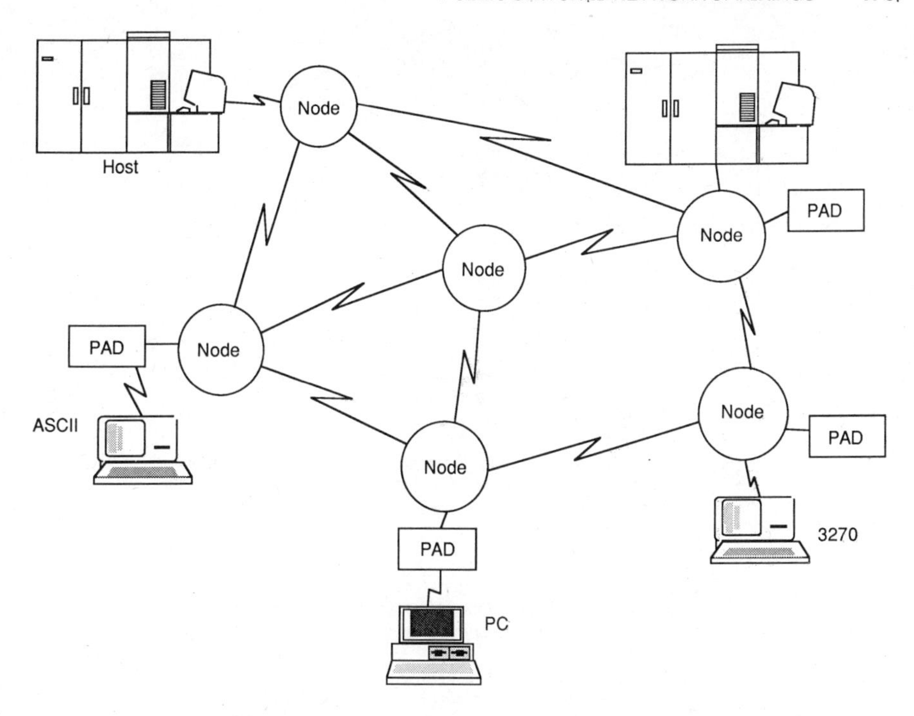

Figure 8.1. X.25 packet switched data network (PSDN).

direct high-speed leased lines to the node or an X.25 HDLC (high-level data link control) connection. Other advantages of packet networks are that local and remote devices may operate at different speeds and protocols. Each device acts independently and is not regulated by the network or the operation of the remote system.

A private leased line can be compared with a packet switched data network. Leased line costs are based on mileage, while packet networks are distance independent. The volume of information transmitted is unlimited on a leased line, whereas packet networks charge for the message length based on the number of 64 octets or characters. Therefore, the amount of data should be considered in evaluating each alternative. In addition, leased lines offer unlimited availability and are independent of the number of users on the network.

Leased lines offer point-to-point connection with fixed responses, while packet networks vary in response times due to network routing, which can be different for two consecutive packets. Terminals cost more to operate at a higher speed, 2400 versus 9600 bps, on a packets network, whereas leased lines can be operated at any speed that the DCE is able to transmit over the line within the quality limits that can be tolerated.

The least expensive arrangement depends on the application, the volume of data to be transmitted, and the response time. For continuous on-line interactive sessions, the inherent delays in a packet network may not be tolerable to some terminal users.

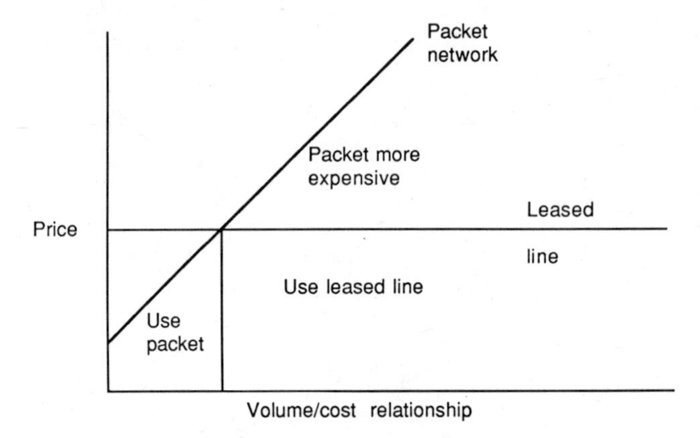

Assumption: Information transfer is between two points, e.g., remote terminal (branch or region) to host.

Consider the following:

- Volume of traffic: Leased lines allow unlimited information. Packets are based on message length.
- Time of day: Usually discounts after 5 p.m., 6 p.m., or midnight for packet network.
- Response time: May be critical for interactive sessions. Leased lines require no initialization. Packet networks have inherent delays due to call setup, packetizing of data.
- Quality of service: Dial may be less expensive, but is it worth the line dropouts and restarts. Packet networks guarantee packet delivery. Leased lines offer better quality than the PSTN.
- Line speed: Small incremental charges for 2.4-, 4.8-, and 9.6-kbps lines in packet networks. Leased lines allow up to rated line speed at no extra cost.
- Can existing equipment be used on the network?: Leased line applications are usually dependent on host protocols. Depending on the country, packet networks allow asynchronous, ASCII, BSC, SDLC, and HDLC connection.

Figure 8.2. Volume/cost relationship—leased line versus PSDN.

Can data be transmitted in off hours? Many packet network providers offer discounts after 5 p.m., 6 p.m. or midnight. The various cost relationships between a leased line and PSDN are shown in Figure 8.2.

8.3. DIVESTITURE OF AT&T —IMPACT ON U.S. NETWORKING

The Department of Justice filed an antitrust suit against AT&T in 1974 asking for the breakup of the Bell System. As a condition for withdrawing that suit, a modification of a 1956 consent decree was entered into between AT&T and the Department of Justice in 1982. That modification, known as the Modification of Final Judgment

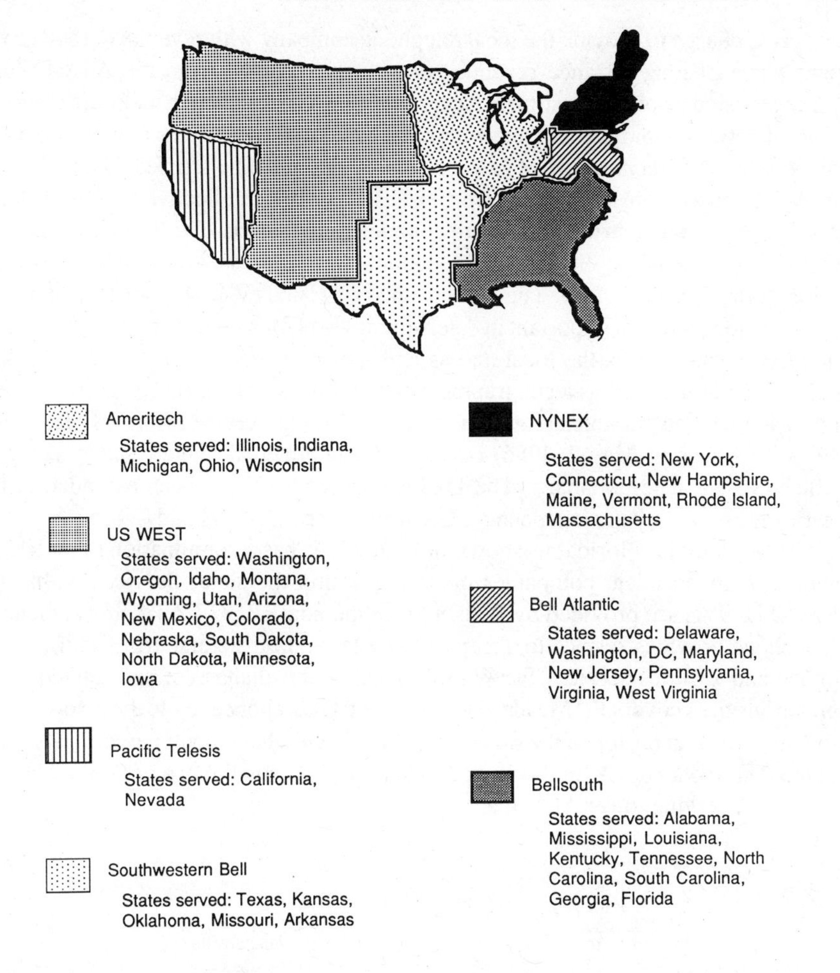

Figure 8.3. Regional Bell Operating Companies and their servicing areas.

(MFJ), required AT&T to divest itself of its operating companies. The 22 Bell Operating Companies were then grouped into 7 fully independent regional holding companies. Figure 8.3 lists the 7 Regional Bell Operating Companies (RBOCs) and the areas they cover.

In accordance with the FCC Computer Inquiry II Decision, and independent of MFJ, AT&T created two fully independent subsidiaries: AT&T Information Systems to serve unregulated markets and AT&T Communications (AT&T-C), a regulated common carrier for interexchange traffic. Local exchange telephone traffic continues to be regulated and is the domain of the RBOCs and the local exchange companies (LECs) or independent telephone companies.

Since the AT&T divestiture on January 1, 1984, the U.S. environment has changed dramatically. Long distance end-to-end circuits normally consist of separate components provided by different carriers. Now users and long distance carriers must pay an access charge to provide the local telephone company with revenue to replace the percentage of long distance revenues previously received from the AT&T Long Lines Division prior to divestiture. Access charges are designed to make efficient use of local networks and prevent uneconomical bypass. As part of the new ruling, each function within a central office or service provided must be tariffed based on cost. AT&T-C must obtain local access from an RBOC or LEC, collectively called Telcos. Therefore, the tariff structure is more complex, and there is a wide variance in end-to-end costs depending on the actual location, street address, area code, and central office code of the customer. The vertical and horizontal (V & H) coordinates or city and town locations are important in determining costs for a given service. A new term that has been added is the local access and transport area or LATA. A LATA is defined under the MFJ as a geographical area within which an RBOC carrier makes provision for and administers various communications service offerings. There are 190 LATAs (as of May 2, 1986) covering the continental United States, some of which cross a state boundary. The LATA concept has since been extended to the territories served by the independent Local Exchange Carriers and Telcos.

As an example, Florida, as shown in Figure 8.4, has a combination of an RBOC and three independent companies operating within the state. Services within the RBOC LATAs are provided by Bellsouth. In the non-RBOC LATA areas, General Telephone provides service to Tampa, Sarasota, and St. Petersburg; Central Telephone and Utilities services Ft. Walton Beach and Tallahassee; and United Telephone Services covers Ft. Meade, Ft. Myers and Okeechobee. Note the arrows in the Pensacola area at the top of the state. Arrow 1 indicates that a small part of Alabama is in the Pensacola LATA, and Arrow 2 indicates that a small piece of Pensacola is in the LATA serving lower Alabama.

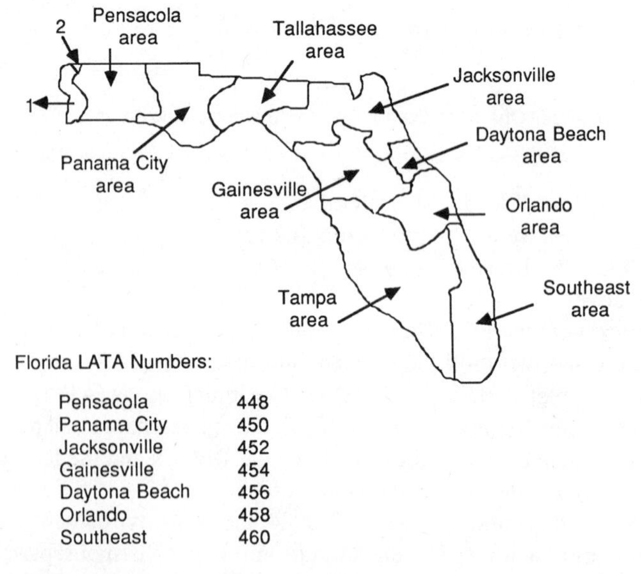

Figure 8.4. Florida RBOC and local exchange carriers.

Intra-LATA refers to services within the same LATA. Telephone calls that originate and terminate within the same LATA are typically handled by the local exchange carrier or local exchange company (used interchangeably) serving that LATA. Intrastate intra-LATA service is within the same state and the same LATA. Telcos provide intra-LATA services, while AT&T-C provides interstate services.

Inter-LATA telephone calls originate and terminate in different LATAs and are transported between the LATAs by an interexchange carrier (IC), AT&T-C, or other regulated carriers.

Two other categories are non-LATA to LATA, which is a non-RBOC company to an RBOC and can be interstate or intrastate, and non-LATA to non-LATA, which is between two non-RBOCs and may also be interstate or intrastate. Local exchange carriers provide connection between customer premises equipment (CPE) and the interexchange carrier network at the point of interface within the interexchange carrier.

Based on current AT&T tariffs, at least three components are required to price a private line end-to-end circuit: interoffice channel, central office (CO) connection, and private line analog local channel. For example, the interoffice channel contains the interexchange mileage between COs, which interconnects channels and other service components at an AT&T central office. The local channel monthly charge includes both the local access and termination charges. Figure 8.5 is an example of the pricing components required to calculate an end-to-end service.

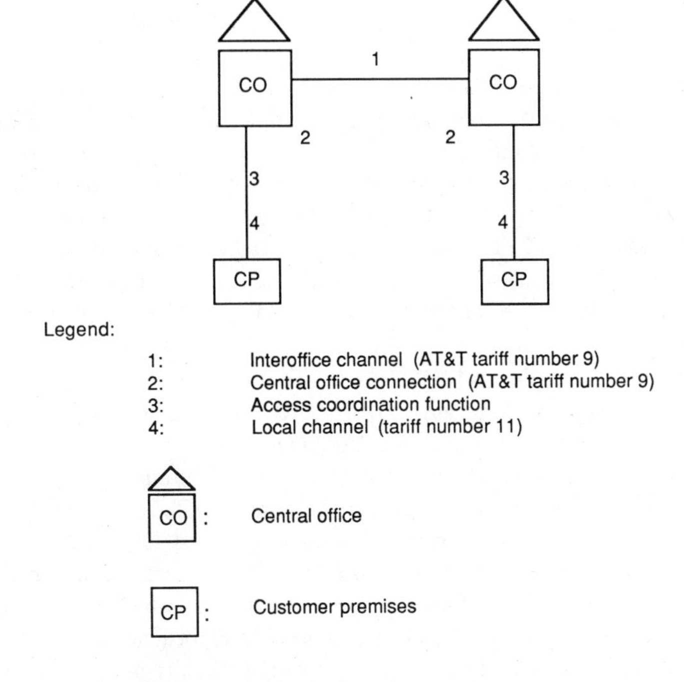

Figure 8.5. Cost components for AT&T private line end-to-end service.

8.4. AT&T PRIVATE LINE TARIFFS

As previously discussed, two major groups determine telephone tariffs and regulatory policies. The FCC has jurisdiction over interstate services, tariffs, and rules and interprets proposed tariffs. The state public utility (or service) commissions have jurisdiction over intrastate traffic and lines. States are concerned with local exchange carrier (LEC) rate changes, possible intra-LATA competition, and bypass problems. A private line, also called a nonswitched or leased line, is a circuit leased by a subscriber from either a local exchange carrier or an interexchange carrier (IC) that will be used to carry only the subscriber's traffic. Long distance calls usually involve an interexchange carrier and a local exchange carrier. Private lines may be intra-LATA intrastate (same LATA serving an area within the same state), inter-LATA intrastate (different LATA serving areas in the same state), intra-LATA interstate (same LATA serving an area between different states), or inter-LATA interstate (different LATA serving areas in different states). Inter-LATA service can be provided only by AT&T-C or other common carriers (OCC). Intra-LATA service is provided by LECs or Telcos although AT&T-C or OCCs may offer this service.

An AT&T-C private or leased line may carry voice or data or both. The monthly cost is based on airline miles and calculated from V&H coordinates between two locations. A leased line ordered from AT&T-C consists of two major components, an interoffice channel and a local channel. In addition to these components, there are central office charges for access coordination, line conditioning, multiplexing, and connections to local channels.

AT&T-C offers three ordering options to private line customers:

1. *Total service circuit:* AT&T-C assumes total responsibility for the circuit and coordinates between the local exchange companies to maintain, test, and guarantee line quality. Total service circuits are available for voice-grade access lines, dataphone digital circuits, and terrestrial digital circuits. These services are furnished within a LATA, inter-LATA, or to an overseas location. If AT&T provides services within a LATA, part of the circuit must be for interstate or overseas communications. The customer receives one monthly bill from AT&T-C for the lease of the entire circuit. AT&T tariff numbers 9, 10, and 11 are used to calculate the monthly bill.
2. *Coordinated service:* AT&T-C provides a transport circuit or a transmission pipeline between two serving offices. AT&T designs, installs, and maintains their components. To complete the connection, the customer must obtain access services from another network provider. The customer is responsible for overall transmission over the line and is billed separately for troubleshooting whenever a problem has to be corrected. Coordinated service is provided for terrestrial digital circuits, overseas circuits, and high-speed digital circuits. The customer receives multiple monthly bills: an interoffice bill from AT&T-C based on tariff number 9 and local access channel bill(s) from the local exchange carriers providing the circuits. Local exchange carriers usually charge different rates.

3. *Baseline service:* AT&T-C provides the interoffice channel portion of a circuit. All coordination, testing, and maintenance are the responsibility of the customer. The customer receives multiple bills: one from AT&T-C and one from each of the local exchange companies involved with the circuit. The customer is billed separately for troubleshooting to solve any problems.

8.4.1. Monthly Charge—New York to Atlanta

The tariffs used in this chapter are for analysis and calculating communications costs between U.S. cities. The rates were effective as of September 1986. The following example demonstrates how to determine the monthly cost for a circuit ordered from AT&T-C.

The monthly cost of a two-way voice-grade private line from New York City (LATA number 132) to Atlanta (LATA number 438) will be based on the following assumptions:

- The distance between New York and Atlanta AT&T-C central offices is 750 miles (taken directly from an interexchange table).
- The Atlanta location is in the same rate center as AT&T-C and 12 miles from the central office. The local channel for a two-way connection includes a fixed and per mile monthly charge.
- The New York City location is in the same rate center as AT&T-C and 10 miles from the central office. The local channel for a two-way connection includes a fixed and per mile monthly charge.
- The entire circuit is procured from AT&T-C using their total service circuit option.

The components are calculated as follows:

1. An interoffice channel (IOC) The interoffice mileage between the two AT&T-C central offices is 750 miles (AT&T tariff number 9):
   A. Fixed rate (applies to channel) = $305.25
   B. Monthly charge
   750 airline miles × 0.30 = 225.00
   $530.25

2. Central office (CO) connection charge (tariff number 11):
   Per central office connection:
   $15.45 per month × 2 offices = $ 30.90

3. Central office access coordination function (ACF) charge, one at each end (tariff number 11):
   $9.00 × 2 ends = $ 18.00

4. Local channel two-way line (tariff number 11):
   Atlanta office:
   $79.87 + (1.63 × 12) = $ 99.43
   New York office:
   $108.61 + (2.07 × 10) = 129.31
   $228.74

   Total monthly charge for a private line = $807.89

8.4.2. Private Line for Boston Converter

A private line for Boston Converter Company was considered in Chapter 1. During the start-up situation, a limited number of home office calls would not be too expensive. However, as the local office grew into a regional office, there would be more traffic to handle additional employees and accounts. Let's assume that Boston Converter would like to evaluate a private leased line from their home office in Boston to the regional office in Los Angeles. In this example, assume that a private leased line will be obtained from AT&T-C and that the distance between their inter-office exchanges is not known. Boston Converter's first step would be to calculate the distance between interoffices using the V&H coordinates supplied by AT&T tariff number 10.

City/Town	V	H
Billings, Mont.	6391	6790
Binghampton, N.Y.	4943	1837
Boise, Idaho	7096	7869
Boone, Iowa	6394	4355
Boston, Mass.	4422	1249
Bridgeport, Conn.	4841	1360
Buffalo, N.Y.	5075	2326
Lawrence, Mass.	4373	1311
Leesburg, Va.	5634	1685
Little Rock, Ark.	7721	3451
Longview, Tex.	8348	3660
Logan, Utah	7367	7102
Los Angeles, Calif.	9213	7878
Louisville, Ky.	6529	2772

Figure 8.6. Selected vertical and horizontal coordinates.

Distance Calculation

The calculation for air miles between Boston and Los Angeles will be used to determine the private leased line monthly charge. AT&T-C and many specialized carriers base intercity airline mileage on a common grid of vertical (V) and horizontal (H) coordinates. The V&H coordinates in Figure 8.6 were selected from over 19,000 rate centers in the United States and correspond to *Category A* rate centers designated by AT&T-C for private line data channels. Other cities are included for practice examples. For a complete listing of exact V&H coordinates, consult AT&T tariff number 10.

The rate mileage formula is used to determine the distance between two rate centers where V1 and V2 are the vertical coordinates and H1 and H2 are the horizontal coordinates. V1 and V2 are the vertical coordinates for Los Angeles and Boston, respectively, and H1 and H2 are the horizontal coordinates for Los Angeles and Boston, respectively. The largest numbers are the V1 and H1 coordinates

$$\text{Mileage Formula} = \sqrt{\frac{(V1 - V2)^2 + (H1 - H2)^2}{10}}$$

1. Take the V and H coordinates from Figure 8.6.
2. Calculate the distance by subtracting the lower coordinate V (Boston) from the larger V (Los Angeles) and the lower H (Boston) from the higher H (Los Angeles):

$$\begin{array}{r} 9213 \\ -4492 \\ \hline 4721 \end{array} \qquad \begin{array}{r} 7878 \\ -1249 \\ \hline 6629 \end{array}$$

3. Square each of the resulting V and H coordinates obtained in part 2, add the two numbers, and divide the total by 10. This results in 6,623,148.2. All fractions are rounded to the next whole number or 6,623,149.
4. Calculate the square root of the result. This number is the rate distance in miles and results in 2573.5 or 2574. Fractional miles are always rounded to the next whole number.

The distance of 2574 will be used to calculate the interoffice channel rate.

1. Interoffice channel (IOC): The interexchange mileage between the two AT&T-C central offices is from AT&T tariff number 9:
   A. Fixed rate (applies to channel) = $305.25
   B. Monthly charge, airline miles (× 0.30):
      From calculation above, 2574 × 0.30 = 772.20

$1077.45

2. Central office (CO) connection (tariff number 9):
   Per central office connection:

   $15.45 per month × 2 offices = $ 30.90

3. Central office access coordination function (AFC) charge, one required per local access:
   $9.00 × 2 ends = $18.00

4. Local channel (LC): This charge includes local access and termination charges AT&T tariff number 11. Assumptions:
   (a) Two-way line.
   (b) Los Angeles state/LATA is 730. The office is 56 miles from the central office.
   (c) Boston state/LATA is 128. The home office is 10 miles for the central office.
   Calculation:
   Two-way Los Angeles local channel monthly charge:
   Rate for 56 miles is $144.15 + $2.40 per mile:
   $144.15 + (56 × 2.40) = $278.55

   Two-way Boston local channel center monthly charge:
   Rate for 10 miles is $110.08 + 2.24 per mile:
   $110.08 + (56 × 2.24) = $235.52

   $ 514.07

Total monthly charge for private line (Boston to LA) = $1640.42

8.4.3. Multipoint Line Example for Boston Converter

The previous example for a direct private line to Los Angeles may cost more than the budget will allow. However, now that Boston Converter has opened up offices in New York and Chicago, they could consider a multipoint line with local loops in New York, Chicago, and Los Angeles. The line could be averaged over three remote offices and could make it feasible for a private line connecting these branches. The advantage of a multipoint line is that the monthly cost is based on the total length of the line. Since additional miles cost less per mile, the result is a lower average cost per mile. Each of the remote offices would share the multipoint line and be in contention for access. Since there is a 1-hour time difference between each remote office, preset times could be arranged to maximize the use of the line.

The major components required for a total AT&T-provided multipoint configuration are shown in Figure 8.7.

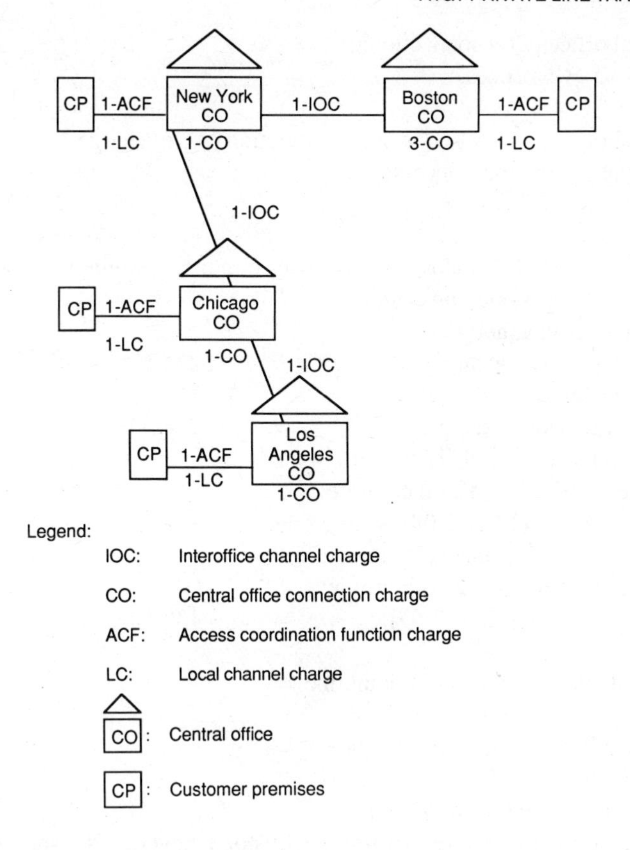

Figure 8.7. Multipoint configuration.

A multipoint private line is available for 24 hours per day, 7 days per week for unlimited use. It may not take many hours of dial-up calls for Boston Converter to justify this arrangement. The following is a calculation for the multipoint line.

- Interoffice channel (IOC) (AT&T tariff number 9):

Distance between Boston and New York	=	185 miles
Distance between New York and Chicago	=	715 miles
Distance between Chicago and Los Angeles	=	1732 miles
Total distance	=	2632 miles

Interoffice channel rate:
$305.25 + 2632 × 0.30 = $1094.25

- Channel conditioning:

C2 for multipoint channels arranged for switching:
$30.00 (per interoffice channel) × 3 = $ 90.00

- Central office (CO) connection:
  $15.45 × 6 offices = $ 92.70

- Central office—access coordination function (ACF) charge, one required per local access:
  $9.00 × 4 offices = $ 36.00

- Local channel (LC) rates—all two-way circuits: Assume Chicago and New York are in the same rate center:

Boston local channel:			
From previous example	=	$235.52	
Los Angeles local channel			
From previous example	=	278.55	
Chicago local channel (12 miles):			
Rate is $94.89 + $2.08 per mile:			
$94.89 + (12 × 2.08)	=	119.85	
New York local channel (15 miles):			
Rate is $108.61 + 2.07 per mile:			
$108.61 + (15 × 2.07)	=	139.66	
			$ 773.58
Total monthly rate for multipoint line	=		$2086.63
Average line cost per remote office	=		$ 695.54

Point-to-Point Versus Multipoint Line

An alternate option to a multipoint line for Boston Converter is a star arrangement where all lines emanate for Boston, the host site, to remote cities. For example, a star network would have three private lines: one from Boston to New York, one from Boston to Chicago, and one from Boston to Los Angeles. The total inter office monthly charge would be calculated as follows:

- Interoffice channel (IOC):

Boston to New York = 185 miles:			
$155.25 + (185 × 0.60)	=	$ 266.25	
Boston to Chicago = 849 miles:			
$305.25 + (849 × .30)	=	559.95	
Boston to Los Angeles = 2574 miles:			
Previous calculation	=	1077.45	
Interoffice channel rate	=	$1903.65	

- Central office (CO) connection:
  $15.45 × 6 offices = $ 92.70

- Central office—access coordination function (ACF) charge:
  $9.00 × 6 offices = $ 54.00

- Local channel (LC) rates—all two-way circuits: Assume Chicago and New York are in the same rate center:

Boston local channel:			
From previous example,			
$235.52 × 3 lines	=	$ 706.56	
Los Angeles local channel			
From previous example	=	278.55	
Chicago local channel (12 miles):			
Rate is $94.89 + $2.08 per mile:			
$94.89 + (12 × 2.08)	=	119.85	
New York local channel (15 miles):			
Rate is $108.61 + 2.07 per mile:			
$108.61 + (15 × 2.07)	=	139.66	
			$1244.62
Total monthly rate for a star or point-to-point line		=	$3294.97

The total monthly charge for the star or point-to-point network is $3294.97 compared with the total monthly charge of $2086.63 for a multipoint arrangement. Comparing three separate direct lines in a star arrangement to a single continuous multipoint line, we find that a reduction of $14,500 per year is possible with a multipoint line. A host-controlled BSC or SDLC protocol with multipoint line control for data traffic is a solution. Another approach is an open contention system where each user may use the line to make voice or data calls when the line is not busy.

Private point-to-point leased lines and a private multipoint leased line are two options available to Boston Converter. Chapter 9 discusses multiple options from multiple locations for a larger company with higher volumes of voice and data traffic.

9

U.S. NETWORK ALTERNATIVES

In this chapter we will

- Compare PSTNs versus leased line networks
- Compare packet networks for inquiry and on-line applications for a PC and 3270 terminal
- Compare the AT&T and MCI digital voice and data network offerings at 9.6 and 56 kbps and 1.544 Mbps

In the United States there are several voice and data networking alternatives from both common carriers and private companies. Some of the larger national network offerings along with their advantages and present restrictions will be considered in this chapter. National networks can be grouped into the following areas of telecommunications services:

1. PSTN or dial-up offerings
2. Private leased line
3. Packet switched data networks
4. Digital services at 56 kbps
5. T 1.5- and 1:544-Mbps services
6. Satellite services

The first five areas will be covered here, while satellite services will be discussed in Chapter 13. The number of satellite networks are growing in order to handle increased traffic, offer bulk data links among two or more locations, extend terrestrial networks, and provide backup for existing networks.

Rate changes will occur monthly for most companies as they position themselves in the marketplace. However, the rate structure and conditions usually remain the same for many years. For current rates, updated rate services are available for both national and international networks.

Note: The tariffs used in this chapter are for analysis and for calculating communications costs between certain U.S. cities. The rates were effective as of September 1986. For a detailed service with periodic updates, contact CCMI/McGraw-Hill, 50 South Franklin Turnpike, Ramsey, N.J.

9.1. PSTN—LONG DISTANCE MESSAGE TELECOMMUNICATIONS SERVICE

The public switched telephone network (PSTN) is used more than any other service for long distance voice traffic. Voice traffic represents the bulk of the telephone companies' revenue. The PSTN is also one of the more widely used networks for transferring data within the United States and overseas. There are several companies offering voice-grade lines within the United States and three companies offering overseas service to Europe. The three overseas carriers are AT&T-C, MCI, and U.S. Sprint. The number of carrier alternatives to the Far East for voice-grade lines is limited today but will increase as satellite circuits and underseas fiber optic cables are installed to meet the growing demand for international telecommunications.

The advantages of the PSTN are that it is accessible to most countries, is easily connected for data calls (acoustic couplers and modems), and offers attractive rates for calls of short duration.

AT&T-C, MCI, and U.S. Sprint offer long distance direct dialing throughout the United States and many countries of the world. Worldwide coverage differs by carriers. AT&T-C has access to most of the countries of the world, while MCI and U.S. Sprint each provide access to over 30 countries. ITT's switched private network system is available only within the United States and to Puerto Rico and the Virgin Islands. The subscriber should expect both MCI and U.S. Sprint to extend country coverage in time. The main advantage to both MCI and U.S. Sprint is reduced charges within the United States and to Europe and other countries.

The tariff or monthly subscriber charge billed by AT&T-C, U.S. Sprint, ITT, and MCI is based on mileage and the duration of the call in minutes. The first minute rate is higher than the rate for subsequent minutes. For AT&T-C subscribers, which are offered more services, there are additional charges for the amount of operator assistance provided, such as assistance in dialing a number or in placing person-to-person calls where the caller does not pay until the party called is on the line. In addition, subscribers are now charged for information calls over some minimum monthly number. MCI and U.S. Sprint do not have the same coverage as AT&T-C, and there may be longer delays in setting up a call due to shared access public telephone network access ports and switching onto and off their national networks, which may consist of terrestrial links, microwave links, and satellite links. ITT offers a measured use service based on a customer dedicated access line or the public telephone network. Other features offered by ITT are long distance directory assistance service and usage volume discounts.

9.1.1. PSTN Long Distance Comparisons

The cost of network offerings will vary by tariff relationships. For example, the public switched telephone network or dial-up network charges will vary by the duration of the call, the amount of assistance an operator will provide, the distance called, and the time of day. If the call to be placed is a long distance call, there are several alternatives based on equal access regulations and tariffs.

AT&T-C rates for long distance calls are based on mileage and on whether a customer dials the call directly, places a credit card call, goes through the operator, or requests a person-to-person call. The surcharge for person-to-person calls is $3.00 over the regular dial rate within the United States. It is obvious that even for the greatest distance, the user can place five calls at the dial rate before a person-to-person call can be justified. The rate table that follows (Table 9.1) is an example of the variances for a normal telephone call within the U.S. mainline based on the AT&T-C long distance message telecommunications service and MCI communications' schedule for interstate traffic. MCI rates are from MCI Tariff Number 1 and apply to interstate telephone calls on MCI's own network. Taxes are not included in Table 9.1.

Rates will vary considerably by carrier due to the AT&T-C umbrella effect. That is, to increase its market share of the long distance message business, a competing interexchange carrier will charge rates lower than AT&T-C or provide additional services for about the same price. Table 9.1 is the rate schedule for both AT&T-C and MCI and is an example of the variances for a normal telephone call within the U.S. mainland.

9.2. PRIVATE LEASED LINE

Private leased lines are used extensively for national and international communications between two points and for multipoint access within the United States. Private lines are available 24 hours per day and 7 days per week, offer unlimited usage, and are of better quality than the PSTN. Their only domestic drawback compared to the PSTN is the higher cost over distance for low usage.

Private leased lines are available for voice, data, facsimile, and image transfer from several carriers. There are several networking alternatives based on point-to-point and multipoint service over various ranges of speed. Private lines can be provided by AT&T Communications, U.S. Sprint's Private Line Service, MCI's Dedicated Leased Line Service, ITT's Interstate Transmission Systems, and Western Union Telegraph Company's Voice Graded Channel Service.

TABLE 9.1. AT&T and MCI Comparisons

	Direct Dial Day Rate		Evening Rate		Weekend Rate	
Mileage	First Minute	Additional Minute	First Minute	Additional Minute	First Minute	Additional Minute
			AT&T Communications			
1-10	$0.36	$0.18	$0.216	$0.108	$0.144	$0.072
11-22	0.43	0.24	0.258	0.144	0.172	0.096
23-55	0.48	0.28	0.288	0.168	0.192	0.112
56-124	0.51	0.33	0.306	0.198	0.204	0.132
125-292	0.51	0.35	0.306	0.210	0.204	0.140
293-430	0.52	0.37	0.312	0.222	0.208	0.148
431-925	0.55	0.38	0.330	0.228	0.220	0.152
926-1910	0.56	0.39	0.336	0.234	0.224	0.156
1911-3000	0.65	0.42	0.390	0.252	0.260	0.168
3001-4250	0.67	0.45	0.402	0.270	0.268	0.180
4251-5750	0.70	0.47	0.420	0.282	0.280	0.188
			MCI Communications			
1-10	$0.1930	$0.1782	$0.1368	$0.1069	$0.0912	$0.0713
11-22	0.2466	0.2313	0.1656	0.1421	0.1104	0.0948
23-55	0.3246	0.2772	0.2140	0.1663	0.1428	0.1109
56-124	0.3510	0.3267	0.2140	0.1960	0.1440	0.1307
125-292	0.3618	0.3465	0.2149	0.2079	0.1449	0.1386
293-430	0.4400	0.3647	0.2149	0.2140	0.1449	0.1435
431-925	0.4698	0.3740	0.2861	0.2257	0.1906	0.1505
926-1910	0.4920	0.3854	0.3129	0.2317	0.2090	0.1544
1911-3000	0.5330	0.4158	0.3465	0.2495	0.2106	0.1663
3001-4250	0.6035	0.4455	0.3560	0.2673	0.2478	0.1782
4251-5750	0.6212	0.4653	0.3759	0.2792	0.2573	0.1861

Source: AT&T and MCI Communications.

9.2.1. AT&T Communications

AT&T-C's major advantages are availability of lines, geographical coverage, and quality of service. AT&T-C can offer the subscriber connection between any two U.S. cities, while most of its competitors are limited to major U.S. cities. However, this advantage is expected to disappear in the mid to late 1990s.

9.2.2. ITT/United States Transmission Systems

ITT's subsidiary offers interstate private line services in 34 major U.S. cities concentrated in the eastern and southeastern part of the country with extensions to the Midwest and Los Angeles. This network is supported by a combination of microwave, fiber optic, and cable facilities. Leased lines are offered for interstate transmission service. Second only to AT&T-C in use by the U.S. government, ITT provides very reliable service to both government and commercial subscribers. ITT has an ample backup of power supplies and triple redundancy for circuits, which results in greater reliability, greater availability, and less downtime.

9.2.3. MCI Communications

MCI offers single-carrier services and provides systems management to subscribers by interfacing directly with other interconnecting carriers. MCIs covers 132 major U.S. cities with voice, analog data, and alternate voice and data types of transmission.

MCI's voice-grade service covers the normal 4000-hertz telephone bandwidth and allows voice, data, or facsimile transmission for speeds up to 9600 bps.

9.2.4. U.S. Sprint

U.S. Sprint is an interexchange carrier providing voice, data, facsimile, telemetering, and other applications over one of the largest private microwave radio relay systems in the United States. Its channels may be shared by any authorized subscriber and are available for leased point-to-point interstate traffic at line speeds up to 9600 bps.

Sprint has expanded its terrestrial facilities with fiber optics and microwave radio as well as satellites. Its rates are lower than AT&T-C for certain cities. Sprint provides voice-grade channel service for speeds up to 9600 bps, although rates above 4800 bps are not guaranteed.

9.2.5. Western Union

Western Union offers interstate services for voice and data for speeds up to 2400 bps. Western Union's terrestrial network is the second largest in the United States. Its Westar satellites supplement the terrestrial system. WU offers telex, the leading worldwide message service, and is the only carrier offering Teletex service to Europe. WU's rates are lower than those of AT&T-C but much the same as those of the other carriers. WU does not offer systems management for circuit connection, and data is limited to 2400 bps over terrestrial circuits.

9.2.6. Private Line Versus PSTN for Voice and Data

There are many cases where a private or leased line is preferred for voice and data calls. Private lines may be point to point (between two locations) or multipoint (among several locations on the same line). Private lines are usually of better quality than the PSTN and provide instant access since dialing is not required. The monthly charge is fixed by tariff and is independent of the time period.

Voice and data calls to the same point will be calculated for both private lines and the PSTN. Figure 9.1 presents the components that make up the tariff for a private line point-to-point connection.

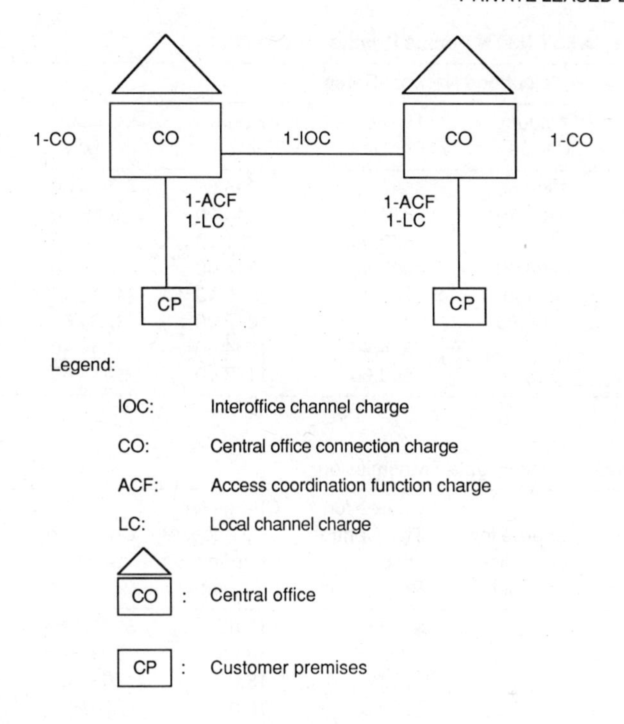

Figure 9.1. Private line point-to-point connection.

Assumptions:

1. Calls via the PSTN are made once a day for 20 working days per month.
2. Five different periods will be calculated for PSTN calls: 1, 1 1/2, 2, 2 1/2, and 3 hours per day. If a call is interrupted, the operator will be contacted and the session will be reestablished for the same call period (1, 1 1/2, 2, 2 1/2, or 3 hours). That is, the call will not start at the first-minute rate but at the each-additional-minute rate. Generally, there is some credit given for a disconnected call.
3. The billing period is 1 month.
4. A private leased line is available 24 hours per day, 7 days a week for unlimited use.
5. AT&T-C dial rates from Table 9.1 will be used. Private line rates are from AT&T tariff number 9.
6. PSTN distances will be calculated from 124 to 5750 miles, which are normal breaks in the tariff schedule. Private leased lines are calculated in increments of 100 miles. All curves are plotted up to a distance of 3200 miles.
7. Voice calls will use the normal telephone handset. Data calls will require two modems, one at each end. It is assumed that the modems will have been

TABLE 9.2. AT&T PSTN Versus Private Line Network

Public Swtiched Telephone Network Rates

Mileage	1 Hour Per Day	1½ Hours Per Day	2 Hours Per Day	2½ Hours Per Day	3 Hours Per Day
124	$399.60	$597.60	$ 795.60	$ 993.60	$1191.60
292	423.20	633.20	843.20	1053.20	1263.20
430	477.00	669.00	891.00	1113.00	1335.00
925	459.40	687.40	915.40	1143.40	1371.40
1910	471.40	705.40	939.40	1173.40	1407.40
3000	508.60	760.60	1012.20	1264.60	1516.60
4250	544.40	814.40	1084.40	1354.40	1624.40
5750	568.60	850.60	1132.60	1414.60	1696.60

Private Line for Voice or Data Transmission

Milleage Between Interoffice Channels	Charge for Interoffice Channel	Charge for Two Central Office Connections	Charge for Two Access Coordination Functions	Charge for Two Local Channels	Total Monthly Charge
100	$ 215.25	$30.90	$18.00	$203.62	$ 467.77
200	275.25	30.90	18.00	203.62	527.77
500	455.25	30.90	18.00	203.62	707.77
1000	605.25	30.90	18.00	203.62	857.77
1500	755.25	30.90	18.00	203.62	1007.77
2000	905.25	30.90	18.00	203.62	1157.77
2500	1055.25	30.90	18.00	203.62	1307.77
3000	1205.25	30.90	18.00	203.62	1457.77
4000	1505.25	30.90	18.00	203.62	1757.77
5000	1805.25	30.90	18.00	203.62	2057.77
5750	2030.25	30.90	18.00	203.62	2282.77

Source: AT&T

purchased or rented. Modem prices are not included in the calculations because they would only shift both curves upwards by the amount of the purchase or rental price.

8. AT&T-C line tariffs were used in the calculations. The results are shown in Table 9.2. All locations are assumed to be in the same rate center and LATA as the AT&T-C central office. Assume that the local channel in each rate center is 8 miles from the AT&T central office. The average cost based on a mix of seven high-density traffic states is $101.81 per month per local channel. Each office requires a local connection charge at each end.

Based on the preceding assumptions and the distances between company facilities, a direct dial call is favorable for distances of less than 150 miles for calls of less than 1-hour duration or less than 800 miles for calls of less than 1 1/2-hour duration. A plot of the monthly cost for a PSTN versus a private line is shown in Figure 9.2.

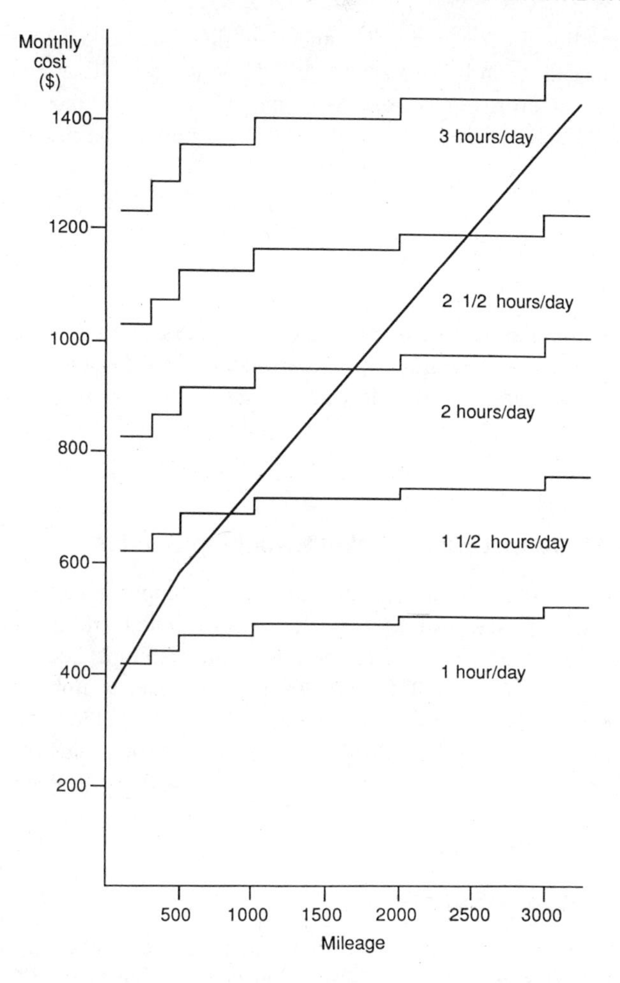

Figure 9.2. Monthly cost for PSTN versus private line.

9.3. PACKET SWITCHED DATA NETWORKS

For data transfers packet switched data networks (PSDNs) are growing faster both nationally and internationally than telex, the PSTN, and private lines. There are many reasons for this growth. First, for most PSDN offerings with broad domestic coverage, a local call to the nearest node will allow access to databases or applications services anywhere in the United States. In some cases, international access is available for an additional charge, but the cost is usually lower than that for a direct call to the remote service bureau in another country. Second, a variety of existing terminals can be used, from ASCII to full-screen 3270 devices and remote batch work stations. Third, a number of different protocols are supported for communications to the host: asynchronous, bisynchronous, SDLC, and HDLC (X.25). Fourth, unlike a private line for international connection, there are few or no restrictions in the types of data that can be sent over a PSDN within a multinational enterprise. Within the same enterprise, electronic mail and electronic document distribution is allowed on a PSDN in Europe. Several PSDNs for public and private use will be presented.

9.3.1. ADP Autonet

ADP offers public dial-up access, private dial-up access, and dedicated access for interstate packet services within the United States, Canada, and over 45 foreign countries. Autonet supports over 150 host systems through over 8000 domestic access ports.

9.3.2. AT&T-C ACCUNET® Packet Switching Service

AT&T-C offers a store and forward message service supporting a variety of analog and digital terminal devices. AT&T-C's offering is regulated, and access is limited to private leased line connection. The leased line connection must be X.25 compatible because the FCC restricts AT&T-C from providing a packet assembly and disassembly (PAD) function. The AT&T-C ACCUNET® Packet Switching Service (APS) is tariffed primarily for companies intending to create their own value added networks for private internal use or to offer their customers value added services.

9.3.3. Telenet (U.S. Sprint)

Telenet is the original domestic public and international public packet switched service. Telenet offers a full range of PSDN services to over 250 U.S. cities and 70 international locations. More than 800 host computers and 100,000 terminals are connected to the network. In addition, Telenet also designs and installs data networks and nodes for foreign PTT administrations and provides international access CCITT X.75 recommended (international interworking of X.25 national networks) equipment and support for many national networks.

In November 1986, Telenet completed the migration of United Telecom's Uninet data division customers onto a single Telenet public data network. The two networks were merged into Telenet and are a part of U.S. Sprint, the joint venture formed by GTE and United Telecommunications, Inc. At year end 1986, Telenet was handling over 16 million calls a month over its domestic network and over 1 million calls a month internationally.

9.3.4. Tymnet McDonnell Douglas

Tymnet, part of the McDonnell Douglas Systems Group, is one of the worldwide leaders in public switching technology and service offerings. Its coverage extends to over 234 U.S. cities and 30 countries. Nationally, Tymnet supports asynchronous, polled asynchronous, 3270 BSC, SDLC, and remote job entry protocols. Internationally, Tymnet provides complete X.25 (packet switched data networks) and CCITT X.75 (international interface between national networks) systems for many of the world's PTTs.

Due to the number of suppliers of X.25 packet networks, PSDNs, or value added networks, as they are called, each network provider is expected to add more value added services to their network structures. In the future, decisions will not be made based on the number of city or country connections but rather on the amount of pertinent value added services that are supported on a national or international basis.

9.3.5. Packet Network Alternatives

There are several configurations of packet switched data networks available. Some of the more commonly used facilities will be considered along with a monthly cost analysis and supporting figures. The charges used in the following calculations were provided by Telenet (U.S. Sprint) and Tymnet McDonnell Douglas.

Inquiry Application Using PCs on Tymnet
Assumptions:

1. Personal computers, intelligent work stations, and word processors with an asynchronous interface are connected to the PAD facility. The connection is by local, public dial-up facilities to a host for periods of 15 minutes per session. There are five sessions during normal business hours, 8 a.m. to 5 p.m., for 20 working days per month.
2. The host has twice as much outgoing activity as a remote office. All host terminals are locally attached.
3. The company has 10 locations, and each site has one terminal.
4. Each terminal uses about 30 screens per session with 250 characters per screen, which results in 750,000 characters per month for five sessions per day.
5. All customer offices are located in high-density cities, and there are no surcharges. All billing is to the host site; remote costs are allocated to each remote office. The following monthly rates apply for the company (modems are not included in each service):

Host Location	**10 Sessions Per Day**
a. General account charge	$ 100.00
b. User name ID (NUI), at $4.00	40.00
c. Private access ports, 1200 bps, 6 ports at $250.00	1500.00
d. Outdial service, 2 ports	450.00
e. Access time, 2 ½ hours at $4.25 per hour	212.50
f. Transmission charge ($0.05 per kilocharacter)	75.00
Total monthly cost for host location:	$2377.50

Remote Locations	**5 Sessions Per Day**
a. User name ID (NUI), at $4.00	$ 20.00
b. Access time, 1 ¼ hours at 4.25 per hour	106.25
c. Transmission charge ($0.05 per kilocharacter)	37.50
Total monthly cost per remote location:	$163.75

Conclusions: The total monthly cost for a packet service would be $4015 (= $2377.50 + $163.75 × 10) for 10 asynchronous terminals and a host based on the assumptions provided. In general, packet networks offer a cost advantage over the PSTN or private leased lines for heavy message traffic, mainly because the subscriber is charged for the amount of traffic over the line and not the distance traveled. Most packet network offerings have a node in most major U.S. cities, resulting in a local call to the node for remote host access.

Figure 9.3 is a layout of an inquiry application using a packet network with associated nodes and PAD facilities for the host and remote terminal connections.

Data Input Using PCs on Tymnet
Assumptions:

1. Personal computers, intelligent work stations, or word processors with an asynchronous interface are connected to the PAD facility. The connection is by local, public dial-up facilities for periods of 1, 3, and 5 hours per day during the normal business hours of 8 a.m. to 5 p.m. for 20 working days per month.

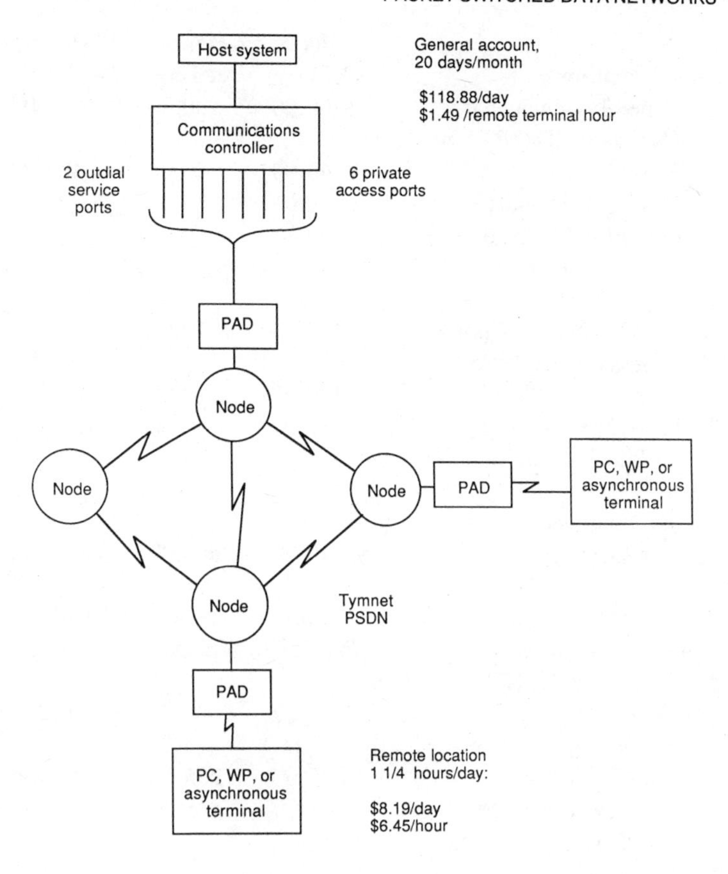

Figure 9.3. Inquiry application using PCs on Tymnet.

2. The host has twice as much outgoing activity as a remote office. All host terminals are locally attached.
3. The company has 10 locations, and each site has one terminal.
4. Each terminal uses about 30 screens per hour with 500 characters per screen, which results in 300,000 characters per month for 1 hour per day, 900,000 characters per month for 3 hours per day, and 1,200,000 characters per month for 5 hours per day. This represents the outgoing traffic to the host for each terminal. Traffic from the host to all terminals is twice this amount.
5. All customer offices are located in high-density cities, and there are no surcharges. All billing is to the host site; remote costs are allocated to each remote office. The following monthly rates would apply for the company (modems are not included):

Host Location	2 Hours Per Day	6 Hours Per Day	10 Hours Per Day
a. General account charge	$100.00	$ 100.00	$ 100.00
b. User name ID (NUI), at $4.00	40.00	40.00	40.00
c. Private access ports, 1200, bps, 3 for 1 hour, 6 for 3 hours, 8 for 5 hours at $250.00	750.00	1500.00	2000.00
d. Outdial service, 2 ports	450.00	450.00	450.00
e. Access time at $4.25 per hour	170.00	510.00	850.00
f. Transmission charge ($0.05 per kilocharacter)	30.00	90.00	150.00
Total monthly cost for host location:	$1540.00	$2900.00	$3590.00

Remote Locations	1 Hour Per Day	3 Hours Per Day	5 Hours Per Day
a. User name ID (NUI), at $4.00	$ 20.00	$ 20.00	$ 20.00
b. Access time at $4.25 per hour	85.00	255.00	425.00
c. Transmission charge ($0.05 per kilocharacter)	15.00	45.00	75.00
Total monthly cost per remote location:	$120.00	$320.00	$520.00

Conclusions: Based on the assumptions provided, a company with 10 remote offices would have a monthly cost of $2740 (= $1540 + $120 × 10) for 1 hour per day, $5890 for 3 hours per day, and $8790 for 5 hours per day.

The total amounts would compare favorably with the PSTN or private leased line, again because the terminals are locally connected to a nearby node and the packet network transfers the information to a remote host. The subscriber is charged only for local calls to the nearest PAD function plus the number of packets. Figure 9.4 is a diagram of the data input system for the terminal and host site.

On Line with 3270 BSC Terminals on Telenet
Assumptions:

1. Ten remote 3274 cluster controllers are connected by leased line to the nearest node at speeds of 9600bps. Once logged onto the host, the sessions are continuous. The terminals are used during normal business hours of 8 a.m. to 5 p.m. for 20 working days per month.

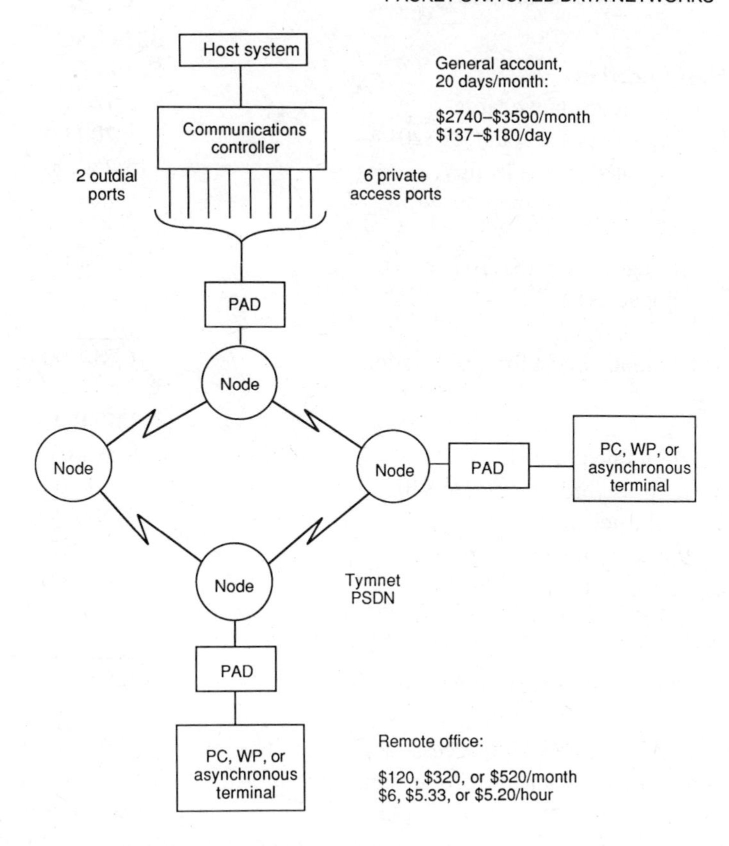

Figure 9.4. Data input for PCs on Tymnet.

2. The host has twice as much outgoing activity as one remote office. All host terminals are locally attached.
3. The company has 10 locations throughout the country.
4. Each remote location has a 3274 cluster controller with eight attached 3278 terminals.
5. Each terminal uses about 30 screens per hour for 8 hours per day, and each screen has 1200 characters. Each remote cluster controller would therefore transmit 46,080,000 characters per month. Telenet packets have a billable length of 128 characters, which results in 360,000 packets or 360 kilopackets.
6. The rates quoted by Telenet include a leased line from the customer's location to their service point and the associated modems.
7. All customer offices are located in high-density cities, and there are no surcharges. All billing is to host site; remote costs are allocated to each office. The following monthly rates would apply for the company (modems are not included in each service):

Host Location	**3270 BSC Session**
a. Regular account charge	$ 140.00
b. ID/password account, at $20	200.00
c. Dedicated access facility, 9.6 kbps, host type, 10 ports	15,250.00
d. Storage charge ($1.70 per kilopacket) at 720	1,224.00
Total monthly cost for host location:	$16,814.00

Remote Locations	**3270 BSC Session**
a. ID/password account, at $20	$ 200.00
b. Dedicated access facility, 9.6 kbps, terminal type	1200.00
c. Storage charge ($1.70 per kilocharacter) at 360	612.00
Total monthly cost per remote location:	$2012.00

Conclusions: A dedicated 3270 access facility with 10 remote 3274 cluster controllers, each with eight attached 3270 terminals, costs $20,120 plus a host charge of $16,814, for a total monthly cost of $36,934. At 8 hours per day, this results in a host charge of $1.31 per hour for each remote terminal and a remote charge of $1.57 per terminal hour, or a total charge of $2.88 per remote terminal hour.

A 3270 BSC application on a packet network may not be the best alternative compared to a private leased line with an upper limit of 34 M bytes a day (8 hours). The main crossover point is determined by the amount of data traffic between remote sites and the host information center. On-line applications with high data rates offset the advantages of shorter data inquiry and message transfers as in the two previous cases. A diagram of the network layout, host system, and remote terminal accounts are shown in Figure 9.5.

9.4. DIGITAL SERVICES AT 56 KBPS

There is a trend to higher-speed lines to meet the growing demand for bulk information transfer and on-line applications. New products such as multiplexers and communications concentrators, which allow several low-speed devices to be attached to one high-speed line (56 kbps or higher) have aided this trend. A 56-kbps line can be multiplexed or shared by several lower-speed devices; for example, four 9.6-kbps, two 4.8-kbps, and three 2.4-kbps devices. Several offerings will be discussed followed by a comparison of two common carriers, tariffed services.

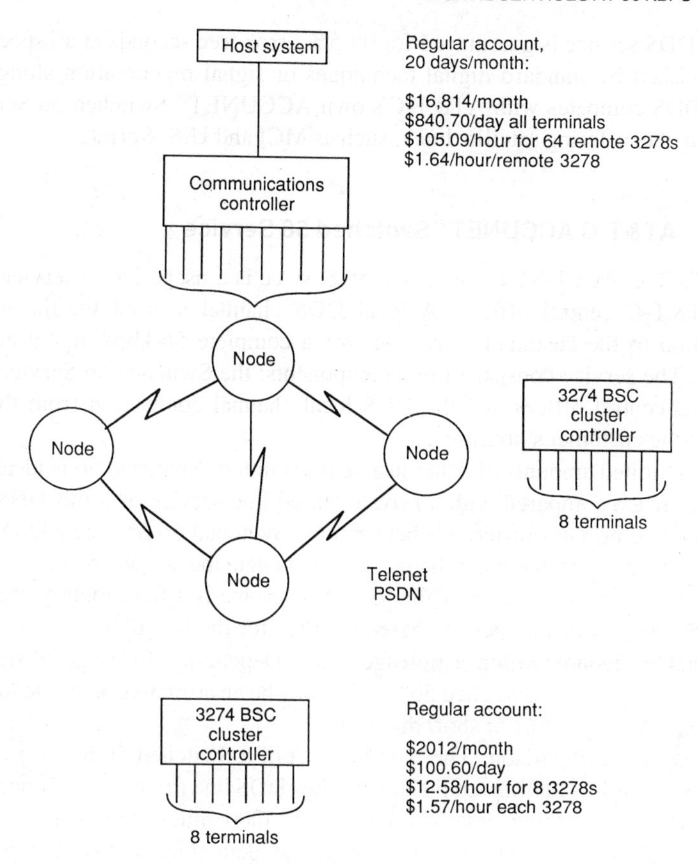

Figure 9.5. On line with 3270 BSC terminals on Telenet.

9.4.1. AT&T-C DATAPHONE Digital Service

AT&T-C DATAPHONE Digital Service (DDS) is a digital end-to-end service for point-to-point or multipoint data communications and is offered over interstate or inter-LATA to over 102 U.S. cities. In addition, there are connections to over 82 Canadian cities through Telecom Canada's Dataroute service. DDS provides communications between terminals and computer host systems at speeds from 2.4 to 56 kbps. DDS rates are based on mileage, transmission speed, and network access charges. Attachment to the network is by a customer service unit (CSU) or data service unit (DSU), available from many vendors. The CSU is a matching device to ensure that there is no harm to the network and is required for customers providing their own digital network attachment. A DSU is required for matching the subscriber's terminal or host output equipment to the network access point. The DSU performs timing and signal control.

The DDS service is guaranteed for 99.5% error-free seconds at all speeds and is accomplished by standard digital techniques or signal regeneration along the network. DDS competes with AT&T-C's own ACCUNET® Switched 56 Service and other carriers offering satellite links, such as MCI and U.S. Sprint.

9.4.2. AT&T-C ACCUNET® Switched 56 Service

The AT&T-C ACCUNET® Switched 56 Service is a usage-based service between two AT&T-C central offices. A local DDS channel is used for the local loop connection to the customer's premises for a complete 56-kbps digital end-to-end service. The service consists of two components: the Switched 56 Service between AT&T-C central offices and the DDS local channel connection from the central office to the customer's premises.

When limited amounts of data must be transmitted, Switched 56 is ideal for long distances when compared with a private leased line service or a full DDS 56-kbps offering. The primary difference between the Switched 56 Service and DDS is the monthly charge for the interchange link between the AT&T-C central offices. Switched 56 is based on use, while the DDS service is a flat monthly charge. The Switched 56 Service charges are based on a fee for the initial 30 seconds plus each additional 6 seconds within a mileage band. Depending on usage, distance, and customer location, the Switched 56 Service may be an attractive alternate to DDS for high-speed data transfers of short duration.

For a quick comparison: The monthly cost of a Switched 56 Service line in the United States is less than an equivalent 56-kbps DDS line for less than 1 hour of traffic per day for 20 working days per month. At 1000 miles, the cost advantage of Switched 56 Service allows up to 2 hours of data transmission compared to a dedicated fixed DDS connection.

9.4.3. MCI Private Line Facilities (Digital Services)

MCI offers a point-to-point private line digital service at transmission speeds of 2.4, 4.8, 9.6, and 56 kbps. The digital service is part of MCI's Terrestrial Digital Service. MCI's digital service includes an end-to-end connection with local access to the customer's premises. MCI will coordinate the local channel attachment.

The emphasis at MCI is to organizations that transmit large volumes of data between two points. MCI does not transmit voice over the same facilities.

9.4.4. Digital Data Service Comparisons

The monthly charges for digital facilities will be calculated for direct private lines from 17 remote locations to Atlanta for 9.6- and 56-kbps lines. Major components for an AT&T DDS point-to-point connection are shown in Figure 9.6. Note the addition of the digital data access multiplexer component for each local connection.

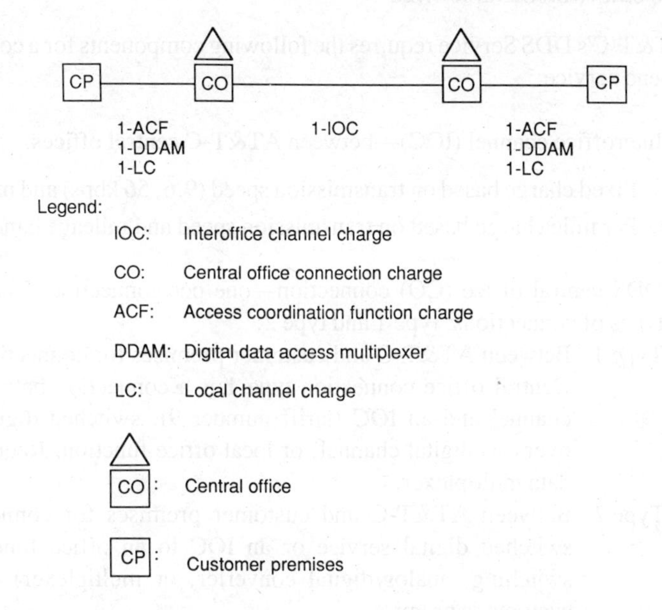

Figure 9.6. AT&T DDS point-to-point connection components.

Assumptions for digital data service:

- AT&T tariff Number 9 applies to interoffice channel rates, and AT&T tariff Number 11 is used for local channel connection rates for AT&T-C's DATAPHONE Digital Service (DDS). MCI's rates are based on its private line facilities for digital data service (DDS).
- Each configuration is a point-to-point line to the host location based in Atlanta, Georgia.
- Taxes, installation charges, and nonrecurring and one-time charges are not included in the calculations.
- All locations will be a distance of 8 miles from the AT&T-C rate center.
- The interoffice mileage from Atlanta is as follows:

Boston	935	Miami	601
Chicago	593	Minneapolis	910
Cleveland	553	New York	750
Dallas	720	Philadelphia	675
Denver	1209	Phoenix	1590
Detroit	594	St. Louis	465
Houston	702	San Francisco	2134
Los Angeles	1930	Seattle	2174
		Washington, D.C.	542

- AT&T-C's DDS Service requires the following components for a complete end-to-end service:

  1. Interoffice channel (IOC)—between AT&T-C central offices:

     a. Fixed charge based on transmission speed (9.6, 56 kbps) and mileage band
     b. Per mile charge based on transmission speed and mileage band

  2. DDS central office (CO) connection—one per connection. There are two types of connections: type 1 and type 2:

     Type 1: Between AT&T-C and customer premises for connection to DDS. Central office connection type 1 is a connection between a local channel and an IOC (tariff number 9), switched digital service, overseas digital channel, or local office function. Requires digital data multiplexer.

     Type 2: Between AT&T-C and customer premises for connection to a switched digital service or an IOC to an office function (e.g., switching, analog/digital converter, or multiplexer) or between customer's premises.

     The type 1 connection applies in this example.

  3. Access coordination (ACF) charge—based on transmission speed. The charge is for end-to-end design, installation, and maintenance.
  4. Digital data access multiplexer (DDAM)—required for type 1.
  5. Local channel (LC)—per central office connection.

- MCI's DDS requires the following components for a complete end-to-end service:

  1. Interoffice channel (IOC)—between MCI central offices:

     a. Fixed charge based on line speed (9.6, 56 kbps) and mileage band
     b. Per mile charge based on air distance between offices within a mileage band

  2. Central office (CO) connection—one per connection.
  3. DDS multiplexer—one per central office connection is required.
  4. Access coordination (ACF) charge—applied because MCI will coordinate the local channel.
  5. Local channel (LC)—assume AT&T rate. Some MCI offices are closer than AT&T, and some are farther. In a 17-office comparison, the rates will average out to about the same monthly charge.

TABLE 9.3. AT&T-C and MCI Interoffice Channel Rates

	AT&T-C Monthly Charges				MCI Monthly Charges			
	9.6 kbps		56 kbps		9.6 kbps		56 kbps	
Mileage Band	Fixed Charge	Per Mile	Fixed Charge	Per Mile	Fixed Charge	Per Mile	Fixed Charge	Per Mile
1-50	$ 75.25	$2.20	$ 285	$9.50	$ 64.50	$1.90	$ 255	$8.00
51-100	135.25	1.00	535	4.50	122.00	0.75	485	3.40
101-500	175.25	0.60	700	2.85	145.00	0.52	580	2.45
501-1000	325.25	0.30	1325	1.60	265.00	0.28	1050	1.50
Over 1000	325.25	0.30	1325	1.60	265.00	0.28	1105	1.45

Source: AT&T and MCI Communications.

Table 9.3 gives the monthly interoffice channel rates for AT&T-C and MCI. Each DDS has a fixed charge and a per mile charge, both based on transmission speed and a mileage band. Table 9.4 shows the AT&T-C 9.6-kbps DDS between Atlanta and 17 cities, and Table 9.5 contains a similar analysis for an AT&T-C 56-kbps service. MCI's 9.6-kbps DDS is shown in Table 9.6, and its 56-kbps DDS is shown in Table 9.7.

For both the 9.6- and 56-kbps digital data services, there is a significant savings on an MCI digital line compared to AT&T. At 9.6-kbps, the AT&T-C total monthly charge is $28,731.77 compared to MCI's $20,454.37, and AT&T-C's 56-kbps service is $72,543.95 compared to MCI's 56-kbps service at $57,879.51.

9.5. T1 OR 1.544-MBPS TERRESTRIAL SERVICES

High-speed services at 1.544-Mbps are now offered between specific locations. One condition should exist for a 1.544-Mbps line to be economical: enough voice and/or data to efficiently use the high bandwidth. T1 can be used as a single high-speed channel, several high-speed channels, or 24 voice-grade channels.

T1 digital service allows several voice channels, bulk information transfer, and video conferencing between two locations. To justify a fixed service connection, large amounts of voice and data traffic is required between two corporate sites on a continuous basis. As an alternative to a continuous direct link, an intermittent service called ACCUNET® Reserved 1.5 is available for bulk data transfers and video conferencing.

T1 was designed to transmit voice frequencies up to 4000 hertz, but with the use of T1 multiplexers, data and voice can be transmitted over the equivalent voice channels. To duplicate speech with quality reproduction at 4000 hertz, the sampling rate is two times the highest frequency component or 8000 samples per second. A pulse-coded technique is used to quantify the pulse amplitude into 8 bits. The result is a

TABLE 9.4. AT&T-C 9.6-kbps DDS

	Inter-Office Channel	Local Channel	Options[a]	Total 9.6-kbps DDS
Atlanta	—	$3,363.62[b]	$5,512.59	$ 8,876.21
Boston	$605.75	216.36	324.27	1,146.38
Chicago	503.15	212.06	324.27	1,039.48
Cleveland	491.15	195.95	324.27	1,011.37
Dallas	541.25	278.72	324.27	1,144.24
Denver	687.95	210.38	324.27	1,222.60
Detroit	503.45	221.86	324.27	1,049.58
Houston	535.85	278.72	324.27	1,138.84
Los Angeles	904.25	212.48	324.27	1,441.00
Miami	505.55	194.42	324.27	1,024.24
Minneapolis	598.25	219.28	324.27	1,141.80
New York	550.25	204.94	324.27	1,079.46
Philadelphia	527.75	174.60	324.27	1,026.62
Phoenix	802.25	193.64	324.27	1,320.16
St. Louis	464.75	279.89	324.27	1,068.91
San Francisco	965.45	212.48	324.27	1,502.20
Seattle	977.45	211.99	324.27	1,513.71
Washington, D.C.	487.85	172.85	324.27	984.97
Total monthly charge—AT&T 9.6-kbps DDS				$28,731.77

[a]Options include the following:

$ 35.00	CO—DDS central office connection
210.00	ACF—Access coordination function charges
79.27	DDAM—Digital data access multiplexing
$ 324.27	

[b]Atlanta has 17 local channel connections and 17 options charges.

Source: AT&T.

channel rate of (8000 × 8 bits =) 64,000 bps or 64 kbps. Each T1 carrier consists of a frame that carries 24 voice channels for a total of 1,563,000 bps. The remaining bits are used for synchronization, signaling, and supervisory information.

A T1 multiplexer will be used to allocate the channel. Each T1 channel is subdivided into subslots that vary by manufacturer up to 64 kbps. If the multiplexer has a minimum slot rate of 4.8 kbps, terminals operating below that rate will take up the whole slot. If the rate is over 64 kbps, e.g., 128 kbps, the device will take up two channels. In practice, the maximum use of the T1 channel is dependent on how many devices are filling up each 24-channel 64-kbps capacity.

TABLE 9.5. AT&T-C 56-kbps DDS

	Inter-Office Channel	Local Channel	Options[a]	Total 56-kbps DDS
Atlanta	—	$4858.94[b]	$6336.92	$11,195.86
Boston	$2821	319.58	372.76	3,513.34
Chicago	2274	324.39	372.76	2,971.15
Cleveland	2210	280.47	372.76	2,863.23
Dallas	2477	319.71	372.76	3,169.47
Denver	3259	315.07	372.76	3,946.83
Detroit	2275	318.75	372.76	2,966.51
Houston	2448	319.71	372.76	3,140.47
Los Angeles	4413	329.75	372.76	5,115.51
Miami	2286	282.01	372.76	2,940.77
Minneapolis	2781	327.30	372.76	3,475.30
New York	2525	291.91	372.76	3,189.67
Philadelphia	2405	267.46	372.76	3,045.22
Phoenix	3869	288.88	372.76	4,530.64
St. Louis	2025	320.78	372.76	2,718.54
San Francisco	4739	329.75	372.76	5,441.51
Seattle	4803	322.12	372.76	5,497.88
Washington, D.C.	2192	251.53	372.76	2,816.29
Total monthly charges—AT&T 56-kbps DDS				$72,543.95

[a]Options include the following:

$ 50.00	CO—DDS central office connection
260.00	ACF—access coordination function charge
62.76	DDAM—Digital data access multiplexing
$ 372.76	

[b]Atlanta has 17 local channel connections and 17 options charges.

Source: AT&T.

9.5.1. AT&T-C ACCUNET® T1.5 Service

ACCUNET® T1.5 Service can be used for voice, data, facsimile, and video transmission in point-to-point private use. The service is based on line configuration and usage charges. A complete service from AT&T-C includes charges for an interoffice channel between two central offices, central office charges, and local connection options. Standard arrangements allow the multiplexing of either 24 or 44 voice channels.

AT&T-C will provide (1) a total service or end-to-end service completely configured and operable; (2) a coordinated service whereby they will design, install, and

TABLE 9.6. MCI 9.6-kbps DDS

	Mileage from Atlanta	Inter office Rate	Local Channel	Options Charges[a]	Total
Atlanta	—	—	$3363.62[b]	$2074[b]	$5,437.62
Boston	935	$526.80	216.36	122	865.16
Chicago	593	453.36	212.06	122	787.42
Cleveland	553	287.56	195.95	122	605.51
Dallas	720	466.60	278.72	122	867.32
Denver	1209	603.52	210.38	122	935.90
Detroit	594	431.32	221.86	122	775.18
Houston	702	461.56	278.72	122	862.28
Los Angeles	1930	805.40	212.48	122	1,139.88
Miami	601	433.28	194.42	122	749.70
Minneapolis	910	519.80	219.28	122	861.08
New York	750	475.00	204.92	122	801.92
Philadelphia	675	454.00	174.60	122	750.60
Phenix	1590	710.20	193.64	122	1,025.84
St. Louis	465	470.75	279.89	122	872.64
San Francisco	2134	862.52	212.48	122	1,197.00
Seattle	2174	873.72	211.99	122	1,207.71
Washington, D.C.	542	416.76	172.85	122	711.61
Total monthly charge—MCI 9.6-kbps DDS					$20,454.37

[a]Options charges:

$ 30 central office connection
65 DDS multiplexer
27 access coordination
$122

[b]Atlanta has 17 local channel connections and 17 options charges.

Source: MCI Communications.

maintain overall service and order the local lines; and (3) a baseline service where the subscriber orders the access line. The service is very reliable and is designed for operation with an error rate of 1 per million bits of information.

9.5.2. MCI Terrestrial Digital Service

MCI's terrestrial digital data service is offered at 1.544 Mbps and consists of a complete end-to-end service. The DDS is available for point-to-point two-way simultaneous data transmission over a dedicated circuit. There are two pricing

TABLE 9.7. MCI 56-kbps DDS

	Mileage from Atlanta	Inter-Office Rate	Local Channel	Options Charges[a]	Total
Atlanta	—	—	$4858.94[b]	$1989	$6,847.94
Boston	935	$2460.75	319.58	117	2,897.33
Chicago	593	1964.85	324.39	117	2,406.24
Cleveland	553	1906.85	280.47	117	2,304.32
Dallas	720	2149.00	319.71	117	2,585.71
Denver	1209	2858.05	315.07	117	3,290.12
Detroit	594	1966.30	318.75	117	2,402.05
Houston	702	2122.90	319.71	117	2,559.61
Los Angeles	1930	3903.50	329.75	117	4,350.25
Miami	601	1976.45	282.01	117	2,375.46
Minneapolis	910	2424.50	327.30	117	2,868.80
New York	750	2192.50	291.91	117	2,601.41
Philadelphia	675	2083.75	267.46	117	2,468.21
Phoenix	1590	3410.50	288.88	117	3,816.38
St. Louis	465	2066.00	320.78	117	2,503.78
San Francisco	2134	4199.30	329.75	117	4,646.05
Seattle	2174	4257.30	322.12	117	4,696.42
Washington, D.C.	542	1890.90	251.53	117	2,259.43
Total monthly charge—MCI 56-kbps DDS					$57,879.51

[a]Options charges:

$ 40 central office connection
50 DDS multiplexer
27 access coordination
$ 117

[b]Atlanta has 17 local channel connections and 17 options charges.

Source: MCI Communications.

components: an interoffice mileage charge and a local access facilities charge. MCI's service does not have the coverage of that of AT&T-C but, where applicable, can be an attractive alternative.

9.5.3. AT&T-C ACCUNET® Reserved 1.5 Service

ACCUNET® Reserved 1.5 Service is a switched version of the ACCUNET® T1 Service and is a good alternative for intermittent applications such as video conferencing (full-motion TV), computer disaster recovery services, or bulk data transfers to backup computer files in another city. Reserved 1.5 is available as a point-

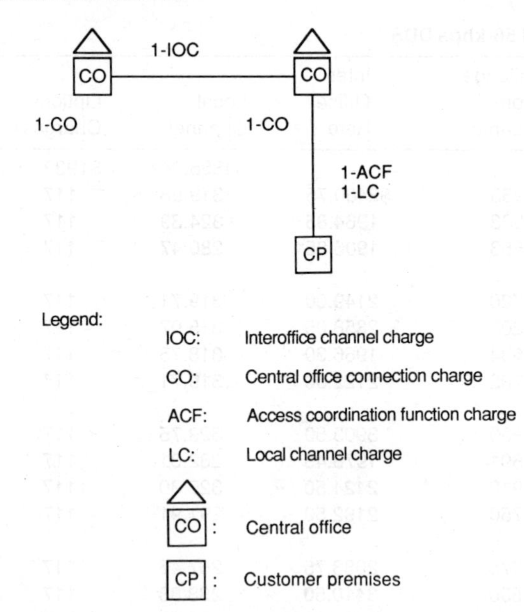

Figure 9.7. AT&T, MCI T1 end-to-end components.

to-point or multipoint service. The multipoint service is available only as a broadcast service, e.g., a headquarters call to all regional or district offices, and is not conducive for normal voice and data information transfer. The point-to-point service can be used for normal voice and data service and is available for interstate channels between two customer locations. As the name implies, the service is reserved in advance and is available 24 hours per day, 7 days per week. The minimum charge is for 30 minutes with additional 30-minute segments based on mileage between the AT&T-C central offices. AT&T-C has extended the ACCUNET® Reserved 1.5 Service to Belgium, France, Germany (FRG), Italy, the Netherlands, Sweden, Switzerland, and the United Kingdom.

9.5.4. T1 End-to-End Connection Comparison

A comparison between AT&T-C's ACCUNET 1.5 Service and MCI's Private Line Terrestrial Digital Service (TDS) will be calculated. Both offer a point-to-point dedicated circuit available for two-way transmission at synchronous speeds of 1.544 Mbps. Each carrier offers their customers end-to-end service and will perform complete design, installation, coordination, and maintenance. The major components for either an AT&T or MCI T1 end-to-end connection are shown in Figure 9.7.

Assumptions for AT&T-C and MCI 1.544-Mbps services:

- Thirteen cities will be compared with a star or point-to-point network arrangement with all lines originating from Atlanta.
- Each office will be in the same LATA at a distance of 8 miles from the central office.

- Customer premises equipment such as T1 multiplexers and line terminating equipment are not included in this calculation.
- Taxes, installation, nonrecurring costs, and line terminating equipment are not included in the calculation.
- AT&T-C's FCC tariff number 9 applies to the interoffice channel between central offices. The local portion and coordination charges are covered under FCC tariff number 11.
- AT&T-C's ACCUNET 1.5 Service requires the following components for a complete end-to-end service:

  1. Interoffice channel (IOC) rate—between AT&T-C central offices:

     a. Fixed charge based on transmission speed of 1.544 Mbps and mileage band
     b. Per mile charge based on transmission speed and mileage band

  2. Central office (CO) connection charge
  3. Access coordination function (ACF) charge—because AT&T-C will provide access coordination.
  4. Local channel (LC)—per central office connection, including:

     a. Fixed charge based on transmission speed of 1.544 Mbps and mileage band
     b. Per mile charge based on transmission speed and mileage band

- MCI's terrestrial digital service (TDS) requires the following components for a complete end-to-end service:

  1. Interoffice channel (IOC)—between MCI central offices.

     a. Fixed charge based on 1.544-Mbps line speed and mileage band
     b. Per mile charge based on air distance between offices within a mileage band

  2. Central office (CO) connection—one per connection.
  3. Access coordination (ACF) Charge—applied because MCI will coordinate the local channel.
  4. Local channel (LC)—The local channel rate varies by the distance from the central office from a low of about 850 to a high of 1200 miles. Assume an average of 1000 miles for each local channel.

AT&T-C ACCUNET T1.5 and MCI terrestrial digital service calculations are shown in the following tables. The mileage rate will be based on the values in Table 9.8. AT&T's ACCUNET T1.5 interoffice, central office connection, local channel, and access coordination function charges are shown in Table 9.9. MCI's terrestrial digital service calculations are given in Table 9.10.

TABLE 9.8. AT&T-C and MCI 1.544-Mbps Mileage Rates

	AT&T-C 1.544 Mbps		MCI 1.544 Mbps	
Mileage Band	Fixed Charge	Per Mile	Fixed Charge	Per Mile
1-50	$ 400.00	$40.00	$ 385.00	$34.50
51-100	900.00	30.00	790.00	26.40
101-500	1400.00	26.00	1100.00	23.30
Over 500	1400.00	26.00	1250.00	23.00

Source: American Telephone and Telegraph Company and MCI Communications.

TABLE 9.9. AT&T-C Accunet 1.5 Total End-to-End Monthly Charge

City	Mileage from Atlanta	Inter-office Channel	Local Channel	Options Charges[a]	AT&T-C Total Monthly Cost
Atlanta	—	—	$15,563.28[b]	$972[b]	$16,535.28
Boston	935	$25,710	1,269.94	81	27,060.94
Chicago	593	16,818	1,537.64	81	18,436.64
Dallas	720	20,120	1,567.32	81	21,768.32
Denver	1209	32,756	861.28	81	33,698.28
Detroit	594	16,844	1,269.94	81	18,194.94
Houston	702	19,652	1,567.32	81	21,300.32
Los Angeles	1930	51,580	1,177.28	81	52,838.28
New York	750	20,900	1,216.78	81	22,197.78
Philadelphia	675	18,950	840.53	81	19,871.53
San Francisco	2134	56,884	1,177.28	81	58,142.28
Seattle	2174	57,924	903.27	81	58,908.27
Washington, D.C.	542	15,492	1,044.13	81	16,617.13
Total					$385,569.99

[a]Options charges:

$60 central office (Co) connection charge
21 access coordination function (ACF) charge
$81

[b]Atlanta has 12 local channel connections at the central office and 12 options charges. The local channel consists of a fixed rate plus a per mile component:

$$\$844.86 + (\$56.51 \times 8 \text{ miles}) = \$1296.96 \times 12 \text{ lines} = \$15,563.28.$$

Source: American Telephone and Telegraph Company.

Conclusion for AT&T-C and MCI 1.544 services: Normally, a T1.5 service is used between two company locations with high concentrations of personnel or between information centers. At $385,570 (AT&T-C) and $338,498 (MCI) per month, a 12-city distribution is very expensive and requires large volumes of data or several voice channels to be cost effective. The comparison has been presented for comparative purposes and was effective for one period in time. Due to competition among carriers, AT&T-C, MCI or another service could be lower for another period of time.

TABLE 9.10. MCI Total End-to-End Cost

City	Mileage from Atlanta	Inter-office Channel	Local Channel	Options Charges[a]	MCI Total Monthly Cost
Atlanta	—	—	$12,000[b]	$732[b]	$12,732
Boston	935	$22,755	1,000	61	23,816
Chicago	593	14,889	1,000	61	15,950
Dallas	720	17,810	1,000	61	18,871
Denver	1209	29,057	1,000	61	30,118
Detroit	594	14,912	1,000	61	15,973
Houston	702	17,396	1,000	61	18,457
Los Angeles	1930	45,640	1,000	61	46,701
New York	750	18,500	1,000	61	19,561
Philadelphia	675	16,775	1,000	61	17,836
San Francisco	2134	50,332	1,000	61	51,393
Seattle	2174	51,252	1,000	61	52,313
Washington, D.C.	542	13,716	1,000	61	14,777
Total					$338,498

[a]Options charges:
$40 central office (CO) connection charge, one at each end
21 access coordination (ACF) charge
$61

[b]Atlanta has 12 local channel connections at the central office and 12 options charges.

Source: MCI Communications.

10

INTERNATIONAL TRAFFIC

In this chapter we will

- Examine the world renumbering system, international overseas channels, and basic networks
- Provide information on international analog, DDS, Datel, and telex channels
- Discuss international packet switched data networks

Countries of the world are classified into two economic categories by the ITU: developed countries and developing countries. The developed countries include the United States, Canada, European countries, Australia, New Zealand, and Japan. The remainder of the world, with one or two exceptions (South Africa), falls into the developing countries category. In telecommunications, developing country offerings vary from basic telephone and telex facilities to some private line facilities and data network offerings. Most of the developed countries offer complete alternatives for voice and data networks.

Telephone message services between the United States and other countries use CCITT recommendations in order to have compatibility of transmission and services for international telephone communications. The CCITT has issued a numbering plan for all ITU member countries whereby each country has a country code followed by the national number of the subscriber's station. At present, the plan provides a maximum of 12 digits, although many countries do not use the full number of digits.

Countries with high requirements have a single-digit number, while countries with the lowest requirements have a three-digit country code.

Many small countries do not have a country code but share a single country code with other countries within a zone. Another consideration is based on the requirements for telephones by the year 2000. Because of higher requirements, North America (Canada and the United States) and the Union of Soviet Socialist Republics have been assigned a single-digit country code, and both have a separate world zone. The code makes it simple to call Canada from the United States. Each province in Canada is assigned an area code, and there is no difference in the procedure for a call between the states and a call to anywhere in Canada. This procedure also applies to the Caribbean countries and Puerto Rico.

Country codes were devised by dividing the world into nine geographic world zones and assigning a number to each zone. All countries within a zone are assigned a country code beginning with that digit. Europe is a special case due to the number of telephones projected for the year 2000. They have two zones and at least a two-digit code for each country. Throughout the world, country telephone networks are unique, and each national telephone numbering system is dictated by the individual country.

Although telephones are considered indispensable for most persons in developed and developing countries, over 80% of the world's telephones are used by 23% of the world's population. The 14 countries in Table 10.1 accounted for the greatest number of telephone lines in 1985. Another useful figure is the number of telephones per 100 inhabitants.

Table 10.1 shows 1985 year-end estimates based on multiple sources including British Telecom of the United Kingdom, the Population Reference Bureau, Inc.; the ITU of Geneva, Switzerland; and the U.S. Department of Commerce.

TABLE 10.1 Distribution—80% of the World's Telephones

Country	Millions of Access Lines	Telephones per 100 Inhabitants
United States	111	79
Japan	46	56
U.S.S.R.	26	12
Germany (Federal Republic)	24	53
France	21	60
United Kingdom	20	53
Italy	16	45
Canada	10	45
Spain	8.8	35
Brazil	6.5	8
Australia	6.0	54
The Netherlands	5.6	58
Sweden	5.1	89
Switzerland	3.2	80

10.1. INTERNATIONAL TELEPHONE WORLD NUMBER ZONES

The world telephone numbering system is grouped by area and listed in Table 10.2. North America includes a number of islands in the Caribbean. Africa includes four major areas that are significantly different in attitudes and cultural and economic ties. The French-speaking countries, although independent, maintain strong economic ties to France, while the English-speaking countries are independent and maintain looser ties to the United Kingdom. This results in telecommunications links to Paris and London and is a key point in planning telecommunications to African countries through international gateways.

Twenty-seven islands in the South Pacific that can be reached by telephone are not included in the listing. Country codes are included in parentheses after each country except for North American countries and the Caribbean Islands, which all have country code 1.

10.2. INTERNATIONAL OVERSEAS CHANNELS

An overseas channel is the component required to extend a private or leased line service from an AT&T-C or other international record carrier (IRC) overseas gateway office to an overseas location. The monthly charges in this section are based on AT&T-C tariffs. See the IRCs' section in this chapter for alternate solutions. Depending on the country, voice-grade lines for voice or data are furnished at either full-channel or half-channel rates. A full channel includes one complete end-to-end line with one tariff paid to AT&T-C that includes the U.S. portion from the gateway city, any intermediary country or countries, and the country PTT where the line terminates. For half-channel service the subscriber contracts with AT&T-C for one half of the channel and with a foreign PTT, overseas company, or foreign administration for the second half channel to complete the end-to-end service. With half-channel service, the subscriber receives two monthly bills: one from AT&T-C or the local exchange company and one from the country PTT where the line terminates. Charges for an intermediary country or countries are worked out by the two end-to-end network providers and are included in both tariffs.

Voice-grade overseas channels are used for voice and data, voice only, or data only. If a private line service to an overseas location involves more than two AT&T-C central offices on the mainland to the overseas gateway location, AT&T-C will assure satisfactory transmission only on their segment or the U.S. mainland. There are five overseas gateway cities established by the FCC: Miami, New Orleans, New York, San Francisco, and Washington, DC.

Overseas channels to the Pacific, South Pacific, and Far East start from the San Francisco overseas gateway. All other nations are reached by the four eastern overseas gateway cities—Miami, New Orleans, New York, and Washington, D.C. The channel rates that follow are based on AT&T-C tariff number 9. Although monthly rates will vary by year, the charges can be used for planning purposes.

TABLE 10.2. Country Codes for International Telephone Calls

Zone 1: North America (all countries have country code 1)

Bahamas	Cayman Islands
Bermuda	Turks and Caicos
Canada	United States
Caribbean Islands:	
Anguilla	Puerto Rico
Antigua	St. Kitts-Nevis
Barbados	St. Lucia
Dominica	St. Pierre & Miquelon
Dominican Republic	St. Vincent
Grenada	Trinidad & Tobago
Jamaica	Virgin Islands, British
Montserrat	Virgin Islands, United States

Zone 2: Africa

North Africa	Southern Africa
Algeria (213)	Angola (244)
Egypt (20)	Botswana (267)
Libya (218)	Lesotho (266)
Morocco (212)	Mozambique (258)
Sudan (249)	South Africa (27)
Tunisia (216)	South West Africa (Namibia)
	Swaziland (268)
	Zimbabwe (283)
French speaking:	English Speaking
Burundi (257)	Ethiopia (251)
Cameroon (237)	Gambia (220)
Central African Rep. (236)	Ghana (233)
Chad (235)	Kenya (254)
Congo (Brazzaville) (242)	Liberia (231)
Dahomey (229)	Malawi (265)
Gabon (241)	Nigeria (234)
Guinea (224)	Sierra Leone (232)
Ivory Coast (225)	Somali Republic (252)
Madagascar (261)	Tanzania (255)
Mali (223)	Uganda (256)
Mauritania (222)	Zambia (260)
Niger (227)	
Rwanda (250)	
Senegal (221)	
Togo (228)	
Upper Volta (226)	
Zaire (243)	

(continued)

TABLE 10.2. Country Codes for International Telephone Calls

Zones 3 & 4: Europe

Albania (355)	Italy (39)
Andorra (33)	Liechtenstein (41)
Austria (43)	Luxemborg (352)
Belgium (32)	Malta (356)
Bulgaria (359)	Monaco (33)
Cyprus (357)	The Netherlands (31)
Czechoslovakia (42)	Norway (47)
Denmark (45)	Poland (48)
Finland (358)	Portugal (351)
France (33)	Romania (40)
Germany (Dem. Rep.) (37)	San Marino (39)
Germany (Federal Rep.) (49)	Spain (34)
Gibraltar (350)	Sweden (46)
Greece (30)	Switzerland (41)
Greenland (299)	United Kingdom (44)
Hungary (36)	Yugoslavia (38)
Iceland (354)	
Ireland (353)	

Zone 5: South Central America

Argentina (54)	Guyana (592)
Belize (501)	Haiti (509)
Bolivia (591)	Honduras (504)
Brazil (55)	Martinque (596)
Chile (56)	Mexico (52)
Colombia (57)	Netherlands Antilles (599)
Costa Rica (506)	Nicaragua (505)
Cuba (53)	Panama (507)
Ecuador (593)	Paraguay (595)
El Salvador (503)	Peru (51)
Falklands Islands (33)	Surinam (597)
French Guiana (594)	Uruguay (598)
Guadeloupe (596)	Venezuela (58)
Guatemala (502)	

Zone 6: South Pacific

Australia (61)	New Zealand (64 or 643)
Brunei (673)	Philippines (63)
Indonesia (62)	Singapore (65)
Malaysia (60)	Thailand (66)

Zone 7: Union of Soviet Socialist Republics

Zone 8: Far East

Bangladesh (880)	Korea (Dem. People's Rep.) (003)
China (People's Republic) (86)	Laos (856)
China (Republic of) (886)	Macoa (853)
Hong Kong (852)	South Korea (82)
Japan (81)	Vietnam (84)
Kampuchea (855)	

TABLE 10.2. Country Codes for International Telephone Calls

Zone 9: Middle East and Southeast Asia

Middle East	Southeast Asia
Bahrain (973)	Afghanistan (93)
Iran (98)	Bhutan (N/A)
Iraq (964)	Burma (95)
Israel (972)	India (91)
Jordan (962)	Maldives (960)
Kuwait (965)	Mongolia (976)
Lebanon (961)	Nepal (977)
Oman (968)	Pakistan (92)
Qatar (974)	Sri Lanka (94)
Saudi Arabia (966)	
Syria (963)	
Turkey (90)	
United Arab Emirates (972)	
Yemen Arab Republic (967)	
Yemen People's Dem. Rep. (969)	

The overseas channels are connected to foreign countries by underseas cable systems or satellites. Underseas cable systems play a part in the economic development of a region and form the majority of circuits for voice and data traffic. After the five INTELSAT VI satellites are launched, satellites will handle a higher portion of overseas traffic, particularly to remote locations where an underseas cable system is not practical from an economic standpoint.

Satellite launchings are more interesting than cable systems. For example, each U.S. satellite launched by the Space Shuttle is viewed by the press. The Space Shuttle countdown, launch phase, orbital tracking, and in-space satellite transfer are closely monitored by the world on television. Suspense builds as viewers await the results. Will the satellite enter its geostationary position and begin functioning? Insurance companies are interested watchers since each satellite is insured for about $100 million. Failure to function becomes a "report card" on the progress of U.S. technology and a total loss to the insurance companies. Europe is interested since Ariane, its launching vehicle owned by a consortium of European administrations, is in competition for satellite launchings.

U.S. cable systems are not front-page news. They are negotiated by AT&T, the IRCs, foreign PTTs, or foreign administrations, and work begins without any fanfare. Once the cable systems are in place, they are integrated into the telecommunications network and may receive some "press" when an administrative person cuts the ribbon at some remote seaside location. A thick cable emerging from the sea is not glamorous. However, one of the greatest threats to the satellite business is the use of ocean floor fiber optic cable. For example, the largest existing underseas cable of the TAT series from the United States to Europe carries 4200 voice circuits, while the planned TAT-8 fiber optic cable will have a 40,000 equivalent voice circuit capacity. Projected estimates of the charge per equivalent voice circuit will be $1500 per year for underseas fiber.

The world's first international fiber optic undersea cable (UK-Belgium No. 5) is the link between Broadstairs, England and Ostend, Belgium. The 112-Kilometer cable uses six single-mode fibers and has three repeater stations.

A current listing of underseas cable systems will be discussed for each of the major areas of the world followed by either the half-channel or full-channel charges to many countries.

10.2.1. Europe

Underseas cables handle a large portion of the voice and data traffic between the United States and Europe, Europe and Africa, and the United States and the Asia/Pacific area. However, satellites are increasing their share of voice and data traffic and handle almost all international television traffic.

Private transatlantic cable link: The use of the transatlantic and submarine routes have been the privilege of common carriers and national PTTs. A privately owned fiber optic cable system across the North Atlantic will introduce competition. The FCC has allowed the construction of a private transatlantic cable under the Cable Licensing Act. Two fiber optic cables offering high-quality digital transmission will extend from England to New Jersey. Each cable will contain at least three fiber pairs, and each pair will operate at a bit rate of at least 280 Mbps or the equivalent of 4000 voice circuits. The northern cable, PTAT-1, is to be in operation by June 1989, and the southern cable, PTAT-2, is planned for June 1992. When operational, the $690 million project for both cables will supply more than 24,000 equivalent transatlantic voice circuits or a mix of high-speed data, video, and voice. The joint operators of the system are Cable and Wireless of the United Kingdom and Tel-Optik of the United States.

There are two major underseas cable systems planned by the public common carriers for 1988. The first cable system (TAT-8) spans the Atlantic Ocean cable, and the second cable system will link Asia, Africa, and Europe.

TAT-8, a new underseas fiber optic cable, will consist of three main cable branches. The U.S. cable will meet with the other two branches, one from Widemouth, United Kingdom and the second from Penmarch, France, in the Atlantic Ocean closer to the European side. TAT-8 is jointly owned by 20 European PTTs, 8 U.S. companies (including AT&T), and Teleglobe Canada, which handles international traffic to and from Canada. TAT-8's total capacity of 40,000 equivalent voice circuits amounts to three times the capacity of all existing transatlantic cables. The more recent TAT series are owned in part by AT&T, the IRCs, and several European PTTs. The percentage of ownership is usually based on each administration's projected needs for traffic.

A second TAT Series fiber optic cable is planned across the Atlantic. The new network, called TAT-9, will be linked directly to a Mediterranean fiber optic network now under development. The network hub will be in Palermo (Sicily), Italy. TAT-9 will cost about $650 million and become operational in 1991.

On the receiving end in Europe, France and the United Kingdom will route their internal traffic and provide transit routing for traffic designated for other European countries.

One of the world's longest underseas cable systems will provide a link between Singapore and France and is owned by a consortium of 21 PTT administrations. The 8000-mile (13,000-kilometer) system will consist of eight end points from Asia to the Middle East to Africa and Europe. The country termination points will be in Singapore, Indonesia, Sri Lanka, Saudi Arabia, Djibuti, Egypt, Italy, and finally France.

The major sources for Tables 10.3–10.10 were AT&T Communications and PTT published reports. Figure 10.1 illustrates the major cable routes between the United States and Europe.

France, Italy, and Spain play a major role in the transit routing of information to the Middle East and Africa due to interconnecting underseas cable systems links. In addition to underseas cable systems, Europe is interconnected by terrestrial cable and satellite systems. The major underseas cable systems from France (except TAT-4, TAT-6, and TAT-8), Spain (except TAT-5), and Italy are listed in the tables. Figures 10.2, 10.3, and 10.4 are a layout of the major cable routes in Europe from France, Italy, and Spain.

Overseas channels to the European mainland from the United States are available from the four eastern gateway cities. Private lines are available from these European countries: Austria, Belgium, Bulgaria, Czechoslovakia, Denmark, Finland, France, Germany (Federal Republic), Greece, Hungary, Iceland, Italy, Luxembourg, the Netherlands, Norway, Poland, Portugal, Romania, Spain, Sweden, Switzerland, Turkey, the USSR, and Yugoslavia. The AT&T half-channel monthly charge from a gateway city is the same to all these countries.

AT&T-C includes line conditioning for data over international leased lines to Italy and Spain to match its high-quality M1020 data lines. France provides an AVD (alternate voice or data) line to AT&T and other IRCs. To calculate the total monthly charge, add the U.S. half-channel rate and the French, Italian, or Spanish half-channel rate. Half-channel prices from France, Italy, and Spain are the same to Canada as the U.S. Taxes are not included.

Notes: 1. Half-channel prices for an alternate voice or data line from three IRCs based in New York are as follows:

MCI	$4120
GE (RCA)	4085
TRT	3130

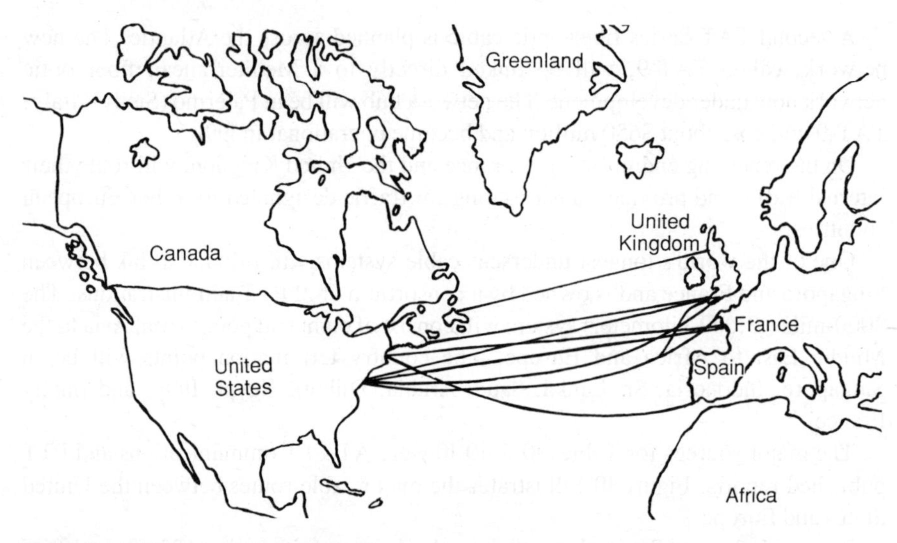

Figure 10.1. Major cable routes in North America.

TABLE 10.3. Major Cable Routes—North American Traffic to Europe

System	Outbound Location	Voice Channel Capacity
CANTAT-1	Grosse's Roche, Canada to Oban, Scotland	—
CANTAT-2	Halifax, Nova Scotia, Canada to Widemouth, England, United Kingdom	—
TAT-3	Tuckerton, New Jersey to Widemouth, United Kingdom	138
TAT-4	Tuckerton, New Jersey to St. Hilaire de Riez, France	138
TAT-5	Green Hill, Rhode Island to Conil, Spain	845
TAT-6	Green Hill, Rhode Island to St. Hilaire de Riez, France	4,000
TAT-7	Tuckerton, New Jersey to Lands End, United Kingdom	4,200
TAT-8	Tuckerton, New Jersey to Widemouth, United Kingdom and Penmarch, France	40,000
TAT-9	Planned between the U.S. east coast and Palermo, Italy	80,000+
Private:		
PTAT-1	New Jersey to west coast of England	12,000
PTAT-2	New Jersey to west coast of England	12,000

Source: AT&T and the U.S. Department of Commerce, NTIA-CR-84-13, Washington, D.C.

2. A current listing of the dollar against foreign currencies can be found in the *Financial Times, The New York Times*, and several major newspapers. Conversion rates of $1.00 U.S. = 6.4 francs, 1340 lira, or 131 pesetas were used to determine the following rates:

Half-Channel Rates in Dollars

Gateway City	U.S.	France	Italy	Spain
Miami	3600	2813	5997	4606
New Orleans	3600	2813	5997	4606
New York	3130	2813	5997	4606
Washington, D.C.	3130	2813	5997	4606

Canada provides a direct route from Australia to Europe. Routing can start in Sydney, Austrailia and go to Aukland, New Zealand to the Fiji Islands to Hawaii to Vancouver, and across Canada and can route either via the CANTAT-1 or CANTAT-2 cable system to the United Kingdom or through the ICECAN cable system through Greenland to Iceland to the United Kingdom. The United Kingdom would then act as a transit routing center for the European mainland.

The United Kingdom plays an important transit role with underseas cable links to the Scandinavian countries (Denmark, Norway, and Sweden) and other countries on the European mainland. Major underseas cable systems, except the TAT Series, are shown in Table 10.7. Figure 10.5 includes the major cable routes emanating from the United Kingdom.

Half-channel data rates apply from the United States to Ireland and the United Kingdom. The United Kingdom includes the Channel Islands, England, the Isle of Man, Northern Ireland, Scotland, and Wales. When compared to France, Italy, and Spain, there is a slight economic advantage to installing telecommunications facilities in Ireland or the United Kingdom. AT&T-C's half-channel charges are listed in the following note for the United States along with the half-channel charges for an M1020 data line in Ireland and the United Kingdom. Add the U.S. half-channel to the Irish or British half-channel to obtain the total monthly line charge.

Note: Conversion rates of $1.00 U.S. = 0.95 Irish pounds and $1.00 U.S. = 0.67 British pounds were used to determine the following rates:

Half-Channel Rates in Dollars

Gateway Cities	U.S.	Ireland	U.K.
Miami	3400	2148.3	2050
New Orleans	3400	2148.3	2050
New York	2925	2148.3	2050
Washington, D.C.	2925	2148.3	2050

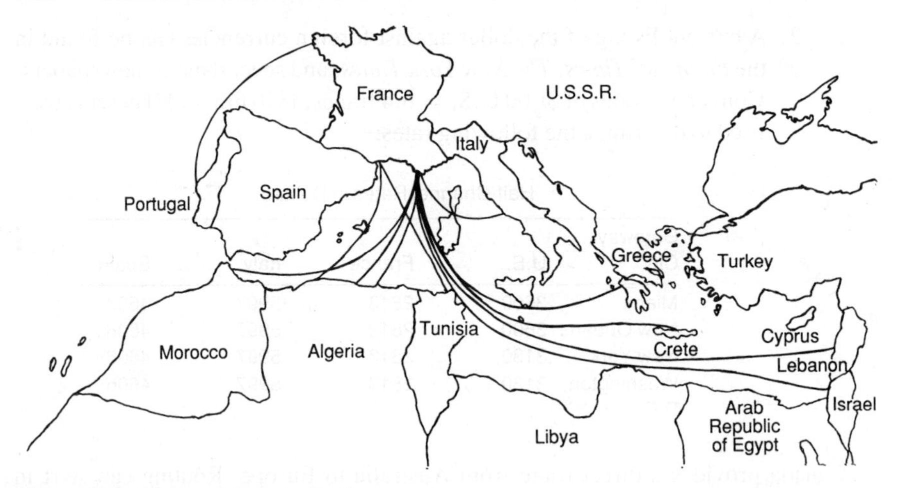

Figure 10.2. Major cable routes in France.

TABLE 10.4. Major Cable Routes Between France and Other Countries

System	France to Foreign Location
ANNIBAL (FR-TUN-2)	Perignan to Bizerte, Tunisia
ARIANE	Marseille to Heraklion, Crete, Greece
AMITIE	Marseille to Tetouan, Morocco
ARTEMIS	Marseille to Lekhaina, Greece
EL FATAH	La Seyne Sur Mer to Tripoli, Libya
FR-ALG-1	Perignan to Oran, Algeria
FR-LBN-1	Marseille to Beirut, Lebanon
FR-LIB	Marseille to Tripoli, Libya
FR-TUN-1	Marseille to Bizerte, Tunisia
FR-TUN-3	Martiques to Bizerte, Tunisia
MARALG-1	Marseille to Bordj el Kiffan, Algeria
MARALG-2	Marseille to Bordj el Kiffan, Algeria
MARTEL	Marseille to Tel Aviv, Israel
MARALG-3	Marseille to El Djamila, Algeria
MARALG-4	Martiques to El Djamila, Algeria
MARPAL	Marseille to Palo, Italy
MOROCCO-1	Carnet Plage to Tetouan, Morocco
MOROCCO-2	Penmarch to Casablanca, Morocco
TAIGIDE	Penmarch to Sesimbra, Portugal

Source: AT&T and the U.S. Department of Commerce, NTIA-CR-84-13, Washington, D.C.

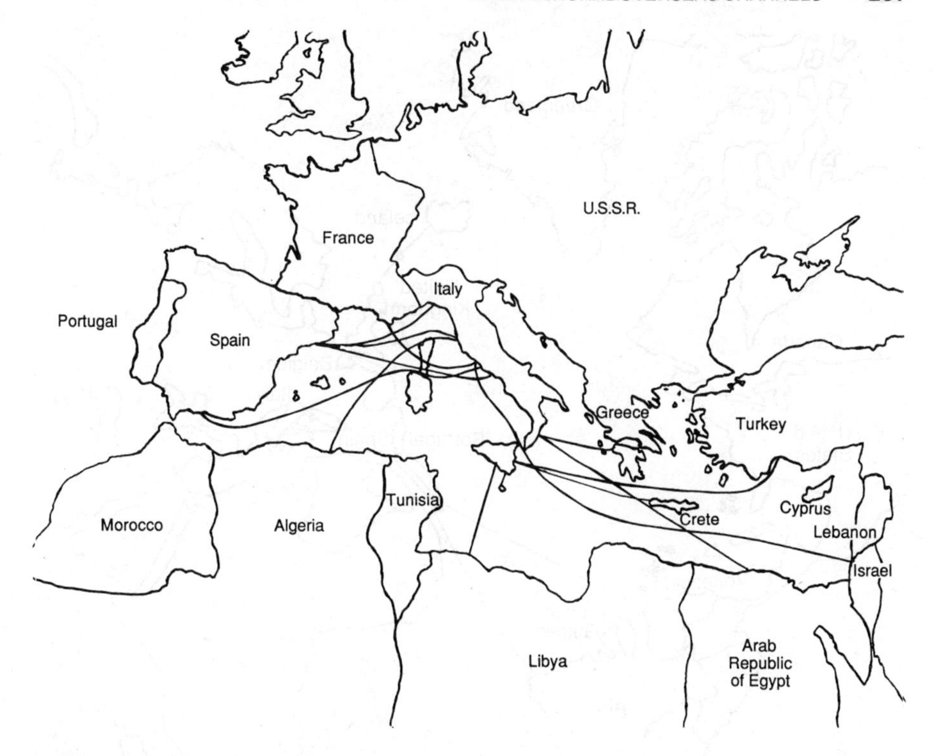

Figure 10.3. Major cable routes in Italy.

TABLE 10.5. Major Cable Routes Between Italy and Other Countries

System	Italy to Foreign Location
BARGEN	Genoa to Barcelona, Spain
BARPIS	Pisa to Barcelona, Spain
BARO	Pomezia to Barcelona, Spain
IT-ALG	Pisa to Algiers, Algeria
IT-EGYPT	Catanzaro to Alexandria, Egypt
IT-LIB	Agrigento, Sicily to Tripoli, Libya
IT-TURK	Catania to Antalya, Turkey
MARPAL	Palo to Marseille, France
MAT-1	Palo to Estepona, Spain
MED-1	Sicily to Malta
TELPAL	Palo to Tel Aviv, Israel
SEA-ME-WI	Marseille, France to Palermo to Alexandria, Egypt

Source: AT&T and the U.S. Department of Commerce, NTIA-CR-84-13, Washington, D.C.

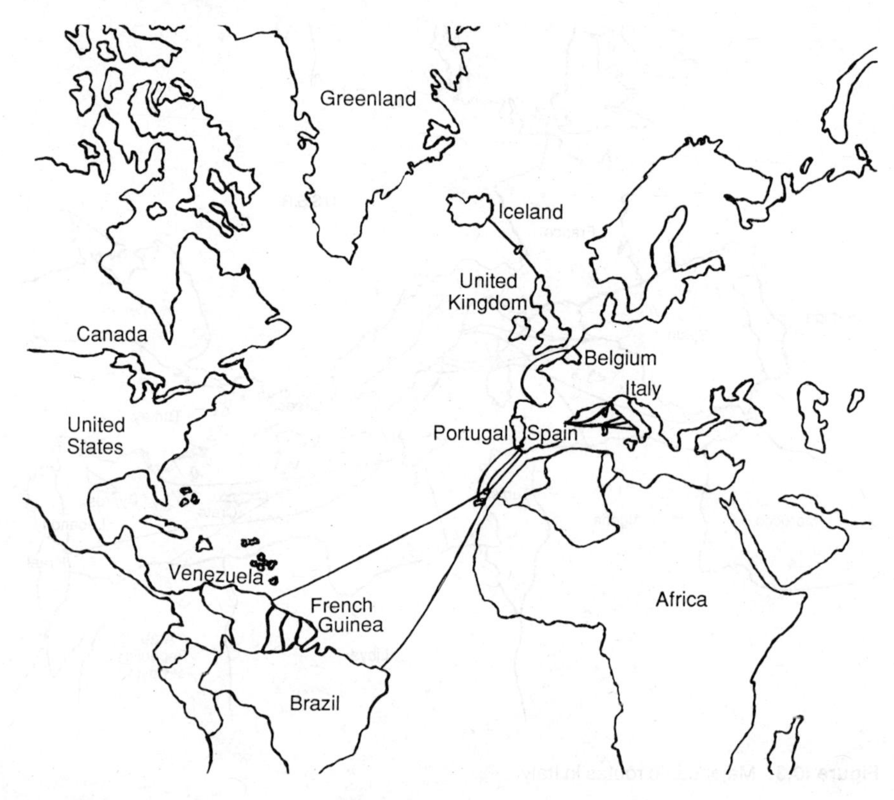

Figure 10.4. Major cable routes in Spain.

TABLE 10.6. Major Cable Routes between Spain and other Countries

System	Spain to Foreign Location
ALPAL	Palma de Mallorca to Bordj el Kiffan, Algeria
BARGEN	Barcelona to Genoa, Italy
BARPIS	Barcelona to Pisa, Italy
BARROM-1	Barcelona to Pomezia, Italy
MAT-1	Estepona to Palo, Italy
MERIDIAN	Rodiles to St. Idesbald, Belgium
PENCAN-1	San Fernando to Canary Islands via COLUMBUS-1 cable to Venezuela
PENCAN-2	Conil to Canary Islands via BRACAM-1 to Recife, Brazil
PENCAN-3	Chipiona to Canary Islands via COLUMBUS-1 cable to Venezuela
SCOTICE	Gairlock, Scotland to Faeroe Islands to Iceland
SLPSL	Palmo to Bordj el Kiffan, Algeria
UK-SP-3	Rodiles to Porthcurno, England

Source: AT&T and the U.S. Department of Commerce, NTIA-CR-84-13, Washington, D.C.

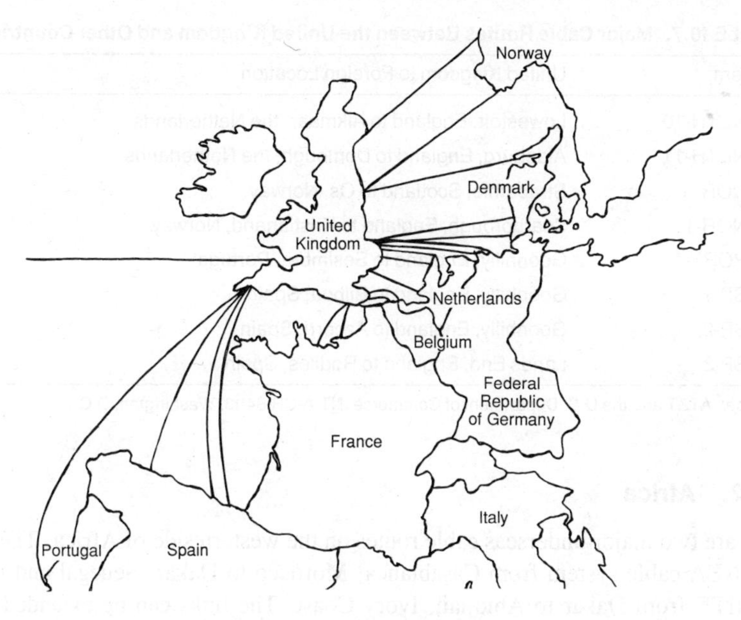

Figure 10.5. Major cable routes in the United Kingdom.

TABLE 10.7. Major Cable Routes Between the United Kingdom and Other Countries

System	United Kingdom to Foreign Location
CANTAT-1	Widemouth, England to Beaver Harbor, Nova Scotia, Canada
CANTAT-2	Widemouth, England to Beaver Harbor, Nova Scotia, Canada
UK-BELG-1	Canterbury, England to Ostende, Belgium
UK-BELG-2	St. Margarets Bay, England to La Panne, Belgium
UK-BELG-4	St. Margarets Bay, England to Veurne, Belgium
UK-BELG-5	Broadstairs, England to Ostende, Belgium
UK-FR-1	Eastbourne, England to Courseulles, France
UK-FR-2	Eastbourne, England to St. Valery en Caux, France
UK-DEN-1	Winterton, England to Made, Denmark
UK-DEN-2	Scarborough, England to Thisted, Denmark
UK-DEN-3	Winterton, England to Romo, Denmark
UK-GER-1	Winterton, England to Leer, Germany (Federal Republic of)
UK-GER-2	Winterton, England to Leer, Germany (Federal Republic of)
UK-GER-3	Winterton, England to Pedderwarde, Germany
UK-NETH-5	Aldeburgh, England to Domburg, the Netherlands
UK-NETH-6	Covehithe, England to Katwijk, the Netherlands
UK-NETH-7	Covehithe, England to Katwijk, the Netherlands
UK-NETH-8	Aldeburgh, England to Dombugh, the Netherlands
UK-NETH-9	Broadstairs, England to Dombugh, the Netherlands

(continued)

TABLE 10.7. Major Cable Routes Between the United Kingdom and Other Countries

System	United Kingdom to Foreign Location
UK-NETH-10	Lowestoft, England to Alkmaar, the Netherlands
UK-NETH-11	Aldeburg, England to Dombugh, the Netherlands
UK-NOR	Strabathie, Scotland to Os, Norway
UK-NOR-1	Scarborough, England to Kristianand, Norway
UK-PORT-1	Goonhilly, England to Sesimbra, Portugal
UK-SP-1	Goonhilly, England to Bilboa, Spain
UK-SP-2	Goonhilly, England to Azcorri, Spain
UK-SP-3	Lands End, England to Rodiles, Spain

Source: AT&T and the U.S. Department of Commerce, NTIA-CR-84-13, Washington, D.C.

10.2.2. Africa

There are two major underseas cable routes on the western side of Africa. They are ANTINEA cable system from Casablanca, Morocco to Dakar, Senegal and FRATERNITE from Dakar to Abidjian, Ivory Coast. The links can be extended with UNION from the Ivory Coast to Lagos, Nigeria. In addition, there are several microwave systems between major cities or towns, with very little overhead or underground cable on the western side of the continent.

Foreign private lines from the four eastern United States gateway cities (Miami, New Orleans, New York, and Washington, D.C.) are available to Algeria, Cameroon, Egypt, Ethiopia, Ivory Coast, Kenya, Liberia, Morocco, Nigeria, Senegal, Somali, South Africa, Southwest Africa, Sudan, and Tunisia. The half-channel monthly rate of $3800 to Africa is the same from each of the gateway cities. The African countries should be contacted directly for current availability and half-channel charges.

10.2.3. Middle East

The Middle East is served by underseas cables from France, Italy, Spain, and other countries in the Mediterranean area. In addition to the cable systems listed in Tables 10.4–10.6, Table 10.8 is a current listing of Middle Eastern links, and Figure 10.6 is a layout of the major cable links in the Middle East. Direct routes from France, Italy, and Spain can be supplemented by transit routing via Greece, Cyprus, and Egypt.

The four U.S. eastern gateway cities provide access to the Middle Eastern countries of Bahrain, Iran, Iraq, Israel, Jordan, Kuwait, Lebanon, Oman, Saudi Arabia, Syria, United Arab Emirates, Yemen Arab Republic, and the Yemen People's Democratic Republic.

The half-channel monthly rate of $3800 is the same to each country from each of the four eastern gateway cities. Middle Eastern PTTs do not publish international tariffs. Each country should be consulted directly.

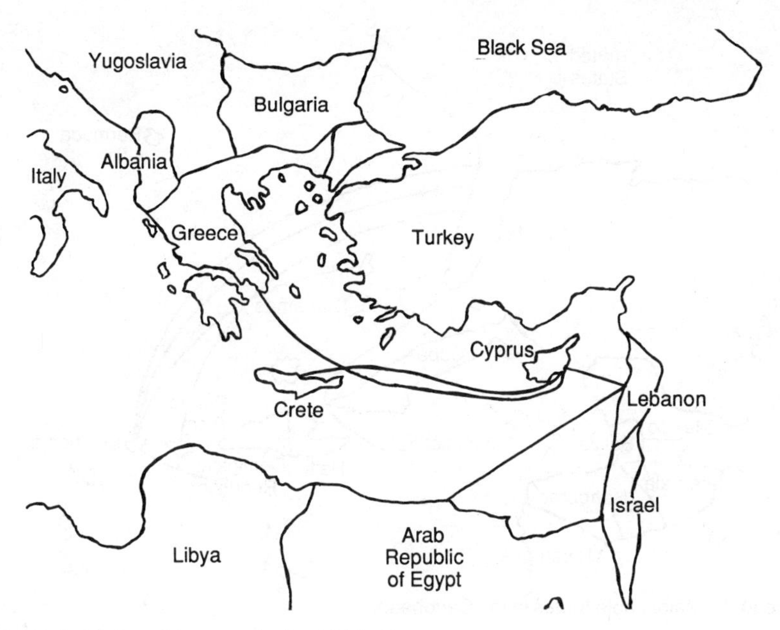

Figure 10.6. Major cable routes in the Middle East.

TABLE 10.8. Major Cable Routes Between the Middle East and Other Countries

System	Middle East Cable Links
ADONIS	Larnaka, Cyprus to Beirut, Lebanon
APOLLO	Lagonis, Greece to Larnaka, Cyprus and via ADONIS to Beirut, Lebanon
ALEX-LEB	Alexandria, Egypt to Beirut, Lebanon
ALEXANDROS	Lagonissi, Greece to Alexandria, Egypt
APHRODITE	Crete, Greece to Cyprus via ADONIS to Beirut
PALMYRA	Heraklion, Crete to Tarous, Syria

Source: AT&T, Morristown, N.J., and the U.S. Department of Commerce, NTIA-CR-84-13, Washington, D.C.

10.2.4. Central America and the Caribbean

Central America is served mostly by land-based cable systems that have been in existence for many years. Central American countries include Costa Rica, El Salvador, Guatemala, Honduras, Nicaragua and Panama.

Central America can be reached from the four eastern U.S. gateway cities. The half-channel monthly rate of $3200 is the same to each country from each of the gateway cities. A listing of the major Caribbean islands is included in Table 10.2.

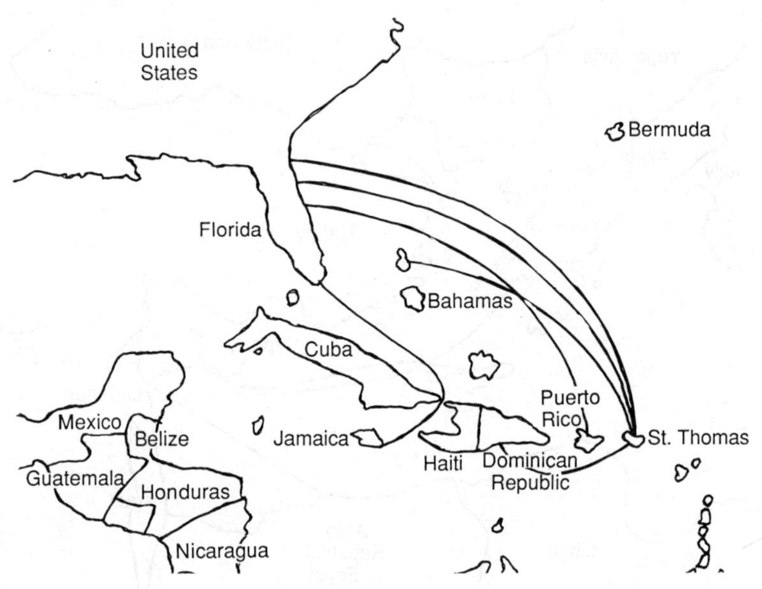

Figure 10.7. Major cable routes in the Caribbean.

TABLE 10.9. Major Cable Routes—U.S. and Caribbean Traffic

System	Outbound Location	Voice Channel Capacity
FLORICO	West Palm Beach, Florida to Puerto Rico from St. Thomas to	48
	Dominican Republic via DOMINICAN	144
JAM-1	Florida City, Florida to Jamaica	144
ST-T-1	Vero Beach, Florida to St. Thomas	142
ST-T-2	Jacksonville, Florida to St. Thomas	845
ST-T-3	Vero Beach, Florida to St. Thomas:	1,901
	From St. Thomas to Dominican Republic via DOMINICAN	144
	From St. Thomas to the Bahamas via BAHAMAS	1,380
—	Transcaribbean (1989)	20,000

Source: AT&T and the U.S. Department of Commerce, NTIA-CR-84-13, Washington, D.C.

The Caribbean can be reached from the U.S. mainland by various underseas cable systems as listed in Table 10.9 and shown in Figure 10.7. The U.S. possession of St. Thomas is the key to cable system extensions to the Bahamas and the Dominican Republic and farther south to Venezuela.

Leased lines to the Caribbean are available from the four eastern U.S. gateway cities. The half-channel monthly rate of $2500 is the same to each country from each of the gateway cities. Appendix B includes the PTTs' addresses for the major islands in the Carribbean.

10.2.5. South America

There are four major underseas cable systems to South America: They include COLUMBUS-1 from Venezuela to the Canary Islands, which in turn provides transit to Portugal or Spain; BRACAN-1 from Brazil to the Canary Islands and then on to either Portugal or Spain; and ATLANTIS, which is a direct link from Recife, Brazil to a drop in Senegal and on to Lagos, Portugal. BRUS is a cable system from Brazil to the Dominican Republic, which in turn can be connected to the United States via the Dominican Republic and St. Thomas.

Several land-based links are available to South America by underground and overhead cables. Future expansions are planned with satellite technology; the total cost will be distributed over a wide area and several countries.

Leased lines are available to South America from the four eastern U.S. gateway cities. The half-channel monthly rate of $3800 is the same to each country from each of the gateway cities. South American countries included in this rate are Argentina, Bolivia, Brazil, Chile, Columbia, Ecuador, French Guiana, Peru, and Venezuela.

10.2.6. Asia/Pacific Area

There are several cable systems from the United States to the Asia/Pacific area that play a strategic economic role as well as acting as a telecommunications network to handle U.S. outbound traffic. These cables are the main reason that the United States is able to operate independently and provide direct links to Asia/Pacific area countries. Underseas cables provide a network to U.S. possessions, trusts, and major trading partners. In 1986, 22 companies and PTTs signed a formal agreement for the construction of two underseas fiber optic cables that will span the Pacific Ocean by mid-1989. The 22 co-owners approved the construction and maintenance agreement to build the Hawaii 4/Trans-Pacific 3 (HAW-4) cable. Another separate agreement was signed by 16 companies and PTTs for a Guam to Philippines (GP-2) cable system. The cable will run from the island of Luzon in the Philippines to Guam. The total investment is $600 million for HAW-4 and $107 million for GP-2.

The geography of the U.S. cable layouts precludes either Japan or Singapore from being a regional center for telecommunications since most of the overseas traffic of the area countries is to the United States. Another reason is that historically most countries have acted independently. There is less cooperation among the various ethnic and religious groups as compared to the European community.

At present, Table 10.10 represents the routes between the United States and the Asia/Pacific area, and Figure 10.8 is a layout of the major cable links.

The full-channel rates to islands in the South Pacific are included from the San Francisco overseas gateway. Overseas channels to the Pacific, the South Pacific, and the Far East start from the San Francisco overseas gateway.

Pacific countries can be reached by a private line from the San Francisco overseas gateway at $4300 per month. The Pacific area zone includes American Samoa, Australia, the People's Republic of China, Hong Kong, Indonesia, Japan, South Korea, Malaysia, New Zealand, the Philippines, Singapore, Taiwan (Republic of China), and Thailand.

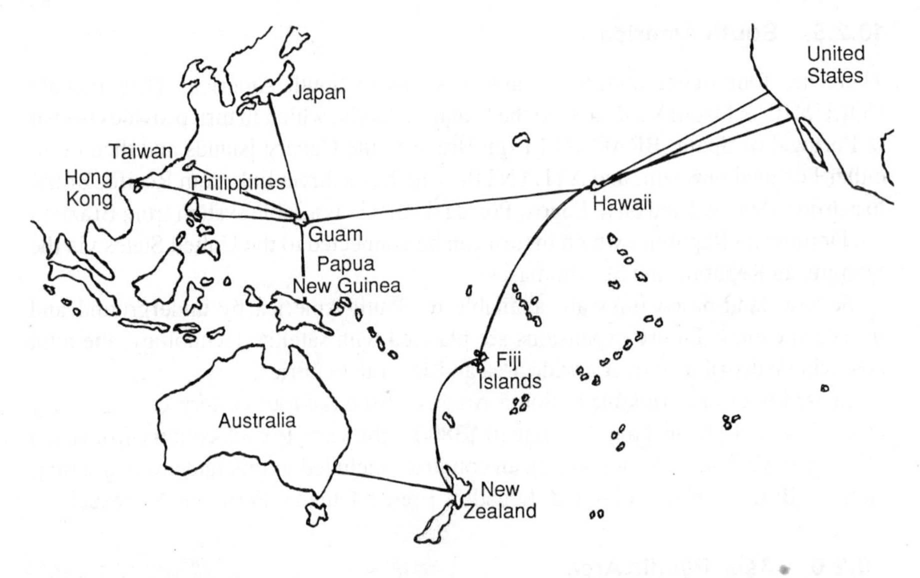

Figure 10.8. Major cable routes in the Asia/Pacific area.

10.3. INTERNATIONAL RECORD CARRIERS

Whenever an international telephone call is placed, an international private line is requested, or an international message is sent, an international record carrier (IRC) or international carrier (IC) is involved. U.S. IRCs and ICs are required to handle all international traffic to and from the United States. ICs handle voice and data traffic, while the IRCs may carry voice, data, telex, and facsimile. In the United States, there are several choices, and the subscriber is free to choose any IRC. Except for the United Kingdom, which has two public carriers to handle international traffic (British Telecom International and Mercury Communications), the remainder of the world deals through their international PTT administration, which handles all international or overseas traffic. In some countries, there is a separate company that handles international traffic. In most cases, the national PTT coordinates the local extension to the subscriber's premises for the international carrier. The exceptions are Australia, Hong Kong, Japan, and some of the smaller countries where there is a separate licensed or designated international carrier, for example, the Overseas Telecommunications Commission in Australia, Cable and Wireless in Hong Kong, and Kokusai Denshin Denwa Company, Ltd. (KDD) in Japan. Cable and Wireless, Ltd. also handles national and/or international telecommunications for many small independent nations.

Depending on the remote country where the call, line, or message terminates, the IRCs use terrestrial links, underseas cables, and satellite links to transfer international traffic. If one or more countries are crossed to reach the final destination, the IRCs and remote PTTs work out the details and settle charges with the intermediary countries. The position of the IRCs in an international link between continents is shown in Figure 10.9.

TABLE 10.10. Major Cable Routes—U.S. and Asia/Pacific Traffic[a]

System	Outbound Location	Voice Channel Capacity
HAW-1	Point Arena, California to Hawaii	51
HAW-2	San Luis Obispo, California to Hawaii	143
HAW-3	San Luis Obispo, California to Hawaii	845
HAW-4	Point Arena, California to Hawaii (1988)	40,000
Hawaii extensions to the Asia/Pacific area:		
COMPAC	Hawaii/Fiji Islands/New Zealand/Australia	—
TCP-1	Hawaii/Midway/Wake/Guam/Philippines	142
TCP-2	Hawaii/Guam/Okinawa	845
Guam extensions to the Asia/Pacific area:		
SEACOM-1	Guam/Hong Kong	—
SEACOM-2	Guam/Papua New Guinea	—
TAIGU	Guam/Taiwan (Republic of China)	—
SEACOM-1	Guam/New Guinea/Australia	—
TCP-1	Guam/Japan/Philippines	142
TCP-2	Guam/Okinawa	845
TP-2	Guam/Luzon, Philippines (1989)	10,000

Notes:

1. Papua New Guinea provides transit routing for two cable systems, APNG and SEACOM-2, to Australia.
2. New Zealand provides transit routing for two cable systems, COMPAC and TASMAN, to Australia.
3. Regional Asia/Pacific cables:

A-PNG	Australia to New Guinea
ASEAN I-S	Indonesia to Singapore
ASEAN M-S-T	Singapore/Malaysia/Thailand
ECSC	Japan to People's Republic of China
IOCOM	Singapore to India
Japan-Korea	Japan to South Korea
JASC	Japan to USSR
Naha-Mujako Jima	Okinawa to Japan
Okinawa-Mujako	Okinawa to Japan
	Okinawa/South Korea/People's Republic of China
OKITAI	Okinawa to Taiwan (Republic of China)
OLUHO	Okinawa/Philippines/Hong Kong
PHILSIN	Philippines to Singapore
TAIGU	Taiwan (Republic of China) to Guam
TAILU	Taiwan (Republic of China) to the Philippines

Source: AT&T, Morristown, N.J., and the U.S. Department of Commerce, NTIA-CR-84-13, Washington, D.C.

Within the United States, if a subscriber requests an end-to-end connection from a local Telco, the Telco will handle all arrangements with the IRCs and the remote country. The subscriber will receive two bills, one from the Telco for all the U.S charges and a separate bill from the country where the call or line terminates.

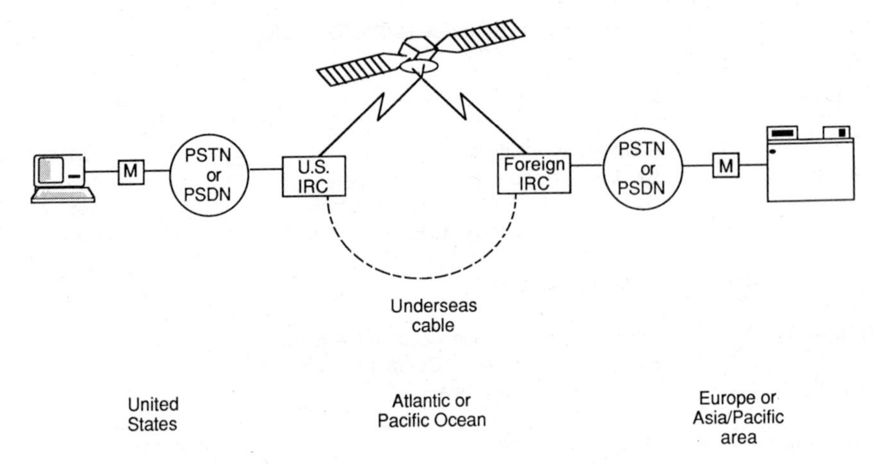

Figure 10.9. Role of international record carrier.

The U.S. IRCs are the following:

1. *American Telephone and Telegraph Company*, 412 Mount Kimbal Avenue, Morristown, NJ 07960: AT&T can provide international voice, data, video, text, and image through terrestrial, underseas cable, or satellite networks.
2. *Comsat, World Systems Division*, 950 L'Enfant Plaza SW, Washington, DC 20024: Comsat is a private company formed by Congress in the Communications Satellite Act of 1962 with a mandate to establish a global satellite communications system. Comsat established and developed the International Telecommunications Satellite Organization or INTELSAT. Comsat is also the U.S. signatory to and the largest shareholder in INMARSAT, the International Maritime Satellite Organization. The main objective of Comsat's space communications is to provide international satellite communications services through INTELSAT and INMARSAT.
3. *FTC Communications*, a Pacific Telecom company, 90 John Street, New York, NY 10038: FTC is an IRC linking the continental United States with major points throughout the world to provide international and domestic telex, international cablegram service, and international electronic mail. Leased channels and alternate voice and data channels are available. FTC offers the following services:

    - A real-time and store and forward telex service from PCs, word processors, or telex terminals.
    - International Business Services (IBS)—a digital satellite offering from 9.6 kbps to 2.048 Mbps with bit error rates of less than 1 in a million.
    - An International electronic mailbox called the Worldwide Electronic Message Service (WEMS), which permits access via telex or overseas packet switched data networks.

- FTC ofers INvoice, a pure voice private line service using the Time Assigned Speech Interpolation (TASI) technique, to France, Japan, and the United Kingdom, and Answer Bank, a worldwide interactive database service.

4. *International Record Carrier, Inc.*, 211 East 43rd Street, New York, NY 10017: IRC offers a worldwide outbound telex service from the United States. Local access in the United States is available from a subscriber's PC with either a 300- or 1200-bps modem to Telenet or Computer Science Corporation's Infonet. The message is sent by IRC via satellite to London where it is then forwarded to its final destination.
5. *International Relay, Inc.* (IRI), 777 Third Avenue, New York, NY 10017: IRI provides store and forward telex for worldwide communications. In addition, it offers 56- and 64-kbps and higher-speed digital circuits including T1 (1.544 Mbps) and CCITT G.703 (2.048 Mbps) over the International Business Services of INTELSAT from the United States to Europe. IRI connects the customer to its earth station across from the United Nations building in New York City.
6. *ITT World Communications*, 67 Broad Street, New York, NY 10004: ITT World Communications offers telex, cablegram, bulk data transfer, leased lines, alternate voice and data, and packet switching in conjunction with its overseas partners. ITT has telex operating agreements with over 200 overseas administrations. A high-speed international communications service is available through satellite stations in New York, San Francisco, and Washington, DC. ITT operates the HOTLINE between Washington and Moscow. The service includes telex and facsimile. Message delivery is guaranteed by either cable or satellite communications.
7. *MCI International Corp.*, International Drive, Rye Brook, NY 10573: MCI offers voice, data, telex, cablegram, Telegram, services, Datel, and alternate voice and data services to over 53 foreign countries and 3 U.S. territories. A new service TalkLine offers telephone access to a single international location. A four-digit extension can be dialed from a regular business telephone to reach the foreign party. MCI's main traffic volumes are outbound voice traffic.
8. *RCA Global Communications* (a unit of GE), 60 Broad Street, New York, NY 10004: RCA Globcom offers a number of international services including telex, store and forward telex, VANS, data, Datel, international leased lines, alternate voice and data, electronic mail, Fax, and wideband data.
9. *Telenet Communications Corp.*, 8229 Boone Blvd, Vienna, VA 22180: Telenet (U.S. Sprint) offers international packet switching and electronic mail (Telemail) from over 250 U.S. cities.
10. *TRT Telecommunications, Inc.*, 1331 Pennsylvania Ave. NW, Washington, DC 20004: TRT provides a wide range of international and domestic telecommunications services including data switching, high-speed data trans-

mission, voice, Fax, and network management. In addition, TRT offers telex, message Telegram, marine radio telex, and telegraph and has direct circuit relationships with over 75 overseas locations. In 1985, TRT entered into an agreement with Comsat for International Business Services (IBS) via the INTELSAT system through jointly owned earth stations at Staten Island, NY and Washington, DC. TRT is the fourth U.S. communications carrier (after AT&T, MCI, and U.S. Sprint) to provide international direct dial telephone service to the United Kingdom, the largest overseas market for voice and data traffic from the United States.

11. *Tymnet Inc. (International)*, 2710 Orchard Parkway, San Jose, CA 95134: Tymnet McDonnell Douglas offers international packet switching and electronic mail (Ontyme) from over 234 U.S. cities.
12. *Western Union Telegraph Company*, One Lake Street, Upper Saddle River, NJ 07458: Western Union Telegraph offers worldwide telex and EasyLink, an electronic mail service. EasyLink has been enhanced with X.PC, an error-correcting service for PC data to WU's packet data network. A binary file transfer program and other electronic data interchange services will be aded in 1987. WU offers international and domestic Teletex, a telex replacement to European countries. In addition, WU will rent, lease, or sell satellite capacity or a transponder on their Westar satellite systems.

10.4. BASIC NETWORKS

There are several choices for international information flow. Key service options and their availability for outbound traffic from the United States will be discussed.

1. *Public switched telephone network* (PSTN): One of the most frequently used services is the PSTN due to the availability of service throughout the world. There is free and open competition for overseas traffic from the United States to Europe. Most of the U.S. carriers will offer lower rates based on volume or discounting on outbound traffic to Europe, because European PTTs proportion their outbound traffic to the United States based on the number of call minutes from American IRCs. For example, if TRT or ITT would like to increase their market share of traffic to Europe, they would lower their rates for U.S. outbound traffic to capture more inbound traffic from Europe, at the full price. On the European side, the PTTs do not discount telephone rates and allocate outgoing traffic based on incoming call minutes to the IRCs.

   International PSTN calls can be made to a number of nations by simply dialing a 1, followed by an international access code, a country code, a city code, and the number of the party the subscriber is calling. The international access code is determined by the U.S. carrier, e.g., AT&T Communications, U.S. Sprint, or MCI International. For example, an MCI subscriber can dial direct by the following procedure:

Dial 1 + 011 + country code + city code + local number

Depending on where the call is placed, it may or may not be necessary to dial the first 1 in this sequence.

2. *Private analog line:* Private analog lease lines for voice or data are available from AT&T Communications as discussed in the section on international overseas channels. MCI also offers international voice and data lines as well as alternate voice or data. Other carrier services are offered via satellite links on a full- or part-time basis. For the world as a whole, 2.4-kbps operation is normal with 4.8 kbps available to some countries. Private leased lines with line conditioning allow operation at 9.6 kbps to Europe and Japan.
3. *Private digital line*: Digital networks are now becoming available in some countries such as Australia, Brazil, Canada, Hong Kong, Japan, and the United States. In Europe, digital networks are available in Austria, Belgium, Denmark, Finland, France, the Federal Republic of Germany, Ireland, Italy, the Netherlands, Norway, Spain, Sweden, Switzerland, and the United Kingdom. Interconnection of private digital leased lines is not common. However, by the 1990s, more countries will offer private digital leased lines for international connection. Today, private digital lines are available to selected countries on a point-to-point basis. Terrestrial and underseas cable links are available at line speeds of 2.4, 4.8, or 9.6 kbps. A 56-kbps digital private line service is available from the United States to Canada and from the United States to the United Kingdom.

   Satellite digital services are available at line speeds from 9.6 kbps to 2.048 Mbps to Belgium, Canada, Denmark, France, Germany (FRG), Ireland, Italy, Mexico, the Netherlands, Norway, Sweden, Switzerland, and the United Kingdom. A T1 satellite connection at 1.544 Mbps is available to some countries and in specific areas where a local loop extension is available (Switzerland and the United Kingdom). The 2.048-Mbps speed conforms to the CCITT G.703 Recommendation but is incompatible with North American standards.
4. *DDS:* AT&T-C International Dataphone Digital Service (DDS) for overseas links consists of one or more components of DDS within the mainland and an overseas cable digital channel.

   International DDS is designed for communications originating or terminating at points directly connected to the International DDS service by cable channels furnished by AT&T.

   The service is a two-point end-to-end service. That is, one point is on the U.S. mainland, and the second is an overseas point. The overseas cable digital channel is furnished in part by AT&T and in part by an overseas PTT or the country's representative international telecommunications organization.

An overseas cable digital channel can be provided by AT&T from New York City to the United Kingdom. Monthly charges vary by transmission speed. The rates shown in the tariff are the half-channel rates or the portion of the overseas cable digital channel furnished by AT&T. The British company (British Telecom International or Mercury Communications) furnishes its portion of the overseas cable digital channel.

Speed, kbps	Half-Channel Monthly charge
2.4	$ 1,800
4.8	2,400
9.6	3,000
56	12,000

5. *Datel:* Datel is a data service for moderate amounts of traffic over voice-grade lines with speeds limited to 2.4 kbps. Some Datel offerings allow users to speak directly with overseas correspondents or to transmit facsimile. Other common exchanges include payroll and billing information, inventory records, engineering specifications, sales orders, and bulk message traffic. Normally, Datel offers the subscriber conditioned international lines, modem compatibility to enable U.S. overseas users with incompatible modems to communicate with each other, and redundant network equipment for backup, which results in greater reliability and less downtime.

   RCA *Global Communications* (a unit of GE) and MCI offer a measured Datel service for voice and/or data transmission to overseas countries. Access to the RCA network is by dial-in to access points at either New York City, San Francisco, or Washington, DC. Rates vary by country from a low of $5.25 to a high of $9.00 for the first 3 minutes and from $1.75 to $3.00 for each additional minute.

6. *Telex:* Message service for international traffic has expanded over the past few years and will continue for global operations. The increase of packet networks with electronic mail offerings has multiplied, but telex will continue for many company operations in countries other than North America, Western Europe, Japan, Australia and New Zealand. Telex is the major worldwide message switching system and offers the greatest number of country destinations of any message service—more than the public switched telephone or dial-up network.

   Telex is a valuable tool for any business with worldwide operations. For example, if you require a hotel room in a distant country, a simple telex message puts it in writing and is a legal document. Also, the originator knows in less than a minute whether or not the telex was received. Sending a message from a telex terminal is simple. Prepare a message offline or online for faster typists. Dial the remote number, press a key or flip a switch, and away it goes. Acknowledgment is included, and the return number and answerback code are printed on the message.

Telex is the most widely used means of sending messages between companies. In many developing countries of the world, except for a direct telephone call, telex is the only other practical means of establishing communications to the outside world. Mail would take days or weeks plus the return time.

Western Union's Telex I is a long-standing service. At less than 40 cents per minute, the sender buys 66 words per minute (based on an average of 6 characters per word and interword spacing) anywhere in the continental United States. Telex II is about 43 cents per minute for 100 words within the same area. However, telex is slow; for domestic use, the transmission speed is 66 words per minute, while international traffic is limited to 60 words per minute.

International telex varies from $1.03 per minute from the United States to the United Kingdom and from $1.54 to France, Italy, and the Republic of Germany to about $2.50 per minute to some countries. Over time, the rates will vary, but the proportions will remain about the same.

Billing on telex is simple: The originator of the telex message pays for the call. Telex machines may be rented for about $60 to $70 dollars per month or purchased for $1000 to $1300.

7. *Telex versus PSTN comparison:* An international telex has two components: an international charge and a domestic charge. International telex messages are based on a per minute charge. Rates vary from a low of 22 cents to Canada to $2.54 per minute for remote countries and continues for the total time it takes to send a message to the destination address. The domestic charge is 0.5308 cent per minute for Telex I and 0.6046 cent per minute for Telex II. After the first minute, additional time is calculated for each one tenth of a second.

   A Telex I 500-word message would cost about $26.50 to many countries of the world. Except for personal contact, a telex puts the message in writing and allows the receiver time to respond to a message.

Many countries may be reached from the United States only by telex. These countries are Belize in South America, Dahomey, Rwanda, and Upper Volta in Africa and Vietnam in Southeast Asia.

In addition to telex, the following countries may be reached by both telex and operator-assisted calls:

Angola	Ghana	Mauritania
Botswana	Gibraltar	Mozambique
Bhutan	Guinea	Nepal
Burma	Laos	Sierra Leone
Burundi	Lesotho	Tanzania
Central African Republic	Macoa	Togo
Chad, Republic of	Madagascar	Swaziland
Falkland Islands	Maldives	Uganda
Gambia	Mali	

The operator establishes a routing procedure for conversation. It may be very difficult for data communications over an operator-assisted line. Any line problem requires an operator to reestablish the line and data to be retransmitted. In addition, some countries may have PTT regulations against data transmitted over telephone lines. The only alternative for data transmission in some countries may be the telex network.

The following countries may be reached by operator-assisted calls but have *no* telex; in these countries, data transmission may be very difficult or next to impossible.

Afghanistan
Bangladesh
Djibuti, Republic of
Mongolian People's Republic
Sri Lanka

Countries where no basic telecommunications services are available from the United States are Albania, Kampuchea (formerly Cambodia), North Korea, and Vietnam (although what was once considered South Vietnam can be reached through operator assistance).

Table 10.11 is a summary of telecommunications to and from the United States. Telephone calls via the PSTN are direct dial without operator assistance. An N symbol in the PSTN column means that a direct dial call is not available to the country from the United States. In Table 10.11, only those countries that have at least two alternatives for data transmission are listed. A direct dial line is considered feasible for some short-duration data calls. A dash represents the lack of a particular service for that country. The number of countries that may be reached by private digital leased lines is expected to grow when new digital networks are implemented. Satellites capable of digital transmission will be in place by the late 1990s to allow private digital circuits to countries that implement a digital network. Although satellites can be economically feasible, the excessive uplink/downlink transmission delays (240–279 milliseconds depending on location) may not be an optimum solution for some interactive or transaction processing applications.

TABLE 10.11. Basic Network Services—Outbound from the United States

	PSTN	Telex	Datel	Private Analog	Private Digital
North America					
Canada	—	X	X	X	X
Caribbean	X	X	—	X	—
Europe					
Austria	X	X	X	X	—
Belgium	X	X	X	X	X
Bulgaria	N	X	—	X	—

TABLE 10.11. Basic Network Services—Outbound from the United States

	PSTN	Telex	Datel	Private Analog	Private Digital
Cyprus	X	X	—	—	—
Czechoslovakia	X	X	—	X	—
Denmark	X	X	X	X	X
Finland	X	X	X	X	X
France	X	X	X	X	X
Germany (Dem. Rep.)	X	X	—	—	—
Germany (Fed. Rep.)	X	X	X	X	X
Greece	X	X	—	X	—
Greenland	X	X	—	X	—
Hungary	X	X	—	X	—
Iceland	X	X	—	X	—
Ireland	X	X	X	X	X
Italy	X	X	X	X	X
Luxembourg	X	X	X	X	—
Malta	X	X	—	—	—
The Netherlands	X	X	X	X	X
Norway	X	X	X	X	X
Poland	X	X	—	X	—
Portugal	X	X	X	X	—
Romania	X	X	—	X	—
Spain	X	X	X	X	—
Sweden	X	X	X	X	X
Switzerland	X	X	X	X	X
United Kingdom	X	X	X	X	X
Yugoslavia	X	X	—	X	—
South and Central America					
Argentina	X	X	X	X	—
Bolivia	X	X	—	X	—
Brazil	X	X	—	X	—
Chile	X	X	—	X	—
Colombia	X	X	—	X	—
Costa Rica	X	X	X	X	—
Cuba	X	X	—	—	—
Ecuador	—	X	—	X	—
El Salvador	X	X	—	X	—
French Guiana	N	X	—	X	—
Guatemala	—	X	—	X	—
Haiti	—	X	—	X	—
Honduras	X	X	—	X	—
Mexico	X	X	—	X	X
Nicaragua	—	X	—	X	—
Panama	X	X	—	X	—
Paraguay	—	X	—	—	—
Peru	X	X	—	X	—
Suriname	X	X	—	—	—
Uruguay	X	X	—	—	—
Venezuela	X	X	—	X	—

(continued)

TABLE 10.11. Basic Network Services—Outbound from the United States

	PSTN	Telex	Datel	Private Analog	Private Digital
Africa					
Algeria	X	X	—	X	—
Cameroon	X	X	—	X	—
Congo (Brazzaville)	X	X	—	—	—
Egypt	X	X	—	X	—
Ethiopia	X	X	—	X	—
Gabon	X	X	—	—	—
Ivory Coast	—	X	—	X	—
Kenya	X	X	—	—	—
Liberia	X	X	—	—	—
Libya	X	X	—	—	—
Malawi	X	X	—	—	—
Morocco	X	X	—	X	—
Namibia	X	X	—	—	—
Niger (Rep. of)	X	X	—	—	—
Nigeria	X	X	—	X	—
Senegal	—	X	—	X	—
Somali Republic	N	X	—	X	—
South Africa	X	X	X	X	—
Sudan	X	X	—	X	—
Tunisia	X	X	—	X	—
Zaire	N	X	—	—	—
Zambia	N	X	—	—	—
Zimbabwe	N	X	—	—	—
Middle East					
Bahrain	—	X	X	X	—
Iran	X	X	—	X	—
Iraq	X	X	—	X	—
Israel	X	X	—	X	—
Jordan	X	X	—	X	—
Kuwait	X	X	—	X	—
Lebanon	N	X	—	X	—
Oman	—	X	—	X	—
Qatar	X	X	—	—	—
Saudi Arabia	X	X	—	X	—
Syria	N	X	—	X	—
Turkey	X	X	—	—	—
United Arab Emirates	X	X	X	X	—
Yemen Arab Republic	—	X	—	X	—
Yemen People's Dem. Rep.	N	X	—	X	—
USSR	X	X	X	—	—
Far East					
China (People's Rep.)	X	X	—	X	—
China (Rep. of)	X	X	—	X	—
Hong Kong	X	X	X	X	X
Japan	X	X	X	X	X
Korea (Rep. of)	X	X	—	X	—

TABLE 10.11. Basic Network Services—Outbound from the United States

	PSTN	Telex	Datel	Private Analog	Private Digital
Southeast Asia					
Afghanistan	X	X	—	—	—
India	X	X	—	—	—
Pakistan	X	X	—	—	—
South Pacific					
Australia	X	X	X	X	X
Indonesia	X	X	—	—	—
Malaysia	X	X	—	—	—
New Zealand	X	X	X	—	—
Philippines	X	X	X	—	—
Singapore	X	X	X	—	X
Thailand	X	X	—	—	—

Source: AT&T, FTC Communications—a Pacific Telecom company, International Record Carrier (IRC), International Relay, Inc. (IRI), ITT World Communications, MCI International Corporation, RCA Global Communications (a unit of GE), TRT Telecommunications, and Western Union Telegraph Company.

10.5. PACKET SWITCHED DATA NETWORKS

International PSDN services are available from several U.S. value added network services (VANS) network providers and IRCs who link foreign networks to domestic packet networks. They are

1. ADP Autonet
2. AT&T ACCUNET® Packet Switch (APS)
3. CompuServe
4. GE Communications and Services
    a. GEISCO Mark*Net
    b. RCA Global Communications
5. Graphnet Freedom Network
6. ITT World Communications
7. MCI International
8. Telenet (U.S. Sprint)
9. Tymnet McDonnell Douglas

VANS providers offer "added value" in terms of store and forward of information, a PAD facility (messages are disassembled into packets at the sending end and reassembled at the receiving end into their original form), or additional services such as electronic mail and electronic data interchange. VANSs link to international and overseas countries through IRCs. To communicate by PSDNs throughout the world, several international transit centers are available at strategic locations that conform to international CCITT recommendations. Over 160 countries have adopted the CCITT X.25 Recommendation.

International packet switched data networks are offered by many IRCs who provide the connection between U.S. VANSs and overseas national PSDNs. Their offerings include a PAD facility that allows terminals of different speeds and protocols to transmit to overseas countries. The nine VANSs listed above would go through the IRCs for connection to other countries. IRCs offering international packet switched data services are FTC Comunications, ITT World Communications, MCI International, and RCA Global Communications (a unit of GE).

For developing countries to acess VANS providers, they need only connect to an international gateway country. Internationally, the interface at the country level is the CCITT X.75 Recomendation. This arrangement was designed by CCITT members to provide a common interface to X.25 networks throughout the world. Both Telenet and Tymnet are suppliers of X.25 node equipment and X.75 interface products to many PTT administrations.

Over 600 databases are available throughout the world via internationally connected PDSNs. Some of the gateway countries are Canada, France, Italy, Japan, the United States, and the United Kingdom. Some of the North American gateway providers are FTC Communications, ITT World Communications, MCI International, RCA Global Communications, TRT Telecommunications, Western Union Telegraph Company, and Canada's Teleglobe. Teleglobe, Canada's international carrier, connects Telecom Canada's Datapac and CNPN's Infoswitch PSDNs to PSDNs in other countries. The North American international gateways connect to other gateways in Europe and Southeast Asia. The international gateways of Europe are connected to North and South America, Southeast Asia, and Africa. For example, Rome is connected to Kuwait, and Paris has a gateway to the Ivory Coast.

Tariffs for intercontinental information are less expensive for data transfers on PSDNs compared to telephone or private analog lines, because PSDN tariffs are based primarily on the amount of packets or characters (8 bits or 1 octet) transmitted. The basic measurement is a 64-octet segment, and transmission time is measured in minutes.

International gateways provide access to smaller countries or territories from the gateway-providing country. For example, France provides packet network links to the French Antilles (the islands of Guadeloupe and Martinique), La Reunion (off the coast of Africa in the Indian Ocean), and French Guiana in South America. The United Kingdom provides access to the Bahamas, Barbados, Oman, and Trinidad through the International Packet Switching Service (IPSS).

Japan has direct access through its international network carrier Kokusai Denshin Denwa (KDD) for international packet switched data traffic to France, Germany (FRG), the United Kingdom, Singapore, and the United States. Access to other countries is provided through these gateway countries. For example, Japan has access to 13 countries through the United States: Canada, Brazil, Austria, Switzerland, Bahrain, Israel, Taiwan, South Korea, Australia, New Zealand, Indonesia, the Philippines, and Thailand. Japan routes traffic through France to 6 countries: Gabon, Ivory Coast, Belgium, Finland, Greece, and Luxembourg. Through the United Kingdom, Japan has access to South Africa, Ireland, the Netherlands, Norway, and Spain. Japan's traffic to Denmark and Sweden is via the German connection; Hong Kong is reached via Singapore; and Indonesia and Thailand are reached through the U.S. connection.

Tymnet offers a transit mode with a PAD facility in Bahrain for asynchronous dial-up at 300 bps or 1.2 kbps to Kuwait, Qatar, and Saudi Arabia. In addition, Tymnet also offers a PSDN connection through a telex gateway. Theoretically, any company in a remote country with a telex that has an account with Tymnet could dial into a telex country gateway and have access to Tymnet services and database accounts.

Table 10.12 lists the countries that carry U.S. traffic. The definitions for the listings in the table are as follows:

AUS: Austria provides the gateway transit interface to Hungary.

GER: Germany provides the gateway transit interface to Austria.

IRC: IRC represents the U.S.-based international record carriers; more than one IRC will provide service to the country.

ITT: ITT World Communications provides the gateway facility to the United States.

MCI: MCI International provides the gateway facility to the United States.

RCA: RCA Global Communications (a unit of GE) provides a gateway.

TRT: TRT Telecommunications provides the gateway facility to the United States.

TYM: Tymnet provides a gateway facility to the United States.

UK: The United Kingdom, via the International Packet Switching Service (IPSS), provides the gateway transit interface to Norway.

TABLE 10.12. PSDN International Gateways

	International Gateway Countries					
	U.S.	Canada	France	Italy	U.D.	Japan
North America						
Canada	X	X	X	X	X	X
United States	X	X	X	X	X	X
South and Central America						
Argentina	X	TYM	ITT	—	—	—
Brazil	X	—	X	—	X	X
Chile	X	TYM	X	—	—	—
Colombia	X	X	ITT	X	—	—
Costa Rica	X	—	TRT	—	X	—
Guatemala	X	—	—	—	—	—

(continued)

TABLE 10.12. PSDN International Gateways

	International Gateway Countries					
	U.S.	Canada	France	Italy	U.D.	Japan
Mexico	X	—	X	X	X	—
Panama	X	—	X	—	—	—
Peru	X	—	—	X	—	—
Venezuela	—	—	X	—	—	—
Africa						
Cameroon	—	—	X	—	—	—
Gabon	X	—	X	—	X	X
Ivory Coast	—	—	X	—	X	X
Kenya	—	—	—	X	—	—
Morocco	—	—	X	—	—	—
Senegal	—	—	X	—	X	—
South Africa	X	X	UK	X	X	X
Tunisia	—	—	X	—	—	—
Europe						
Austria	X	X	GER	—	X	X
Belgium	X	X	X	X	X	X
Denmark	X	X	X	X	X	X
Finland	X	X	X	X	X	X
France	X	X	X	X	X	X
Germany (Fed. Rep.)	X	X	X	X	X	X
Greece	X	X	X	—	X	X
Hungary	X	—	AUS	—	X	—
Iceland	X	—	—	—	—	—
Ireland	X	X	X	—	X	X
Italy	X	X	X	X	X	—
Luxembourg	X	X	X	X	X	X
The Netherlands	X	X	X	X	X	X
Norway	X	X	UK	X	X	X
Portugal	X	—	X	X	X	—
Spain	X	X	X	X	X	X
Sweden	X	X	X	X	X	X
Switzerland	X	X	X	X	X	X
United Kingdom	X	X	X	—	X	X
Yugoslavia	—	—	X	—	—	—
Middle East						
Bahrain	X	X	IRC	—	X	X
Dubai	—	X	RCA	—	—	—
Israel	X	X	MCI	—	X	X
Kuwait	—	X	—	X	X	—
Oman	—	—	—	—	X	—
Qatar	—	X	—	—	—	—
Saudi Arabia	—	X	—	—	—	—
United Arab Emirates	X	X	RCA	—	X	—

TABLE 10.12. PSDN International Gateways

	International Gateway Countries					
	U.S.	Canada	France	Italy	U.D.	Japan
Far East						
China (Taiwan/ROC)	X	X	ITT	—	X	X
Hong Kong	X	X	UK	—	X	X
Japan	X	X	X	X	X	X
South Korea	X	X	ITT	—	X	X
South Pacific						
Australia	X	X	UK	—	X	X
Indonesia	X	—	ITT	—	X	X
Malaysia	X	—	—	—	X	X
New Zealand	X	X	ITT	—	X	X
Philippines	X	X	MCI	—	X	X
Singapore	X	X	X	—	X	X
Thailand	X	—	MCI	—	X	X

Source: AT&T; British Telecom International, London; Deutsche Bundespost, the Federal Republic of Germany; Direction Generale Telecommunications, Paris; FTC Communications, a Pacific Telecom Company; International Record Carrier; ITT World Communications; Kokusai Denshin Denwa Co. (KDD), Tokyo, Japan; MCI International Corporation; Radio Austria, Vienna, Austria; RCA Global Communications (a unit of GE); TRT Telecommunications; Tymnet McDonnell Douglas; and Western Union Telegraph Company.

11

VALUE ADDED NETWORK SERVICES

In this chapter we will

- Discuss VANS providers, key legislation, and procedures
- Look at VANSs in the United Kingdom, Japan, and the United States
- Cover existing and new VANS applications and restrictions and provide a listing of international VANSs.

What is a value added network service or VANS? There are many definitions—those used by the U.S. FCC and each PTT. Some VANSs provide only a value added service and therefore have been called a value added service or VAS. However, the term that has prevailed is VANS, which encompasses both a networking and service offering or a service only offering. Value added providers build their services over packet switched data networks, circuit switched data networks, or private analog or digital leased line facilities. Connection to the VANS' nearest node is usually by private leased line or the public switched telephone network. Depending on the country, VANSs may be provided by private companies, only the PTT, or both private companies and the PTT.

In general, most definitions of VANSs state a requirement for computer processing, and if information (or a message) is to be distributed to a local device or retransmitted from a computer to a remote device, the information (or message) must be stored in the computer before it is forwarded to its final destination, either to the locally attached device or over communications facilities to a remote device.

From a PTT business policy point of view, VANSs should be based on the use of public switched networks. VANSs based on switched networks make use of the transmission capacity (circuits) and the switching capacity (switching centers) of the PTTs. VANSs on switched lines are welcome by the PTT as they produce traffic in the network. The more successful a private VANS provider becomes, the more revenues increase due to higher traffic volumes.

VANSs based on leased lines make use only of the transmission capacity (circuits) of the public networks. The VANS provider interconnects the leased lines in its own switching nodes.

A major concern expressed by many PTTs is that as soon as VANSs are provided via leased lines (dedicated connections), a private communication system can perform switching functions at the same time and can then trade off switched connections against dedicated connections.

VANS network providers use their leased circuits and switching nodes. The public networks are used as access networks for customers in remote and rural areas only where circuits provided by the VANS carrier will be unprofitable.

In many countries, there is no formal definition of VANSs because many of the PTT administrations are not clear on the subject due to the rapid changes in technology, interconnection possibilities, and global marketing potential.

11.1 VALUE OF VANSs

VANSs offer a minimum investment to users. The customer pays a flat monthly fee to connect their host computers and work stations to the network and a variable monthly amount based on actual usage.

If a company is designing a network based on leased lines and is not willing to invest in more than one redundant leased line, reliability can be greater with a VANS provider. VANSs have several lines for backup and offer automatic alternate routing in the case of a line failure or node processor failure. Most nodes or central concentrators and network managers are connected to at least two other nodes to ensure that the message will go through to its final destination. In addition, if a company requires greater reliability on its access to the packet data network, multiple paths via leased line from its host to one or more nodes will increase uptime. In the case of a line failure, at least one line is still connected. Also, if leased line modems are used with a switched network backup (SNBU) feature, the leased line can be bypassed by the public switched telephone network if the line goes down.

For a medium- or small- sized company, VANSs offer a degree of reliability that it could not afford by designing its own networks. The company requires a minimal capital investment to use larger, more powerful computer systems, specialized applications software, and large databases—all supported by professional staffs—which can provide an international networking capability for a price that is less than the company's cost in supplying its own network. The number of new terminals can be reduced because VANSs usually support a variety of terminals with different protocols.

A company may set up its own international network. However, there are many problems to overcome. A technical staff will be required to work with each national PTT. Language barriers can be overcome with local personnel. Remember, all nations operate differently and independently, and there may be little transfer of technical knowledge among personnel.

11.2. VANS PROVIDERS

VANS providers offer a *network* or *service*. In general, a network provider offers a networking capability in addition to a VANS service. The difference is that the network provider normally obtains national and international leased lines from Telcos and PTTs and either offers services over its leased line facilities or implements its own packet switched data network. The network provider contracts with the international record carriers (IRCs) on the U.S. side and the PTTs or RPOAs on the remote side to establish its networking facilities. A service provider may use the various national packet data networks, where allowed, or use international networks provided by network providers such as Telenet and Tymnet. As network providers, both Telenet and Tymnet offer access to hundreds of remote computing and remote database services throughout the world. In addition, Tymnet offers remote computing and database facilities.

The subscriber accesses the nearest network node by either a switched telephone call or a private leased line. Depending on the VANS implementation within a country, the private leased line may be either analog or digital. In general, the subscriber contracts with the Telco or PTT and pays the monthly access charge to the nearest node. Line attachment equipment such as modems may or may not be included in the monthly charge from the network or service provider. It depends on country restrictions and the network provider's marketing approach. Most national PTT networks require that the modem be provided by the PTT for packet or circuit switched data networks.

Most VANS network providers offer some or all of the following functions as part of their services:

- End-to-end communications consisting of a national and international packet switched data network or leased line network over Telco and PTT facilities. For overseas connections, VANS providers integrate national networks with IRCs for connection at both ends of the ocean
- Message routing to the final destination
- Guaranteed message delivery to its final destination
- On packet switched data networks, a packet assembly/disassembly (PAD) facility at the nearest node to facilitate multiple use of existing terminals and remote job entry (RJE) work stations. Connections usually allowed are asynchronous ASCII terminals, asynchronous block-type terminals such as the IBM 3101 and DEC VT-100 terminals, 3270 BSC terminals, and 2780/3780 BSC, HASP, and other RJE work stations

- Host communications protocol support at the receiving end to work with the various terminal and RJE work station protocols
- A processing capability or access to remote computing and remote database facilities
- Protocol conversion between incompatible data terminals:

  3270 BSC to X.25
  3270 BSC or 2780/3780 BSC to X.25/HDLC
  ASCII to asynchronous Teletypewriter (TTY)
  ASCII TTY to X.25/HDLC
  Telex conversion

Most VANS service providers include one or more of the following functions as part of their offerings:

- Remote computing, remote database, or applications software packages over the Telco and PTT packet switched or circuit switched data networks on a national basis or internationally via a network provider's international network
- A contract with the VANS network provider to provide a special terminal protocol at the nearest node to handle their specific application, for example, a banking terminal, special card reader, credit card input terminal, or point of sale (POS) terminal
- Support of different protocols on their host to operate with different types of remote terminal and RJE work stations.

The distinction between VANS network providers and service providers will diminish in time, because there is little profit in the reselling of leased lines or traffic on a packet or leased line network. Therefore, network providers now offer more and more services, and in time the service providers will use national and international packet switched data networks and private leased lines to design, develop and integrate their own networks to provide end-to-end connectivity or turnkey solutions for specific customer problems.

11.3. NETWORK MANAGEMENT

As part of their normal operations, VANSs assume all the responsibilities of network management and reduce the cost required to operate complex communications systems over national and international networks. Normally, they have experienced communications professionals located on two or more continents who provide network management, interface with the Telcos and PTTs and the IRCs, and provide 24-hour, 7-day-a-week service between continents. Normally, VANSs have network hardware and/or software supervisor programs that monitor and control line ports, modems, lines, and nodes in the network. Network node controllers and programmed supervisors are used for automatic routing over alternate paths in case of a problem between nodes.

VANSs usually provide automatic code conversion, speed matching, and protocol conversion that allows a customer to use existing equipment to protect its investment. In addition, many VANSs provide personal computers that emulate 3270 BSC terminals. The intent is to offer enhancements to PCs so that they may operate as a PC or as a remote work station to a host.

VANSs offer security as a key feature that starts when the subscriber logs into the system. The log-on sequence requires a user name and a password that is checked against encoded files for validation before a circuit is completed. In addition, a closed user group can be set up to limit access to specific persons or applications.

On a national basis several VANSs offer a permanent virtual circuit (PVC) for dedicated access to a port on the host. For example, if a customer needs to access only one system, a PVC connection may be requested to provide automatic access to only that system.

Some VANSs also provide monthly accounting data including detailed usage information. A bill back feature is available from some VANSs that allows the monthly bill to be sent to the individual user or offices that originate the call. Other information that can be provided in monthly reports can include the time and date of each connection, originating entry-port node and line port numbers, the number of minutes that the session lasted, the number of incoming and outgoing characters, the network user name, and caller information.

11.4. VANS PHILOSOPHY IN THE UNITED STATES

The term VANS has no regulatory meaning in the United States. What are called VANS in other countries are encompassed within the term enhanced services in the United States.

The dividing line between telecommunications and data processing is not very clear, but to provide some definition, the FCC developed a policy of drawing dividing lines in its *Computer Inquiry-I Decision* of 1971. The decision defined three categories of services: regulated telecommunications services, nonregulated data processing services, and hybrid services. The hybrid services were to be regulated or remain unregulated from case to case, depending on whether they were more telecommunications or more data processing.

Unfortunately, technical developments created a number of marginal cases and carriers and prospective competitors argued before the FCC as to whether particular services were or were not telecommunications and should or should not be regulated. This led to the Computer Inquiry II Decision of 1980 by which the FCC created two new terms to define the difference between *basic services* and *enhanced services*. A *Basic service* which remains subject to common carrier regulations, is defined as the transport of information from one place to another with the information remaining unchanged. An *Enhanced service* offers additional features or service attributes (e.g. storage, processing, code or protocol conversion). They may be defined as offerings whereby a subscriber interacts with stored information or is provided additional, different or reformatted information. All inquiry and retrieval services, electronic banking services, electronic mailbox services, and Videotex are included in the definition. Enhanced services are not subject to common carrier regulations. Common carriers are not allowed to cross-subsidize their customer premises equipment and enhanced service offerings with revenues from basic services. Only AT&T (and the RBOCs) were limited to separate subsidiaries for the offering of enhanced services and customer premises equipment. AT&T created a separate subsidiary: AT&T Information Systems.

The FCC has since determined in Computer Inquiry III (1986) that the costs of structured safeguards outweigh the benefits to the public. It removed the separate subsidiary requirements for AT&T and the RBOCs, subject to the implementation of effective non-structural safeguards. Like other carriers, they can provide basic and enhanced services as well as customer premises equipment.

11.5. VANS IN THE UNITED KINGDOM

In the United Kingdom, discussions about VANS services began in 1980. In 1982, the "General Licence" for VANS was issued allowing private providers to offer VANS services (which included value added services over leased circuits). The VANS license program has been well subscribed to in the United Kingdom. As of January 1987, more than 300 companies received over 800 VANS licenses to provide a variety of new services in the following categories.

VANS Services

Automatic ticket reservation
Conference calls
Customer's data bases
Deferred transmission
Long-term archiving
Mailbox services
Multi-address routing
Protocol conversion between incompatible computers and terminals
Secure delivery services
Speed and code conversion between incompatible terminals
Store and retrieve message systems
Telephone answering services
Telesoftware storage and retrieval
Text editing
User management packages (e.g. accounting, statistics)
Viewdata
Word processor and facsimile interfacing

British Telecom (BT) and Mercury Communications, the two Public Telecommunications Operators (PTOs or common carriers) compete with independent suppliers in the provision of VANS under the General Licence. Some of the BT VANS are:

- Prestel—Interactive Videotex services
- Electronic Yellow Pages
- Interactive cable services (video games)
- Telemarketing (direct marketing over telecommunication networks)
- Telecom Tan (information and acceptance of orders by computers)
- Telecom Red (security services)
- Dialcom (an electronic mail service purchased from ITT)
- Telecom Gold (Electronic Mail and other electronic office services)
- Telecom Violet (conference calls assisted by facsimile and the slow-scan transmission of pictures).

The General License for VANS left several grey areas in the definitions and applications of VANS. In view of the limitations in the original General Licence applied to VANS, the U.K. government (Department of Trade and Industry) decided revision was necessary. In 1987, the United Kingdom decided to broaden the scope of competition to include not only VANS, but also managed data network services (MDNS) through a new class license for value added and data services (VADS) open to all except the PTOs and associated firms. The Office of Telecommunications (OFTEL) has negotiated amendments to the PTO licenses to permit them to compete in the offering of VADS, subject to fair competition obligations. Firms associated with PTOs may offer VADS under the terms of a separate class license.

The VADS licenses allow competitive provision of value added services both nationally and internationally. Data services may be offered nationally to closed user groups. Internationally, the simple resale of leased circuits is not allowed because of the risk of cream-skimming. In addition, the following licensing conditions apply to major VADS suppliers (i.e., those with VADS revenues in excess of $1.5 million or total group revenues in excess of $75 million.

- fair competition conditions (e.g., no cross-subsidies by other business activities)
- obligation to provide access via OSI interfaces

11.5.1. OSI Requirements for VANSs in the United Kingdom

The government has a clear policy encouraging the introduction of OSI standards: There is a 12-month compliance period that applies to the standards in the initial license as well as to those added later. The use of proprietary standards is permitted, but they should not be implemented in such a way as to discriminate unfairly against the use of OSI standards. Licensees may be excused from supporting OSI if they can show that there is no actual or potential demand. In all other cases the licensees will be required to provide OSI compatibility and meet conformance testing requirements.

11.6. VANS STATUS IN JAPAN

Since the new Japanese Telecommunications Business Law became effective on April 1, 1985, Japanese telecommunications carriers have been divided into two classes:

Class 1 carrier: Carriers that are network operators and have their own telecommunications circuit facilities and provide telecommunications services at the same time.

Class 2 carrier: Carriers offerings services on the basis of leased lines from class 1 carriers. Class 2 carriers, in turn, are divided into *special class 2* and *general class 2*. Special class 2 carriers are large-scale VANS providers and must satisfy the telecommunication demand of any customer at the national or international level. General class 2 carriers are small VANS providers and are active only at the regional level and for specific groups of customers.

Class 1 carriers are licensed by the Ministry of Posts and Telecommunications (MPT), and each license is contingent on the following conditions:

There is a regional demand for the service.

There are no local and general excess capacities.

The potential carrier has the necessary technical and financial resources available to provide the service.

The concept of services offered is "resonable" and there is a "public interest" to license the carrier.

The licensing criteria provide the MPT with a flexible structure. In addition, class 1 carriers are subject to conditions such as the obligation to serve and to offer a universal service. Class 1 carriers compete with Nippon Telegraph & Telephone Corporation (NTT). When the existing monopolistic structures were erased, NTT became active in the field of new telecommunications services and VANS activities without any restrictions.

The tariffs of type 1 carriers are subject to the approval of the MPT, while type 2 service providers face fewer restrictions. Over 300 companies have registered for or plan to offer general VANS. Foreign-owned companies face no restrictions as type 2 carriers, but a limit of one-third foreign ownership is applied in the type 1 area.

Telenet (U.S. Sprint) joined with Intec and Sumimoto to form a joint multinational electronic messaging service called Ace-Telemail. The Ace-Telemail system will handle both English and Japanese messages and will contain interfaces to telex as well as facsimile. Future plans include X.400 interconnections with other Telemail systems. At start-up, Ace-Telemail had a proprietary gateway to the United States through the Japanese packet network.

GEISCO (a unit of GE) has joined with Japan's NEC for international VANS and other international network-based business. AT&T joined 16 Japanese companies to start Japan Enhanced Network Services Corp., a networking system to permit communications between different computers.

11.7. VANSs IN THE FEDERAL REPUBLIC OF GERMANY

Communication services for third parties are possible as long as there is no exclusive or predominant transmission of messages for third parties or between other subscribers.

Private VANSs in Germany are usually restricted to special networks to satisfy the needs of closed user groups with special applications. Examples are the following:

SWIFT banking network
Reuters Company news network
GEISCO, a VANS provider
Various international networks for data teleprocessing
Networks for press agencies
START network for the travel business
EARN (European Academic Research Network) for universities and research institutes
DATEV network for the tax advisers within Germany
DIMDI network for the medical databases

Source: Deutsche Bundespost, the Federal Republic of Germany.

11.8. VANS APPLICATIONS

VANS offer network management, remote computing, remote database, electronic mail, and software application packages. Applications include data entry, word processing, electronic mail, program development, and industry applications packages.

The European perspective is that VANSs must add a service to the basic telecommunications network. The definition of VANS is different; it varies by country but generally involves store and forward of data or some modification or processing of the data sent through the network. Unlike the United States, which has completely deregulated VANSs including the resale of circuits, Europe is moving at a slower pace. The prime leader in Europe for VANSs is the United Kingdom, due to the liberalization policies of the government. Japan is now opening VANSs to competition.

Some of the more common VANS offerings are the following:

1. Remote database applications
2. Remote computing applications
3. Electronic mail
4. Electronic document distribution
5. Electronic data interchange
6. International networking—multiple enterprises
7. Managed data networks

11.8.1. Remote Database Access

A remote database access service is offered by most international VANSs to their own or complementary service bureaus. International databases satisfy a need that cannot be provided in the local country. Some of the key businesses and services that use VANSs for remote database access are manufacturers, banks, oil companies. security firms, consultants, legal firms, agricultural departments, publishers, libraries, and research and government agencies.

Some of the more popular databases are the Dow Jones Service, CompuServe's applications, Merrill Lynch Economic Services, and the Official Airlines Guide.

International access is important to developing nations. For example, DACOM of South Korea is offering a service within South Korea that provides its customers with access to over 33 external databases throughout the world.

11.8.2. Remote Computing

All international VANSs offer remote computing, either on their own facilities or to different mainframes offered by other companies. Each network provider will usually publish a list of mainframes and applications that can be accessed through its network.

Some of the remote computing applications include marketing information, payroll, scientific and engineering programs, government and financial services, hospital, manufacturing, brokerage, banking, budget control, and treasury management systems.

Table 11.1 lists the major international companies offering remote computing and database services in North America, Western Europe, and Japan. The major U.S. VANSs provide services to both North America and Western Europe, while two French firms are concentrating on Western Europe. General Electric Company has strengthened its position for end-to-end connectivity with GEISCO and RCA Global Communications (a unit of GE), an international record carrier that can carry both voice and data.

Future additions to the list could include Electronic Data Systems (EDS)/General Motors Corp., AT&T-Communications, Boeing Computer Services, MCI International, and major banks such as Citicorp and Bank of America. Boeing is expected to acquire additional clients via its manufacturing and marketing distribution base, MCI International via its users of voice and data services, and Citibank and Bank of America via their banking clients.

Table 11.1 includes Canada and the United States, Austria, Belgium, Denmark, Finland, France, the Federal Republic of Germany, Ireland, Italy, Japan, the Netherlands, Norway, Portugal, Spain, Sweden, Switzerland, and the United Kingdom. The advantage of working with an international VANS provider is that it coordinates information flow and end-to-end connectivity.

TABLE 11.1. List of Major Remote Computing and Database Services

Company Name	US	CANADA	JAPAN	AUSTRIA	BELGIUM	DENMARK	FINLAND	FRANCE	GERMANY	IRELAND	ITALY	NETHERLANDS	NORWAY	PORTUGAL	SPAIN	SWEDEN	SWITZERLAND	UK
ADP	X	X	—	—	X	—	—	X	X	—	X	X	—	—	—	—	—	X
Comshare	X	X	—	—	X	—	—	X	X	—	X	X	X	—	—	X	—	X
CDC (Control Data)	X	X	—	X	X	X	X	X	X	—	X	X	X	—	X	X	X	X
CISI	X	X	—	—	X	—	—	X	X	—	X	—	—	—	X	—	X	X
GEISCO	X	X	X	X	X	X	X	X	X	X	X	X	X	—	X	X	X	X
GSI	—	—	—	—	X	X	—	X	X	—	X	—	X	X	X	X	—	X
IBM IN/INS	X	X	X	X	X	X	X	X	X	—	X	X	X	—	X	X	X	X
McDonnell Douglas	X	X	X	X	X	X	X	X	X	X	X	X	X	X	X	X	X	X
Telenet	X	X	X	X	X	X	X	X	X	X	X	X	X	X	X	X	X	X

[a]*Notes:*

1. CISI is Companie Internationale de Service en Informatique and is owned by two commercial banks (CEA and BNP) with headquarters in Paris.
2. GSI is General de Service Informatique of Paris.
3. IBM's international capability includes its Information Network in the United States and its Information Network Services in Europe and Japan.
4. McDonnell Douglas includes Tymshare, the service provider, and Tymnet, the network provider.

Source: Company brochures, publications, and press releases.

11.8.3. Electronic Mail Offered by VANSs

Electronic mail is the transfer of information to one or more destinations over an internal network or externally via PTT networks. Information is a message that can be in the form of digitized voice, a note, a data file, text, or image. The mail is the delivery system that transmits messages to one or more destinations. If an electronic mail service uses the PTT facilities, delivery requires a store and forward capability. That is, the message must be stored in the computer before it is retransmitted to its final destination.

A mailbox feature is usually found in intraenterprise store and forward electronic mail systems. A mailbox system can be characterized by access from an asynchronous terminal, communicating word processor, or PC; an editing and filing capability for the creation and storage of messages; and a simple command structure to send and receive messages and notify of receipt of a message. The advantage of a mailbox system is that recipients receive the messages after logging on the system to check their mail. There are no internal delays or waiting for delivery of mail, telex, or Teletex. The mail is received by the users on their terminal, word processor, or PC at their work location.

Some of the advanced features in mailbox offerings include the following:

Electronic file folders: Messages can be kept for secure, long-term storage. Any message can be filed, be sent, or remain in any number of file folders.

Electronic bulletin boards: Messages of general interest can be posted.

Predefined address lists: Persons with related interest can establish a group. Each group can be centrally defined or privately defined for one mailbox. The subscriber controls who may access each centrally defined group.

Delegated message receipt: A single user at a location can easily obtain copies of all incoming messages under password control. Incoming messages are sorted for easy distribution.

Telex and TWX access: A person's mailbox may communicate with any telex or TWX terminal in the world.

On-line assistance: By typing HELP, assistance and directions are brought up on a screen.

11.8.4. Trends in Electronic Mail Services

The Yankee Group predicts that 50% of all U.S. business transactions will be handled electronically by 1990. Recent developments have contributed to expanding international electronic mail. British Telecom is offering its Dialcom service in 16 countries, including Australia, Canada, Germany (FRG), Hong Kong, Ireland, Israel, and the Netherlands. Dialcom has over 120,000 mailboxes throughout the world. Most of the PTTs that use Dialcom software for their public mailbox services are connected to an international network, which allows subscribers to send messages to one another. In addition, two VANS private electronic mail services are connected to public electronic mail services offered by a PTT. Western Union's EasyLink and MCI's Mail subscribers can send messages to subscribers of the French Missive system. Missive is operated by France Cables et Radio, a subsidiary of the French PTT.

Geonet, an electronic mail service operating in the Federal Republic of Germany, has 20 nodes in Europe, including Paris, Marseille, and London. The network has linked 7000 terminals, with projections for up to 800,000 by 1989.

There are several electronic mail services offered througout the world by international VANS suppliers:

VANS Supplier	Product
ADP Network Services	Automail
CompuServe	CompuServe
British Telecom (U.K.)	Dialcom
GEISCO	QUIK-COMM
Geonet (FRG)	Geonet
IBM International Business Services	ScreenMail, PROFS
MCI Communications	MCI Mail (MailLink)
Telenet	TELEMAIL (Ace Telemail in Japan)
Tymshare McDonnell Douglas	On-Tyme
Western Union	EasyLink

11.8.5. Electronic Mail Offered by PTTs

Table 11.2 shows the services that are offered by European PTTs within their country, although there is an attempt by the European Economic Community (EEC) to establish a pan-European electronic mail service by the PTTs. A connection to one of the preceding VANS suppliers would be required for an international electronic mail service. A partial list of national PTT mail/mailbox services is given by country.

11.8.6. Electronic Document Distribution

Electronic document distribution (EDD) software applications programs provide for the creation, distribution, storage, retrieval, processing, archiving, and printing of documents and messages. In addition, an EDD interchange software program offers display, printing, and redistribution of received documents and document access protection.

In all cases, the creation and distribution of information is processed in the computer and distributed to recipients listed in the control segment of the program. The amount of message sending time or information transfer is small compared to the document preparation time. Most EDD offerings include either a simple editor, text editor, or word processing capability.

TABLE 11.2. PTT Electronic Mail/Mailbox Services

Belgium: DCS Mail	Licensed from Telenet and run by the Belgium Telecommunications Authority.
Denmark: Databooks	Part of the BT/Dialcom network run by the Danish PTT.
France: Missive	Based on COMET software licensed from Service Systems Technology; run by France Cable et Radio.
Ireland: Eirmail	Based on BT's Telecom Gold (Dialcom software).
Ireland: QUIK-COMM	The Irish Post Office, a separate entity, is offering the GEISCO product.
Israel: Aurec Goldnet	A service offered as part of the BT/Dialcom network.
Germany: Telebox	Part of the BT/Dialcom network run by the DBP.
Netherlands: Memocom	Part of the BT/Dialcom network run by the Directoraat Commerciele Zaken Telecommunicatie.
Norway: Teleboks	Teleboks is Telenet's Telemail under software license.
Switzerland: Datamail	This is the Comet System provided by Radio Suisse.
United Kingdom: BT Gold	A public service based on BT/Dialcom available in the United Kingdom
United Kingdom: Prestel	Mailbox—a mailbox service to users of Prestel.

Source: PTT published reports.

11.8.7. Electronic Data Interchange

Electronic Data Interchange (EDI) is the standard exchange of business documents between computers, an alternative to the movement of large amounts of paper between companies for buying or selling of goods and services. EDI consists of two components: telecommunications and standardization of information to facilitate the exchange of information to reach as many companies as possible. Standardization of information is achieved using translation software that changes the information in a company format into a standard format. For example, in an international trade and transport transaction, over 20 different enterprises can participate in the transaction. Each firm would have its own set of documents. The forms can be print-outs from computer systems that are sent by normal mail to the next company in the chain and rekeyed into another computer. EDI would make it possible to replace documents and conventional mail by computer to computer communications. Existing standards and draft proposals for business data interchange transactions include purchase order, purchase order acknowledgement, invoice, remittance/payment advice, request for quote, planning schedule, shipping notice, receiving advice and price/sales catalog.

11.8.8. Other VANS Services

The following is a list of VANS services that are offered by PTTs and VANS providers.

1. Services for computer communication (normally by a PAD facility):
   Transmission speed conversion
   Reduction of error rate
   Code conversion
   Protocol conversion
   Format conversion

2. Distribution and storage services:
   Audio distribution systems (voice and data, storage, and forwarding)
   Retrieval services
   Voice (voice mailbox, automatic answering machine)
   Text (interactive Videotex)
   Special dialing facilities (abbreviated dialing, repeat last call)

3. Information services:
   Database inquiry services
   Entertainment services
   Information retrieval
   Advertising through Videotex services
   Electronic telephone directories
   Facsimile or Bureaufax service

4. Security services:
   Monitoring (on-line remote sensing)
   Telecommand (turns remote devices on/off)
   Teleadjusting (varies remote controller based on predetermined values)
   Teleindication (remote sensing and indication)
   Encoding (provides encryption/decryption of data)

5. Booking and money transfer services:
   Transactions between banks (SWIFT)
   Electronic funds transfer (EFT)
   Point of sale (POS)
   Booking (travel business for hotels, railroads, and rent-a-car services)
   Orders for goods (shop by Videotex service—United Kingdom)
   Home banking
   Self-service banking
   Electronic money

6. Terminal equipment services:
   Storage facilities (e.g., automatic answering machine)
   Information services (information telephone, databases)
   Point-of-sale devices
   Special dialing facilities (abbreviated dialing, repeat last call)
   Teleconferencing (audio and video conferencing)

11.8.9. International Networking—Multiple Enterprises

International data exchange or an any-to-any connection is the desirable telecommunications goal of most companies—that is, the ability to access and have an on-line dialog with any terminal or mainframe, in any enterprise connected into the network, anywhere in the world.

International networking for related or unrelated businesses is a growing need and a requirement for doing business around the globe. In fact, it is one of the stated requirements for doing business with either the Boeing Company or the Ford Motor Company. Both companies have stated that international suppliers must be tied into their international network to qualify as a supplier. They will not accept previous methods of information transfer such as telex, the mail service, or the PSTN. In many nations, this requirement is a conflict between PTT policy and the national interest (additional employment and investment). In many cases, there will be a negotiated accommodation to protect jobs and investment in the country.

11.9. MANAGED DATA NETWORK SERVICES

Managed data network services (MDNSs) are commonplace offerings by many VANS and system houses in the United States. The only other country where they are allowed is the United Kingdom. The new VADS licensing agreement allows managed data networks. Essentially, an MDNS provider will receive a contract to provide end-to-end communications. In this case, one enterprise or group of related enterprises require only terminals and a host system. The network provider will contract for the PTT lines and provide network adapters, communications equipment, network management controllers, operating system software, or a turnkey solution. Network providers such as the international VANSs will be able to obtain a license to provide a managed data network service within the United Kingdom.

11.10. RESTRICTIONS ON VANSs

Depending on the country where a message originates, electronic mail or information transfer between enterprises may not be allowed. However, there are several exceptions for a group of related businesses. Other than in the United Kingdom, there is no licensing agreement for VANSs in Europe. In most countries, the PTT does not have a crisp, clear definition of a VANS, and each application for a private leased line is evaluated based on the location of the two end points. If the lines are between two enterprises, there must be some business reason for the connection or both companies should be in related businesses. Electronic mail may be allowed in Spain and Switzerland if the final destination is outside the country. Internal mail would be in competition with their national offering. France will allow a VANS offering only on TRANSPAC, their packet data network for national and international messages. Austria, Denmark, Finland, Italy, the Netherlands, Norway, and Sweden require negotiations for a competing national or international service. In all cases, if a VANS provider offered the service on a switched data network, the service would probably be allowed.

Electronic document distribution as a VANS offering is allowed in Belgium, Ireland, the Netherlands, and Norway. Austria, Denmark, Finland, France, Germany, Italy, Sweden, and Switzerland require negotiations if the service is offered on a leased line. The chances of obtaining a contract are increased if the service is offered over a packet switched or circuit switched data network.

Interconnection of packet data networks has been accomplished throughout Europe. However, the PTTs do not offer international services over the packet data networks because there is no electronic messaging standard for packet networks. Other than telex, each country implements a different standard. This should change with the CCITT X.400 Recommendation for a message handling system.

In some countries, VANSs may not transfer information between enterprises if the purpose is other than a remote data service bureau or database application. Normally, these functions are supplied over the VANS providers leased line network. The VANS provider obtains the lines from the PTT, provides network attachment devices, performs network routing, provides alternate paths, and guarantees reception of information.

Concentrators and multiplexers on a private leased line are allowed in all countries if the customer accesses a remote computer or remote database. They are not allowed if the VANS offers switching between customers, which is usually within the scope of the PTT monopoly.

A PAD facility is allowed over a private leased line in most countries if the customer accesses a remote computer or remote database. A PAD facility is not allowed to switch between two customers. In the United States, the FCC considers a PAD function as an enhanced service and would not allow AT&T-C to offer a PAD facility on their "basic" packet switched data network offering. PADs were left in the domain of the VANS providers.

Protocol converters are allowed on private leased lines in all countries if the application is to a remote computer or remote database.

In the United States, subscriber line charges are being phased in to reduce the subsidy of local rates by long distance traffic. It is likely that European countries will also reduce long distance tariffs in the same way over the next 10 years.

11.11. VANSs—A MAJOR GROWTH AREA

There are several opportunities for enhanced services where the PTT can not provide an adequate international solution. One example is in international Videotex. There are five major systems in the world: Prestel in the United Kingdom, Teletel in France, CEPT in the Republic of Germany and other European countries, CAPTAIN in Japan, and NAPLPS in North America. There is an opportunity for VANSs to offer interfacing programs to match different Videotex country standards. The PTTs may allow international Videotex since they will receive additional revenue within their own country and usage and time charges for international traffic.

Britain, which has the largest potential in Europe, will lead the way. Frost and Sullivan projects a $5.7 billion European market by 1990 with more that 50% going to private companies and the remainder to PTT administrations. Based on the Frost and Sullivan assumptions, the United Kingdom will have revenues of $1.4 billion and Germany and France will have $1.1 billion each.

In 1985, there were two major acquisitions concerning international service industries. The largest was the acquisition of Electronic Data Systems (EDS) by General Motors for $2.5 billion, and the second was the purchase of Tymshare by McDonnell Douglas for $307 million. The two largest purchases in the service industry history indicates the potential growth in the information processing industry. In explaining its entry into the industry, GM stated that it has an annual DP budget of $6 billion a year spread out over a worldwide organization. It would make sense for GM to develop, test, and integrate its internal worldwide network and then offer that capability to other companies.

11.12. INTERNATIONAL VANSs IN BANKING

Citibank has a network called Citiswitch that offers worldwide service through four nodes (London, Bahrain, New York, and Hong Kong). Barhain is the telecommunications center for some of the gulf countries such as Qatar, Saudi Arabia and Kuwait. Citibank has purchased five transponders at $11 million each on the Western Union Telegraph Westar satellite. One of its major holdings is Diners Club, which uses the Citibank network for international credit card authorization and accounting.

Bank of America, Chase Manhattan, Citibank, and Manufacturers Hanover Trust are among the many U.S. banks offering electronic services in Europe.

For European banks to meet U.S. competition, they have signed up with commercial worldwide networks, such as GEISCO and ADP, to provide time sharing offerings for the electronic banking market. For example, all the major U.K. banks are using a service bureau to provide cash management services on these networks. Table 11.3 lists several treasury management systems on VANS networks.

Normally, banks that use a time sharing service bureau to provide cash management services provide software that performs formatting tailored to their banking procedures and distribute the information over the time sharing network. New banking applications have been added to provide customers with an "electronic window" into the bank files, enabling them to see their data in a format of their own choice.

TABLE 11.3. Major U.K. Treasury Management Systems

Banks/Nonbanks	System	Network
Banks:		
Bank of America (B of A)	Micro Star	B of A's network
Barclays Bank	Bar Cam	Packet Switching Services, GEISCO
Chase Manhattan	Global Microstation (Infocash)	Chase Data Network & IDC
Chemical Bank	Chemlink	GEISCO
Citibank	Citibanking	Infospool
First Chicago	International First Cash	GEISCO
Lloyds Bank	Cashcall	ADP
Manufacturers Hanover Trust (MHT)	Transend	Geonet (MHT's network) and Telenet
Midland Bank	CMS	ADP
Morgan Guaranty Trust	MARS and MORCOM	GEISCO, Global Data Network, Tymnet, and Telenet
National Westminister	NETWORK	GEISCO
Royal Bank of Canada	Royal Command	GEISCO
Nonbanks:		
Automatic Data Processing (ADP)	Cash Express	ADP
Interative Data Corp.	IDC	IDC
National Data Corp.	NDC	GEISCO

Source: The Financial Times, London, United Kingdom.

11.12.1. SWIFT

The Society for Worldwide Interbank Financial Telecommunication (SWIFT) was established in 1973 by 239 banks in 15 countries as a nonprofit bank-owned cooperative society. SWIFT is dedicated to meeting a number of specialized service needs relating to interbank financial transactions including electronic funds transfers (EFTs) and message services. See Table 11.4 for a listing of member banks by country.

At year end 1986, SWIFT had over 2164 member banks throughout the world using its money message service. However, 35 banks accounted for about 50% of the 700,000 daily transactions in the network. SWIFT operates a pure peer-to-peer telecommunications network for message transfer between member banks located in 60 countries. The network operates on 9600-bps international leased lines. The system software allows retrieval of up to 4 months of data. Messages of up to 4000 characters can be sent through the system at a rate five times faster than that of telex. EUROCLEAR, a clearinghouse for bank transfers, was added in 1982, and CEDEL, an international clearinghouse, was later added for securities transactions.

TABLE 11.4. SWIFT's Worldwide Link—1984

Country	Member Banks	No. of Banks Connected to SWIFT	Processed Financial Transactions
Andorra	2	3	28,549
Argentina	33	38	266,758
Australia	9	11	1,844,517
Austria	41	46	4,945,144
Belgium	28	46	7,250,470
Bermuda	3	3	117,065
Brazil	25	25	247,391
Canada	8	22	2,392,344
Channel Islands	—	1	—
Chile	13	16	186,334
China (People's Rep.)	1	—	—
Colombia	16	—	—
Cyprus	4	—	—
Czechoslovakia	1	1	211,593
Denmark	35	37	2,670,141
Ecuador	11	14	100,738
Finland	9	11	1,935,367
France	74	89	8,787,783
Germany	118	154	14,441,036
Greece	5	14	488,168
Hong Kong	18	64	2,008,922
Hungary	4	3	131,811
Iran	1	—	—
Ireland	2	9	484,889
Israel	12	14	704,529
Italy	150	146	9,649,456
Japan	65	100	3,412,646
Jordan	1	—	—
Liechtenstein	3	3	110,108
Luxembourg	9	26	1,845,169
Mexico	14	13	362,450
Monaco	1	2	19,686
Morocco	7	—	—
Netherlands	22	34	6,627,198
New Zealand	2	4	492,321
Norway	25	24	2,721,813
Peru	8	—	—
Philippines	10	13	78,078
Portugal	11	13	529,395
Singapore	9	65	1,480,248
South Africa	12	18	1,160,099
Spain	35	47	1,682,680
Sweden	15	17	3,359,559
Switzerland	63	79	11,167,168
Taiwan	12	—	—
Thailand	11	—	—

(continued)

TABLE 11.4. SWIFT's Worldwide Link—1984

Country	Member Banks	No. of Banks Connected to SWIFT	Processed Financial Transactions
Tunisia	4	—	—
United Kingdom	38	136	12,987,308
United States	157	273	22,251,766
Uruguay	11	21	166,351
Venezuela	20	—	—
Total	1188	1656	129,953,693

Source: SWIFT, as reported in *The Financial Times*, London, United Kingdom.

11.13. INTERNATIONAL VANSs

Most of the larger European service bureaus operate within their own country. However, there are two French computer service bureaus that provide their services to other European countries: CISI (Compagnie Internationale de Service en Informatique) and GSI (Generale Service Informatique), both based in Paris.

11.13.1. Automatic Data Processing

Automatic Data Processing (ADP) Autonet is the networking division of ADP, based in Ann Arbor, Michigan. Through Autonet, more than 100,000 subscribers in over 250 U.S. locations and 55 countries may access remote database, remote computing applications, and AutoMail.

ADP is oriented toward specific industries and offers propriety solutions with emphasis on packaged solutions.

ADP Autonet is the networking portion for ADP services to Europe. Any user via a PC or 3270 terminal in Europe has the ability to access a U.S. host. Autonet offers VANSs, including AutoMail, facsimile, and Videotex support.

Remote computing in the United States from France: A customer in France logs into TRANSPAC, the French national packet switched data network. The French PTT passes the information through one of the IRCs. In the United States, the IRC passes the information over to the U.S. Autonet network, which can be tied to another host computer or database supplier that accesses the Autonet network. Autonet negotiates the complete link with the IRC and TRANSPAC.

AutoMail is not available for customers in Europe through the Autonet network due to current PTT regulations. To access ADP's AutoMail, the customer has to dial into the national packet switched data network, which passes the message through the IRC to ADP's mailbox in Ann Arbor.

11.13.2. AT&T Information Systems

AT&T-Information Systems is a subsidiary of AT&T based in Parsippany, New Jersey. AT&T is a new provider of international packet switching and VANSs but has the resources and capacity to eventually match competition if it chooses to become a major international VANS provider. It is in a strong position to offer managed data network services to companies or related user groups.

11.13.3. Boeing Computer Services

Boeing Computer Services (BCS) provides information processing and communications services both internally and to over 1500 commercial customers throughout the world. BCS manages over $600 million worth of equipment from over 100 different vendors. One of the software products offered by Boeing is the Technical and Office Protocol (TOP). TOP will link technical and office environments by allowing systems developed for the office to communicate with products from other vendors that are designed for the factory.

Boeing is offering an electronic mail service in Europe. The service is based on IBM's Professional Office System (PROFS) and is located in a computer in London.

11.13.4. British Telecom

British Telecom (BT), a major common carrier in the United Kingdom, is 49.2% privately owned by individual shareholders and 50.8% government owned. Although the telecommunications environment in the United Kingdom is regulated, BT is able to offer its products and VANSs on a worldwide basis.

BT has acquired 51% of Mitel, a Canadian telecommunications equipment manufacturer with 6000 employees in 14 countries, and Dialcom, an electronic mail service licensed in 16 countries. Prestel, the U.K. national Videotex hardware and software system, is marketed throughout the world, either directly to the United Kingdom or on a licensing basis by private or public groups. Other VANS offerings include MEDIAT (a broker package for insurance contracts) and DELEGAT (handles user inquiries on the insurance policies of more than 20 companies).

11.13.5. Comshare

Comshare offers remote database and remote computing in the United States on many major VANSs such as Autonet, Telenet, and Tymnet, and in Europe from its London office.

Comshare operates four data centers, two in the United States, one in Canada, and one in London, to provide network access from over 380 cities in 32 countries. It offers both time sharing and applications software.

11.13.6. CompuServe Network Services

CompuServe is a wholly owned subsidiary of H&R Block and is based in Columbus, Ohio. CompuServe has passed the 250,000 active subscribers mark. Along with each subscription, a customer has a free mailbox service on the Easy Plex electronic mail system.

CompuServe is a VANS provider using public and private networks for large volumes of data traffic. Its network service is a virtual switched data network based on X.25 packet switching protocol. CompuServes's network transmits data packets between customers' host computers and remote terminals. An added feature of its network is the following: When a session to the host is created, the transmission path is maintained throughout the session. This method provides faster response because the network does not require routing information after the first packet is sent.

11.13.7. Control Data Corporation

Control Data Corporation (CDC) offers remote computing and database services to international accounts from Minneapolis, Minnesota.

CDC, through its Cybernet, operates 20 centers worldwide to over 250 cities including most of Europe, Canada, the United States, and Japan. Cybernet is a large international network with high reliability that provides large data banks, an extensive software library, and turnkey systems offerings.

11.13.8. Electronic Data Systems/General Motors

Electronic Data Systems (EDS), acquired by General Motors (GM) in 1984, has assumed responsibility for all DP operations within GM. The magnitude of GM's operations, along with GM's purchase of Hughes Aircraft (a U.S. common carrier, satellite manufacturer, and major electronic systems house), provides EDS with the resources to construct a sophisticated private global telecommunications network. In addition to GM's operations, EDS is offering its clients a total integrated DP and networking solution for global information processing.

11.13.9. GEISCO

GEISCO is part of General Electric Company's communications and services organization. GEISCO is a pioneer in remote computing and time sharing and claims to have the world's largest commercial network. The company is a VANS provider offering database, information management, and industry applications programs in North America, Europe, the Middle East, and the Asia/Pacific area (Australia, Hong Kong, Japan, and Singapore). GEISCO is focusing on office communications, clearing house services, supplier and dealer systems, international trade, and electronic data interchange (EDI).

GEISCO has introduced new EDI services throughout the world. MOTORNET, which originated in the United Kingdom, is an EDI service for the automobile industry that links suppliers with customers for the exchange of orders, invoices, inventory, etc. Express is a generalized EDI service offered in Europe for linking various trade partners.

GEISCO offers QUIK-COMM, a worldwide electronic mailbox service, as well as remote database and remote computing applications. In Europe, QUIK-COMM is offered in Denmark, Finland, France, Germany, Italy, Luxembourg, the Netherlands, Norway, Sweden, and the United Kingdom. In three other countries, Belgium, Spain, and Switzerland, the customer must dial into the IRC. The QUIK-COMM service in Europe is allowed to access only GEISCO computers.

11.13.10. Generale de Service Informatique

Generale de Service Informatique (GSI) is a French VANS provider owned by two banks and Alcatel, a major shareholder. Alcatel, in turn, is part of CGE (Companie Generale d'Electricite), a French holding company with interests in undersea cables, power plant control systems, nuclear reactors, and telephone switching equipment. GSI's international VANSs are primarily focused in Europe.

11.13.11. Graphnet Freedom Network

Graphnet, a subsidiary of Graphic Scanning Corporation, bases its national and international services in Teaneck, New Jersey. Graphnet is a VANS offering a facsimile service that accepts, stores, and forwards graphic images in digital form. The image is scanned at the sending station, moved throughout the network, and reconstructed at a registered facsimile device at a subscriber location. The store and forwarding feature allows distribution lists and multiple addressing.

Graphnet also offers a digital record service that uses its existing network at speeds up to 1200 bps. Speed and code conversion is offered on the system to accept information from terminals, telex, and TWX. Delivery of the image is to a remote facsimile machine.

11.13.12. Telenet (U.S. Sprint)

Telenet of Vienna, Virginia was one of the first commercial VANS providers. Telenet is the largest data communications network of its kind in the world and provides a broad range of data transmission services as well as private network systems and messaging information services.

On a typical day, companies transmit about 50 million packets of information over the Telenet data network or the equivalent of 1 million typewritten pages. Telenet is used to communicate business data such as invoices, orders, inventory levels, and financial records.

Telenet is offering a complete end-to-end transaction processing service, called TeleCARD, that will allow a retailer to check a major credit card transaction and then have the information sent automatically to its bank for reimbursement. The Tele-

CARD service uses GTE's Micro-Fone II credit card terminal to capture credit card and transaction information and sends it to the TeleCARD computer. From there, the transaction is relayed to either a VISA or MasterCard host computer to check for authorization. After checking the account data, TeleCARD sends back an authorization or rejection. Later, it forwards the transaction information to the bank or credit card company for reimbursement.

Retailers have two choices for processing their credit card transactions. They can send their slips directly to a bank, or they can send them to VISA or MasterCard, who have their own clearing networks. TeleCARD can work with VISA or MasterCard for nationwide clearing or can be operated directly by a bank. The service has two major features: instant payment settlements and no rekeying of credit card data. Normally, retailers get paid after the credit card slips are received by the bank, a process that can take days or weeks.

Remote Access to Telenet
International service to a number of VANS service providers is available from Telenet. For example, access to a CompuServe database via the Telenet international packet switched data network can be accomplished by the following procedure from Argentina to the United States:

1. The customer in Argentina establishes an account with CompuServe for remote services to be supplied in the United States through Telenet.
2. The customer establishes a connection with ARPAC, the Argentina national packet switched data network.
3. The customer enters a password for either a dial call or direct access facility. ARPAC connects to Telenet, who in turn passes the information to CompuServe. After logging into CompuServe, the customer has access to its subscribed database in the United States. The same customer could have access to Telemail, Telenet's electronic mail service.

   Depending on the services subscribed to, a customer receives either one or two bills; one from ARPAC in Argentina and a second from CompuServe.

   Another option is a collect call service to CompuServe. A customer in Argentina would receive one bill from CompuServe. An advantage of this type of account is that dollars flow to the calling PTT. The customer in Argentina pays CompuServe, who pays Telenet its share; Telenet, in turn, pays the national PTT its share. Seven other nations have collect call agreements with Telenet: Canada, Chile, the Dominican Republic, Israel, Mexico, Panama, and the Philippines.

11.13.13. IBM International Business Services

IBM International Business Services provides a number of international services including transaction processing, Videotex, electronic office systems support, EDI, and business professional support.

The electronic office system is based on PROFS (Professional Office System) and offers international communications between offices in the same enterprise. PROFS includes document preparation, calendar maintenance, document search and storage, and time management. Business professional support is based on Applications Systems (AS), an integrated system for information retrieval, business graphics, business planning, statistics, and forecasting.

Data communications is available via IBM's international data communications network within North America, Europe, and Japan. The network service is called the Information Network (IN) in the United States and Information Network Services (INS) in Europe and Japan.

11.13.14. Tymnet McDonnell Douglas

Tymnet is a subsidiary of Tymshare, which is part of the McDonnell Douglas Company. Each month, Tymnet establishes over 10 million sessions and sends 72 billion characters, for over 2 million hours of transmission time over their national and international networks.

Tymnet offers local distribution using microwave broadcasting, microwave links, and satellite earth stations to supplement its national leased line network. Tymnet is in a position to bypass local Telcos in the United States.

Tymnet provides data communications for over 600 organizations with more than 1100 host computers in the United States to more than 1400 nodes and 10,000 public access ports in over 500 cities. In addition, Tymnet provides international access to over 50 countries through five IRCs. Tymnet is also the principal supplier of packet networking hardware and software used by the IRCs. Additionally, countries may access Tymnet via the telex network. On-tyme, Tymnet's electronic mail service, is available internationally.

11.13.15. Western Union Telegraph Company

Western Union of Upper Saddle River, New Jersey has a national network consisting of 23 nodes with links to Canada, France, Switzerland, and the United Kingdom. Western Union, through its Easylink message switching service, is offering InFact, a subscriber service, to over 630 databases. InFact was designed for users with limited knowledge. Log-on is simple; the customer needs only to know the topic of interest, and the system responds with questions to narrow the search. In addition, the system is backed up by librarians, who offer subscriber assistance.

12

TELEMATICS

In this chapter we will

- Present and evaluate telex and Teletex, a new interoffice mail service
- Discuss the new CCITT X.400 Recommendation for message handling systems
- Review the various types of facsimile offerings
- Discuss Videotex and video conferencing

Telematics or the interconnection between computers and communications is like electricity in that it allows instant access from one isolated area to another linked by networks. Whereas electricity transmits energy, telematics transmits information.

Telematics will eventually bring about substantial increases in productivity, the decentralization of autonomous basic operating units, and interaction between social groups. Certain organizations, CEPT in Europe and CCITT on a worldwide scale, form the traditional structure for developing international telematics or information services.

Telematics is a set of European services that was later picked up by Japan and the United States. Viewdata, now Videotex, was developed in the United Kingdom. Teletex was developed in the Nordic countries and the Federal Republic of Germany. High-speed circuit switching data networks and cellular radio were pioneered for commercial use in the Nordic countries. The results are the Nordic Datex Network and Nordic Mobile Telephone System (NMT). Both Nordic networks are successful (NMT had over 260,000 subscribers at year end 1986) and offer complete network integration among four of the five Nordic countries (Denmark, Finland, Norway, and Sweden). The Datex network is also integrated with Germany's (FRG) Datex-L network. The NMT provides access into the PSTN, which allows a direct telephone call from an automobile to anywhere in the world.

Telematics has been expanded by the European PTT administrations and private companies, where allowed (notably the United Kingdom and the United States), into four broad categories:

- Electronic mail/messaging services

  Teletex
  X.400 message handling systems
  Telefax
  Videotex
  Mailbox (voice or text)
  Facsimile

- Inquiry

  Credit card checking
  Electronic directory

- Transaction processing

  Home banking
  Home shopping
  Booking services
  Smart Card

- Teleconferencing

  Audio conferencing
  Video conferencing

Four telematic services are CCITT recommendations: Teletex, the X.400 Message-Handling System, facsimile, and Videotex. They will be discussed in greater detail.

12.1. TELETEX

Teletex is a high-speed public electronic mail service that enables the communication of text between Teletex-compatible terminals on a memory-to-memory basis.

Teletex is a new form of office-to-office telecommunication that uses a new type of terminal with an extended typewriter character set to prepare and transmit text. Teletex integrates with the office environment and is a logical supplement to existing telecommunications services such as telex (message exchange service) and Telefax (remote facsimile or Fax copying).

The international service is defined by the CCITT F.200 Teletex Recommendation. The objective is to provide compatibility between users in transmitting text information. The word *compatibility* is important since there are many different communicating systems (including word processors and PCs) that offer text transfer capability.

The existing features of a Teletex service can be summarized as follows:

- International service for any CSDN, PSDN, or PSTN.
- Twenty-four hour availability.
- Reproduction of the transmitted document at the distant end.
- Identical format—European A4 (210 × 297 millimeter) and North American (216 × 280 millimeter or 8 ½ × 11 inch) standards.
- Identical layout—The standard number of characters is 10 characters per inch (10 pitch) with standard options of 12 and 15.
- Identical line spacing—The standard is six lines per inch; a standard option is eight lines per inch. It is possible to use partial line up or down and one and one-half and double line spacing.
- Memory-to-memory transmission.
- Guaranteed permanent copy.
- Fully automatic reception.
- Essentially error-free transmission.
- Automatic destination verification.
- Page-by-page transmission.
- Journal of send/receive activity.
- Character transmission rate 45 times faster than telex.
- High transmission rate (2400 bps).
- Unique customer identification.
- Uninterrupted local operation while receiving a message.
- The 309 character set consists of the Latin alphabet with lowercase and capital letters, accents, digits, currency symbols, and mathematical symbols.
- Interworking with 1.7 million worldwide telexes.

Some of the enhancements under consideration within the CCITT are the following:

- Interworking with other telematic services such as Telefax and Videotex.
- National option for conversational mode to establish a dialog with computers and databases.
- Message handling features within the network for delayed delivery, multiple addresses, and a mailbox feature.

12.1.1. Teletex Tariff Comparison

National charges are usually based on charges applicable to a particular network. For example, PSTN charges will be the same as a telephone call of the same duration. There are a large number of variables (time of day, number of documents) that affect charges, but in general the Teletex charges should be a fraction of other comparable services such as telex, letter, or facsimile. One feature of all Teletex offerings is that they will be able to send telex messages. A telex is charged at the normal telex rate.

Table 12.1 offers a comparison of Teletex versus telex in the Federal Republic of Germany (the amounts shown are in deutsche marks). For more than 10 Teletexes per day, there is a 30 times price improvement for Teletex over telex. Teletex is priced attractively in Germany.

12.1.2. Teletex Terminal Configurations

The terminal is structured into two areas: user facilities and communications requirements. Terminals may include both user facilities and integrated communications, a low-cost terminal with a stand-alone Teletex adapter, or a cluster controller and a more intelligent communications adapter.

Teletex standards relate mainly to communications, printing, and assurance of delivery. The manufacturer may provide its own user facilities (e.g., keyboard, display, printer, and storage).

Each Teletex terminal has a separate subscriber number and a unique terminal identification. The terminal identification consists of a network code, the subscriber number, and a mnemonic abbreviation. The terminals automatically exchange terminal identifications before the transmission of a document, verifying to the calling terminal that it has established connection with the correct terminal. The two terminal identifications and the date and time of the transmission remain on the document and can be printed. (Figure 12.1)

12.1.3. Communications

The communications adapter has transmit and receive memory, a communications controller, and a line interface. The memory size will depend only on the number and length of the documents the user wishes to transmit and receive; memory must be

TABLE 12.1. Teletex Versus Telex Tariff (Federal Republic of Germany)

	Teletex (Deutsche Marks)	Telex (Deutsche Marks)
Monthly charge	170	65
Per call charge	0.05	—
Usage charge/minute, peak rate, long distance	0.84	0.60
Incremental usage charge per 1000 characters	0.05	1.50

Source: Deutsche Bundespost, the Federal Republic of Germany.

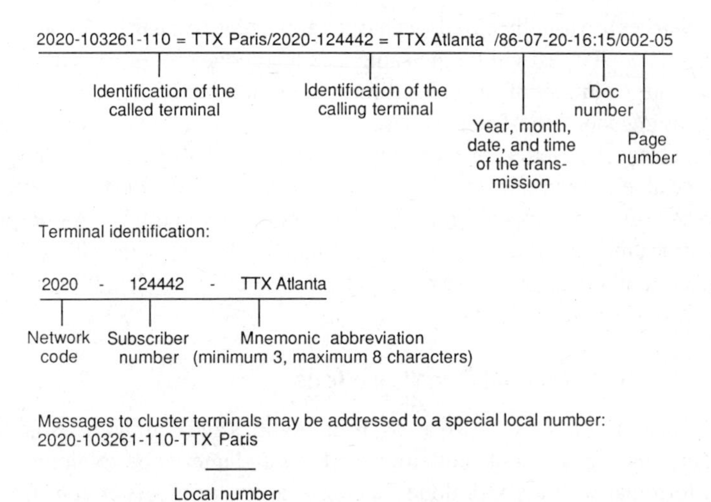

Figure 12.1. Teletex Call Identification Line (Courtesy of Nordic public data networks).

large enough to avoid interruption to local operation and to reject messages due to overloading. Typically, the minimum memory size is 32K characters. The communications controller handles the Teletex protocols, which are structured by the following CCITT recommendations:

Recommendation T.60 defines functional characteristics of the terminal including the presentation layer requirements (the Teletex standards combine the presentation and application Layers into a single document Layer).

Recommendation T.61 defines repertoire and codes for the basic graphic characters and basic control functions. The code is based on an 8-bit structure providing 256 combinations.

Recommendation T.62 defines network-independent control procedures for the session and document Layers. The calling terminal effectively "controls" the call through a series of commands to and responses from the called terminal.

Recommendation T.70 defines the network-independent transport Layer procedures and the interface to level 3. Levels 1, 2, and 3 are network dependent, and the procedures are based on existing standards as follows:

Packet switched data network	X.25 interface
Circuit switched data network	X.21 interface
Public switched telephone network	V Series modems

12.1.4. Interworking with Telex

The CCITT Teletex Recommendation requires access to the telex network. Teletex interworking with the telex network requires a conversion facility provided centrally between the two networks for speed and code conversion. To send telexes, a Teletex terminal must restrict its message to the telex character set and line length. The

TABLE 12.2. Teletex Network Attachment

Country	Network
Austria	CSDN[a]
Belgium	PSDN[a]
Canada	—
Denmark	CSDN
Finland	CSDN
France	PSTN[a]/PSDN
Germany	CSDN
Hungary	CSDN
Italy	CSDN
Japan	CSDN
Netherlands	PSDN
Norway	CSDN
Portugal	PSDN
South Africa	CSDN
Spain	PSDN
Sweden	CSDN
Switzerland	CSDN
United Kingdom	PSTN/PSDN (future)
United States	Leased line to nearest node

[a]CSDN = circuit switched data network.
[a]PSDN = packet switched data network.
[a]PSTN = public switched telephone network.

Source: PTT published reports.

conversion facility is usually a store and forward device that looks like a telex terminal to the telex exchange and looks like a Teletex terminal to the Teletex exchange. To ensure proper delivery of all documents (or notification in the event of nondelivery), special procedures involving the use of control documents have been agreed upon by the CCITT and are contained in the T.90 Recommendation.

12.1.5. International Connections

Two countries that use the same network will have relatively few problems between Teletex data interchanges; e.g., two X.21 networks will work using X.71 protocol, and two X.25 networks that use the X.75 international gateway protocol will work. However, interconnection of different types of networks (e.g., X.71/X.75) will cause problems. A gateway device is required in order for a national PSDN network to communicate with a CSDN network. Table 12.2 shows the Teletex services available on existing national networks.

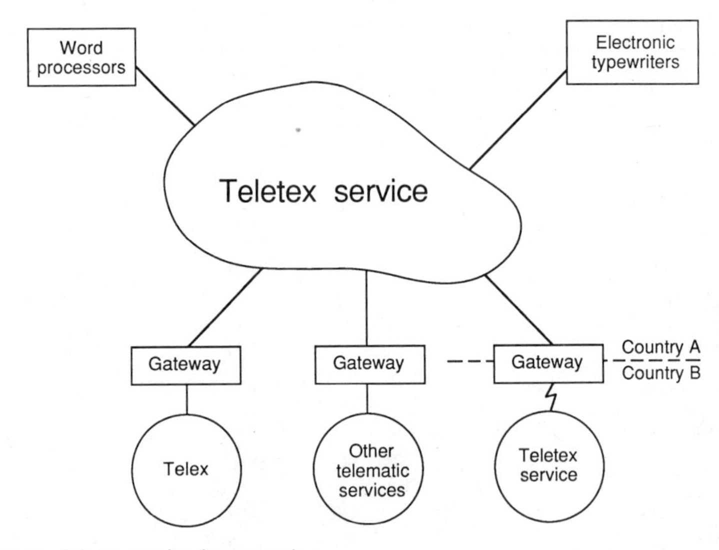

Figure 12.2. Teletex service (gateways).

To have Teletex compatibility between countries, each national PTT must implement a gateway or interface to another country's network to provide a full worldwide Teletex service. For example, France would require a gateway to move messages from their PSDN (Transpac) to Germany's CSDN (Datex-L) network. As part of every Teletex offering, telex interworking is usually provided by a gateway into the telex network. In addition, some countries plan to offer gateways to other telematic services such as Videotex and Telefax. For example, the Federal Republic of Germany allows interaction between their Videotex and Teletex systems. Figure 12.2 gives a sampling of gateways between countries and Telematic services.

Teletex in Japan. NTT, the national Japanese PTT, implemented the Japanese Teletex Service in 1984. The service is based on the Japan Unified Standards for Telecommunications (JUST). This Japanese Teletex standard is able to process about 7,000 characters including KANJI, alphabets, numerals, and other types of characters.

12.1.6. Teletex Versus Telex Comparison

At Teletex speeds of 2400 bps, an A4 page (European standard) or 8½ × 11 page (North American standard) can be sent in 7 seconds, or 30 times faster than telex.

In the following example, assume an average document size of 1500 characters.

Teletex (per document). Teletex uses an 8 bit code at 2400 bps. Assuming a 25% overhead, the net character rate is

$$2400 \div 8 \times 0.75 = 225 \text{ net characters per second}$$
$$\text{transmission time} = 1500 \div 225 = 6.67 \text{ seconds per document}$$

Telex (per document). Telex uses a 7.5-bit code at 50 baud:

$$\text{transmission time} = 1500 \times 7.5 \div 50 = 225 \text{ seconds per document}$$

If an additional 10 seconds is allowed for error correction on retransmission and answerback exchange, nominal transmission time becomes

Teletex = 17.7 seconds
Telex = 235.0 seconds

Dial time is approximately 5 seconds, resulting in a call and dial time of

Teletex = 22.7 seconds
Telex = 240.0 seconds

Maximum messages per 24-hour day are 3800 for Teletex and 360 for telex.

Table 12.3 lists several features of telex and Teletex.

Western Union Telegraph Corp. (WU) offers Teletex in 25 U.S. cities with international connections to Canada, Sweden, and the Federal Republic of Germany. Teletex has not had broad acceptance on a worldwide basis. For example, there were less than 57,000 Teletex terminals installed in Europe and less than 1000 in North America as of year end 1986. Teletex has not had the impact that many PTT administrations originally planned and may give way to the X.400 Message Handling System, a more comprehensive offering including message, document, and text handling.

12.2. X.400 MESSAGE-HANDLING SYSTEM

To improve communications within an establishment or an enterprise or between enterprises, there must be a method of linking persons through an electronic mail or messaging system. While telex and Teletex are normally restricted in most companies to a fixed location within a building, electronic mail is a terminal-to-terminal or desk-to-desk service.

In addition to telex and Teletex, there are a number of different systems capable of generating messages including text, image, voice, facsimile, and a number of public and private voice and data electronic mailboxes.

The CCITT X.400 Recommendation is an attempt to have one standard for internal, national, and international electronic mail. Series X.400 specifies the network architecture, protocol, message transfer, and interpersonal messaging services for interconnecting electronic mail networks.

The message-handling network will allow any combination of message content including text, voice, Fax, graphics, or binary data structures. Message transfer will provide for a store and forward service allowing transparent voice, data, or image transfer.

TABLE 12.3. Teletex Versus Telex Comparison

	Telex	Teletex
Characters (depends on country)	Upper- or lower-case	Full office capability
Character set	57	309
Character per second	6.6	300
Transmission rate	50 baud	2400 bps
Time to transmit A4 page (1500 char)	3 minutes, 45 seconds	6 seconds (approx.)
Page format	Continuous roll	A4/A4L pages or 8 1/2 × 11
Guaranteed format	No	Yes
Code	5 bit	8 bit
Local text preparation while receiving	No	Yes
Guaranteed (auto-acknowledged) delivery	No	Yes
Conversation mode	Yes	Future
Message transfer	Usually keyboard or paper tape, usually manual	Memory to memory, automatic
Document control procedures	No	Yes
Session control procedures	No	Yes
Error detection/correction	No	Yes
Capability for interworking with other services	Limited	Limited

The message-handling system (MHS) model is shown in Figure 12.3. In this model, a user is either a person or a computer application that is either an originator (sends a message) or a recipient (receives a message). Message-handling service elements define the set of message types and the capabilities that enable an originator to transfer messages of those types to one or more recipients.

The originator prepares a message with the aid of its user agent (UA), an application process that interacts with the message transfer system (MTS) to submit messages. The MTS delivers the message to the recipient or UAs. Each MTS may have a number of message transfer agents (MTAs) that relay messages and deliver them to the intended recipient or UAs. UAs and MTAs make up a message-handling system (MHS). The MHS and all its users are referred to as the message-handling environment. The basic structure of a message includes an *envelope* that carries information to be used when transferring the message and the *content*, which is the information that the UA sends to one or more recipient UAs.

The MTS provides an application-independent, store and forward message transfer service by which UAs can exchange messages. There are two basic interactions between MTAs and UAs.

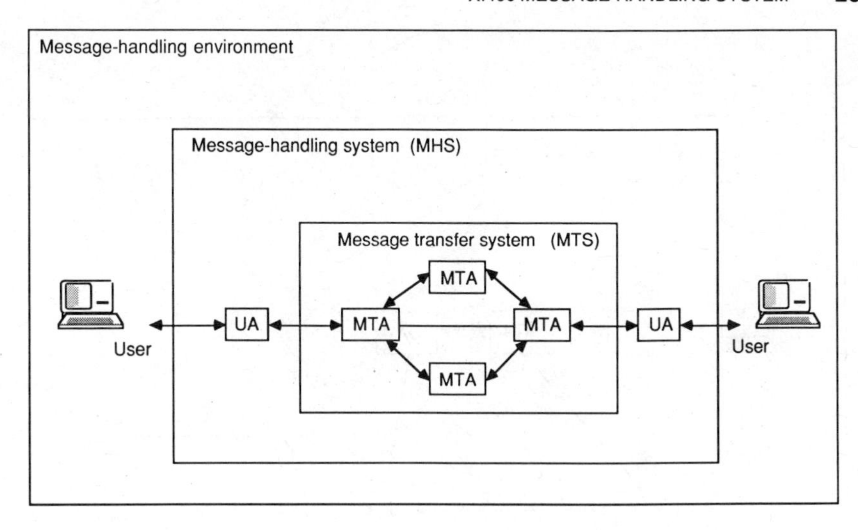

Legend:

↔ : An interaction

UA : User agent

MTA : Message transfer agent

Figure 12.3. X.400 message-handling environment.

The originating UA submits or transfers to an MTA the content of a message plus the submission envelope containing the information required by the MHS to provide the requested service. The MTA delivers or transfers the message and delivery envelope to the recipient UA through other MTAs.

MTAs relay messages to another MTA (content plus relaying envelope) until the message reaches the recipient's MTA for delivery to the recipient UA. The relaying interaction is the method used by an MTA to transfer the contents of a message plus the relaying envelope. The relaying envelope contains information related to the operation of the MTS plus the service elements requested by the originating UA.

MTAs transfer messages that may contain any type of binary coded information. MTAs do not interpret or alter the content unless the UA requests a specific service element. A UA is a set of computer application processes that contain functions necessary to interact with the MTS. A functional view of the MHS model is shown in Figure 12.3.

The Interpersonal Messaging System

The interpersonal messaging system (IPMS) provides individuals with services to assist them in communicating with other individuals. The IPMS uses the MTS. Users of other CCITT Telematic and telex services may gain access to the X.400 message-handling system through specialized user agents, e.g., a Teletex access unit. IPMS interrelationships are shown in Figure 12.4. The IPMS should include a message preparation facility and deliver or receive the message with the MTS.

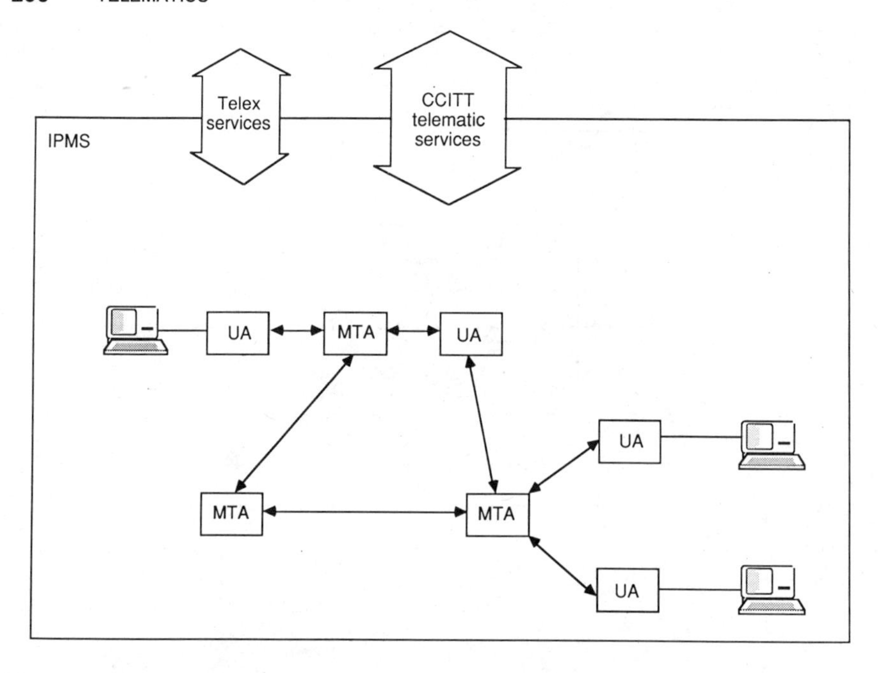

Figure 12.4. Interpersonal messaging system (IPMS).

Messaging-Handling Boundaries

A collection of at least one MTA and UA(s) owned by a PTT or a private or noncommercial organization constitutes a management domain (MD). The MD managed by a PTT is called an administration management domain (ADMD). The MD managed by any other organization is called a private management domain (PRMD). A PTT may provide three types of access for its subscribers:

1. A user may connect to a PTT-supplied end user device (UA).
2. A private end user device (UA) may connect to a PTT node (MTA).
3. A private node (MTA) may be connected to a PTT node (MTA).

The subscriber may have a telex or telephone and may interact with the PTT-supplied end user device (UA) or a PTT-supplied intelligent stand-alone terminal containing the UA functions. In Figure 12.5, an intelligent terminal user may communicate through the PTT's administration management domain (ADMD) to an externally connected input or output device.

Private End User Device (UA) to PTT-Supplied Node (MTA)

The subscriber can have a private stand-alone UA (PC or word processor) to interact with the PTT supplied node (MTA). From his or her terminal, the subscriber would use a standard set of delivery procedures to obtain an MTS service. In this case, a private, stand-alone user (UA) is not in a management domain but is associated with a PTT management domain as shown in Figure 12.6.

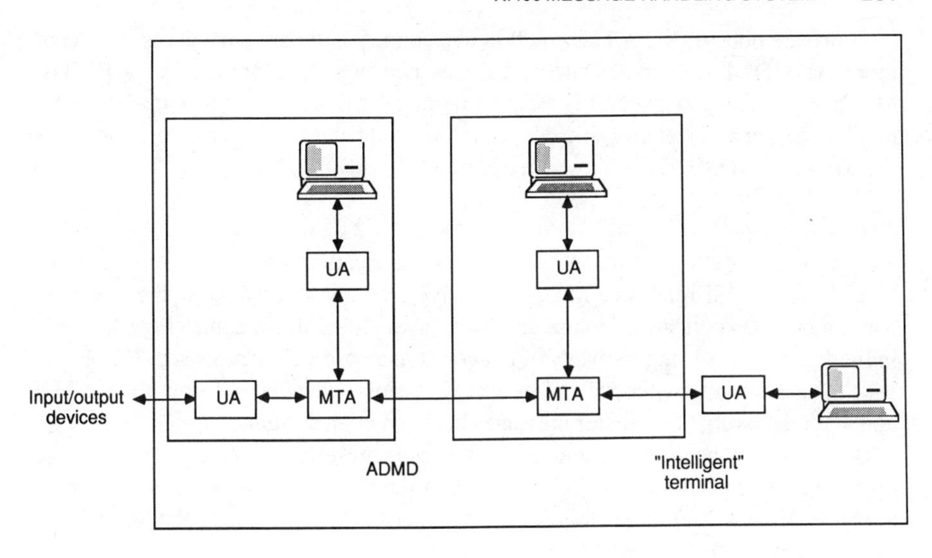

Figure 12.5. PTT-supplied end user device (UA).

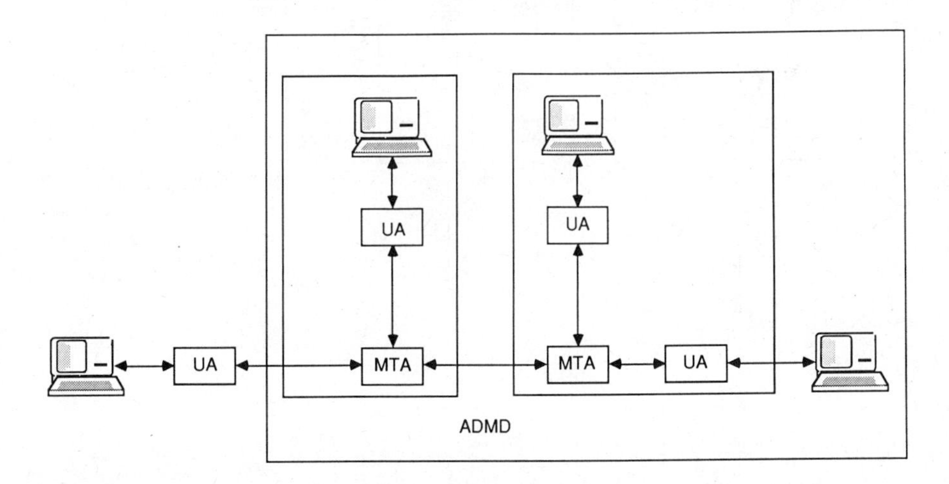

Figure 12.6. Private end user device (UA) to PTT node (MTA).

Private Node (MTA) to PTT Node (MTA)

A PTT's subscriber may have one or more nodes (MTAs) and one or more user agents (UAs). These subscriber UAs may be stand-alone or integrated into the same system. The subscriber's MTA(s) and UA(s) form a private management domain (PRMD) and may interact with the PTT's administration management domain (ADMD) on a node-to-node or MTA-to-MTA basis.

A private node or MTA exists within one country and may have access to one or more ADMDs but cannot switch messages between PTT domains or ADMDs. Message switching between the PTT domains or ADMDs is kept within the PTT monopoly. The interrelationship between the private and PTT management domains (PRMDs and ADMDs) are shown in Figure 12.7.

X.400 and OSI

The message-handling entities and protocols are located in the application layer or layer 7 of the OSI reference model. The layers below layer 7 allow the message-handling (MH) applications to use the lower layers to establish connections between individual systems using a variety of network types, e.g., packet, leased, telephone, LANS, and circuit switched, and establish session connections to permit the MH applications to reliably transfer messages between open systems.

The message-handling functions in the application layer are divided into two layers: the user agent layer (UAL) containing the UA functions associated with the message contents and the message transfer layer (MTL) containing the MTA functions to provide the message transfer service.

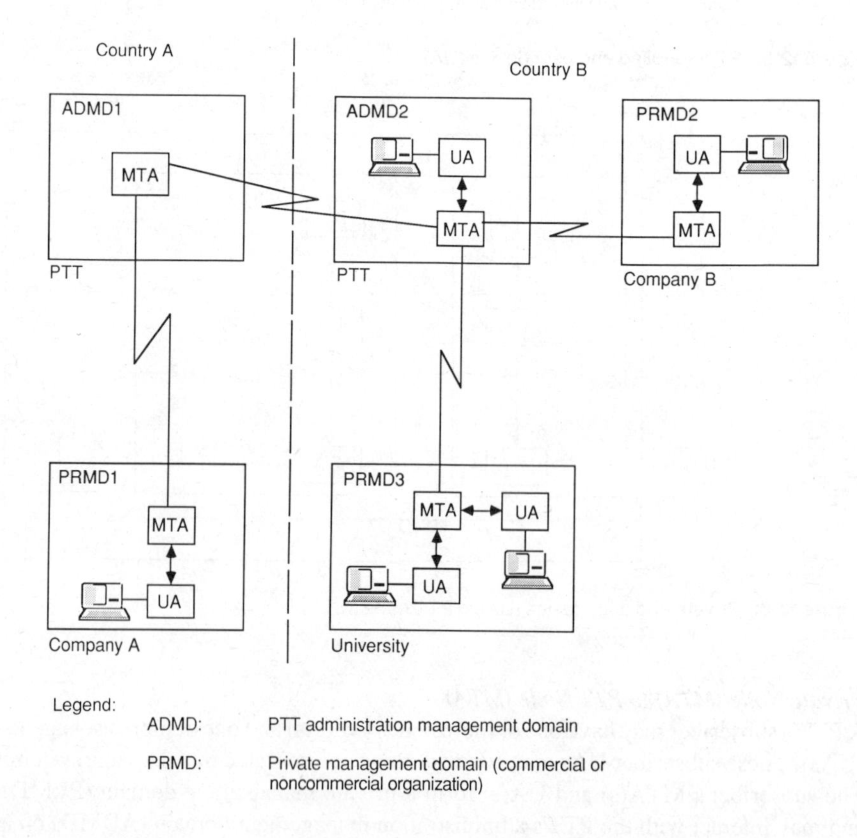

Figure 12.7. X.400 administration and private management domains.

There are three types of systems that are based on the functional model:

1. Systems that contain only end user (UA) functions (S1)
2. Systems that contain only node (MTA) functions (S2)
3. Systems that contain both end user (UA) and MTA functions (S3)

The three functional entities are shown in the OSI model in Figure 12.8. The UA entity (UAE) consists of the end user functions (UA) that represent the message's content and other associated UA functions. The MTAE or the MTA entity provides the functions required to support the layer services of the message transfer layer (MTL) used with other MTAEs (nodes). The submission and delivery entity (SDE) makes the services of the MTL available to a UAE through the MTL boundary. The SDE does not provide message transfer services but interacts with the peer MTAE to provide access to the MTAE. There are three peer protocols in the layered model:

1. The message transfer protocol (P1) defines the relaying of messages between MTAs and other interactions necessary to provide MTL services.
2. The submission and delivery protocol (P3) allows the SDE in an S1 system to provide its UAE with access to the MTL services.
3. Pc is a range of protocols for defining the syntax and semantics of the message content being transferred. A Pc protocol is associated with a class of end users (UAs).

Figure 12.8 contains the items in the message-handling system and shows their position in the OSI reference model. All items above the presentation layer (6) are contained in the applications layer (7).

CCITT recommendations that apply to the X.400 message-handling system are listed in Table 12.4.

12.3. TELEFAX

Telefax, facsimile, or Fax is a primary method for transmitting hard-copy documents on an instant basis between two sites. Fax units can transmit alphanumeric and graphic images, which makes it convenient for transmitting charts, drawings, photos, graphs, and letters with signatures. Fax units produce a copy of the original document at the remote end.

The advantage of Fax is that it is simple to use and easy to install and there are no restrictions on the page content to be transmitted. As in many forms of electronic mail, rekeying is not required. Generally, most Fax traffic is within the same enterprise and can be justified for urgent requests.

A Fax terminal is classified as either analog or digital based on its design for handling electrical information. Essentially, a scanning device illuminates the original document and focuses reflected light from the copy onto a photoelectric transducer or photocell. The purpose of the photocell is to convert the image patterns of the original based on a light to dark scale gradient into equivalent current for

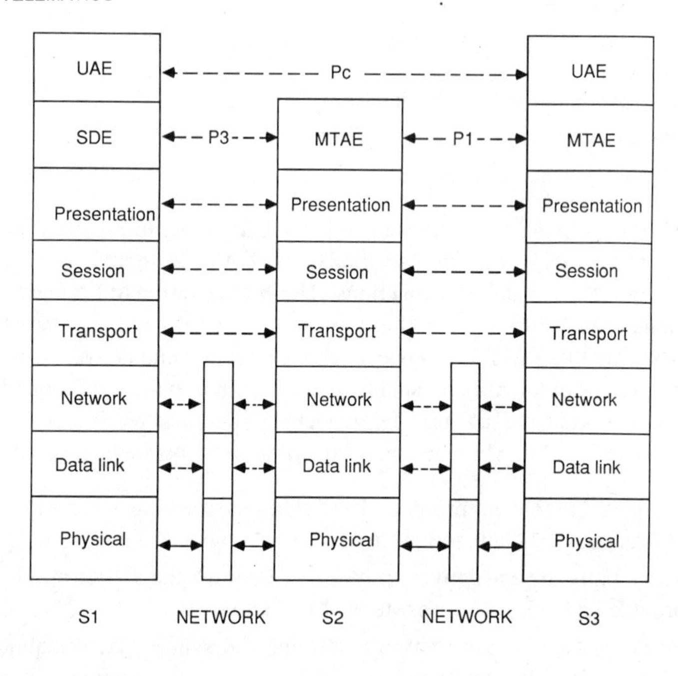

Figure 12.8. X.400 Message-Handling System and OSI.

TABLE 12.4. CCITT X.400 Series Recommendations

X.400	System model of user agents and message transfer agents, the message transfer and interpersonal messaging services, and protocol layering
X.401	Conformance requirements for basic service elements and optional user facilities
X.408	Conversion from one type of encoded information to another
X.409	Standard notation for describing complex data structures and the transmission bit sequences for the protocols
X.410	Remote operations and relationship with OSI model to support MHS applications
X.411	Message transfer sublayer protocol elements of the Application layer (PC)
X.420	Content protocol for the Interpersonal messaging services and formats for memo headers and multipart body types
X.430	Describes how Teletex terminals access the MHS and message exchange between Teletex and message handling users

transmission. New techniques for scanning include lasers and fiber optics. Analog Fax units range from 2 to 6 minutes of transmission time per page, while digital is faster at less than 2 minutes a page.

An analog system scans every portion of an original including characters, space between characters, spaces between lines, and margins. This process is relatively slow but also allows for greater image quality with an appropriate high-quality printer. The electrical signal for transmission is a direct analog representation of the document.

A digital unit analyzes a document in terms of its actual picture content. It only spends time processing real picture elements and considers the page to be composed of a matrix of white and black spaces. As the unit analyzes a scan line, it recognizes a sequence of white and black spaces, which in turn are converted into binary codes of 1s (black) and 0s (white). The digital process is more compact than the continuous process by an analog Fax machine, making it quicker to communicate the digital signal over the telephone line. Fax can be transmitted by either the public switched telephone network or private leased line.

At the remote site the incoming signals are converted, if required, to digital form for output to a printer. Printing devices vary by design based on the type of paper used, the printing speed, and the resolution or quality of the reproduced copy. The key to Fax machines is the quality of the image resolution or reproduction at the remote end.

To achieve compatibility between units, Fax machines are separated into four CCITT groupings, three of which are in use and a fourth (Group 4) under study.

Caution: Not all units within a group are compatible. Either purchase two similar units or request a demonstration with an existing unit.

12.3.1. Fax Groups

There are four CCITT groups for Fax machines; two are analog and two are digital designs. Standards were written and adopted by CCITT Study Group XIV in 1981. The number of copies per hour and the time per copy along with the Fax design type and applicable network of each group is presented in Figure 12.9. Groups 1 and 2 are analog designs and usually work with the public switched telephone network. Groups 3 and 4 are digital types. Group 3 uses the telephone switched network or private leased line, while Group 4 units will use packet or circuit switched data networks.

12.3.2. Telefax Services

Telefax services are generally operated by common carriers. Depending on usage, this can be a cost-effective alternative to designing a Fax network. The carrier provides the switching and control hardware, and the user's only requirement is to provide a compatible Fax terminal. Some of the current Telefax services are the following:

FAXPAK: ITT World Communications, Inc. provides store and forward communication between compatible and incompatible facsimile terminals as well as between

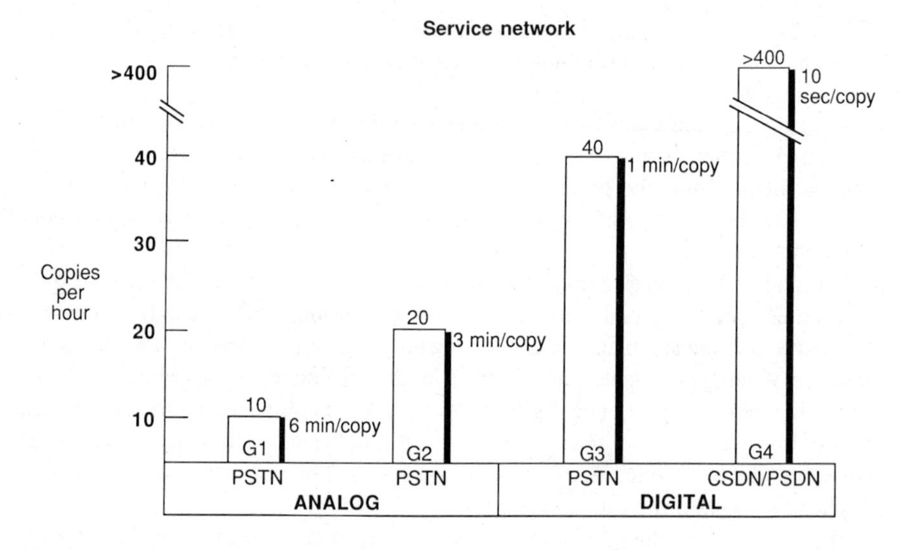

Figure 12.9. Telefax groups by CCITT recommendations.

300-bps ASCII data terminals and facsimile units throughout the United States Subscribers can access FAXPAK switching centers via analog or digital private leased lines or the public switched telephone network.

Graphnet Freedom Network: Graphnet is an authorized special common carrier with a tariffed service based on a VANs that accepts, stores, and forwards digitized graphics and messages to any type of Fax, terminal, or telex registered for use with the network.

MCI International: MCI offers an international Fax service from the United States. MCI has five public facsimile centers located in New York City, Washington, D.C., San Francisco, Miami, and New Orleans. MCI can transmit material to over 20 overseas locations. After the message arrives in the foreign country, MCI arranges for its delivery by Fax, mail, or messenger.

Network Facsimile Service: Network Facsimile Service operates a network of independent Fax centers to serve the business community. All centers use Group 3 Fax equipment and operate 24 hours a day, 7 days a week. They provide pickup, delivery, and proof of delivery to a shipper within 30 minutes of delivery.

Q-fax: RCA (a unit of GE) offers a service for international Fax communications. Documents may be delivered to one of the six RCA Q-fax operation centers via Fax or in person. Documents can be sent to more than 25 overseas locations, including Argentina, Bahrain, Bermuda, Hong Kong, Japan, and Switzerland. The message can either be picked up or delivered via Fax or messenger.

Syndifax: Executive Telex/Syndifax provides Fax and telex sending and receiving services to businesses. Contracts, briefs, affidavits, and other documents can be sent via telecopier anywhere in the world to a network office or to equivalent private Fax machines.

TABLE 12.5. Bureaufax Service in Europe

Country	Service Name	CCITT Group	Domestic Only
Austria	Telepost	G2, G3	Yes
Belgium	Telefax	G2, G3	Yes
Finland	Post office	G2	Yes
France	POSTECLAIR	G2, G3	Yes
Germany (FRG)	Telebrief	G2, G3, G2/3	Yes
Greece	Private	As required	—
Ireland	Private	As required	—
Italy	Telefax	G2, G3	No
Luxembourg	Post office	G2	Yes
Netherlands	FAXPOST	G2, G3	1200 locations
Norway	Postfax	G2	Yes
Portugal	CORFAC	G2	No
Sweden	Post office facsimile	G2	Yes
Switzerland	Radio Suisse	G2	Yes, 39 cities
United Kingdom[a]	Faxpost	G2	

Note: [a]There are several private offerings in the United Kingdom for both national and international service.

Source: PTT published reports.

12.3.3. Telefax Status

Group 1 and 2 are declining in total units shipped, while Group 3 is increasing. According to EMMS, over 160,000 Fax machines were delivered in the United States in 1986. British Telecom International (United Kingdom) states that Britain had over 48,000 installations at year end 1985 compared with 850,000 units in Japan and 550,000 in the United States.

Facsimile in Japan. There are several reasons for Japan's heavy use of Fax. Japanese text processing is heavily character based with over 7000 characters required for a good working business vocabulary. Consequently, keying in the complex Japanese characters is usually a burdensome and time-consuming task. In addition, telex or personal computer terminals have fixed character sets. Also, the heavy use of Japanese word processors that use English-language software operating systems are difficult for the end user.

12.3.4. Bureaufax

Bureaufax is a Telefax service provided by many postal administrations of the PTTs in Europe. Typically, customers bring their documents to the local PTT office for transmission to a remote post office. The remote post office has similar Fax equipment and delivers the reproductions the same or next working day by letter carrier. Table 12.5 is a listing of Bureaufax services operated by some of the PTTs in Europe, mainly within their country. A few PTTs offer international Telefax, and some allow private companies to provide their own Fax units.

12.4. VIDEOTEX

Videotex is a particular type of on-line, real-time transmission medium that is used in the storage and retrieval of databases. It incorporates graphics that transmit pages or frames of text and/or graphics. The pages are edited on a keyboard or generated from a computer-stored database. The database design is such that it permits rapid access and retrieval of specific items of information and the billing for customers using the system. Interaction between the user and the computer is through the public telephone switched (dial-up) network.

Figure 12.10 includes the major components of a Videotex system. The user at home or business requires a telephone, TV receiver, keypad, and modulator unit that interfaces to the public switched telephone network.

Most Videotex systems use a modified TV receiver or separate external box attached to a TV set to translate the data and build up the images on the TV screen. Pages for transmission are selected by the user on a numeric keypad or an alphanumeric keyboard. Calls are made through the PTT facilities to a remote database. In some countries, notably France, the PTTs are supplying a specially designed terminal for their Videotex offerings.

In the early stages of Videotex in the United Kingdom, databases were located on the PTTs premises or often as stand-alone private in-house Videotex systems. Today, databases are offered by companies to the general public through a service bureau or via the national Videotex system. Information providers (a person, company staff, or separate service company) create the frames and interactive programs that make up

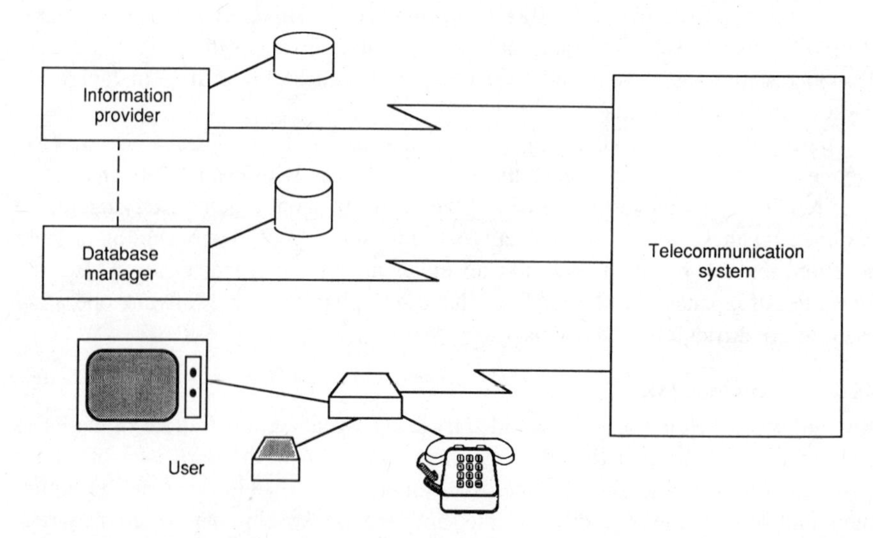

Figure 12.10. Major components in a Videotex system.

the database or service offering. Videotex has progressed from being used as a fixed retrieval system for train and airplane schedules to interactive on-line applications for update, order entry, billing, and storage. New offerings include home banking, home shopping and travel bookings for hotels, trains, and airline services.

There are many incentives for using Videotex. Basically, there are benefits to end users, the PTT, and the information provider:

1. End user
   Low-cost terminals or television set
   Ease of use, simple menu and help commands
   Variety of applications for shopping, news, and directories

2. PTT
   Increased use of its lines and telephone service
   Attractive modern service
   Revenue generated from additional services

3. Information provider
   Inexpensive information distribution
   Widens range of users and customers
   Opens up new marketing channel

12.4.1. Videotex Standards

Videotex was originally developed by the United Kingdom post office. Due to different national requirements and to varying television standards, five major Videotex standards have emerged in the world:

- *Prestel:* The original and United Kingdom standard allows full-duplex connection to a remote computer over a normal dial-up telephone line. It includes a 64-character set, seven colors, and simple *alphamosaic* graphic capability. The main objective of Prestel was to make maximum use of existing television and telephone technology. Many PTTs in a number of countries run public systems based on the Prestel standard.
- *Teletel:* This standard was developed by the French PTT. Its primary objective is to provide inexpensive access to many information systems. Both black-and-white and color terminals called Minitel are available. It is similar in principle to the Prestel standard and is based on an alphamosaic graphic capability. The French PTT network, called Teletel, consists of concentrators, called *Videopads*, through which end users can access various Videotex systems offered by public or private enterprises.
- *CEPT:* The first implementation of the CEPT standard was in September 1983 by the German Bildschirmtext system. The current implemented level provides an almost unlimited range of color and improved facilities for producing pictures using advanced alphamosaic techniques. Approximately 500 different

characters are available for text and graphics. Up to 16 foreground colors and 16 background colors allow more than 4000 color combinations. The standard uses high-resolution graphics by using a dynamically redefinable character set (DRCS). Graphic characters are displayed in a matrix of 10 × 12 points. Each point can be programmed with a different color.

- *NAPLPS:* The North American Presentation Level Protocol Syntax evolved in part from Telidon, a Canadian system. Table 12.6 lists the differences in Videotex terminals and line speeds for the North American and European standards.
- *CAPTAIN:* The Japanese standard, the Character and Pattern Telephone Access Information Network, was designed to accommodate the Japanese Kanji characters. All points on the screen are addressable, allowing unique character generation. The near photographic-type patterns are formed by mosaics or geometric forms. CAPTAIN uses a pattern recognition approach that requires more time to form a frame or display. The transmission rate is 3200 bps to speed up the frame buildup.

Most PTT Videotex systems provide a public database of pages contributed by a variety of information providers. The databases in the Prestel and Bildschirmtext systems each contain over 300,000 frames. The trend is for PTT systems to provide a gateway through which Videotex terminal users can be attached to computers external to the national Videotex system, enabling organizations to offer data processing services and information retrieval on their private systems to Videotex users. An external computer is usually connected to the public PTT system using the packet switched data network. Three different national networks are shown in Figure 12.11.

TABLE 12.6. Differences Between North American and European Videotex Systems

	North America	Europe
Videotex terminals		
Frequency	30	25
Lines	525	625
Visible	388	480
Vertical pixel	200	240
Horizontal pixel	250	400
Screen characters	40 × 20	40 × 24
Line speed		
Receive	1200 bps	1200 bps[a]
Transmit	1200 bps	75 bps

[a]Additional speed combinations of 1200/1200 bps and 2400/2400 bps are available in a few countries.

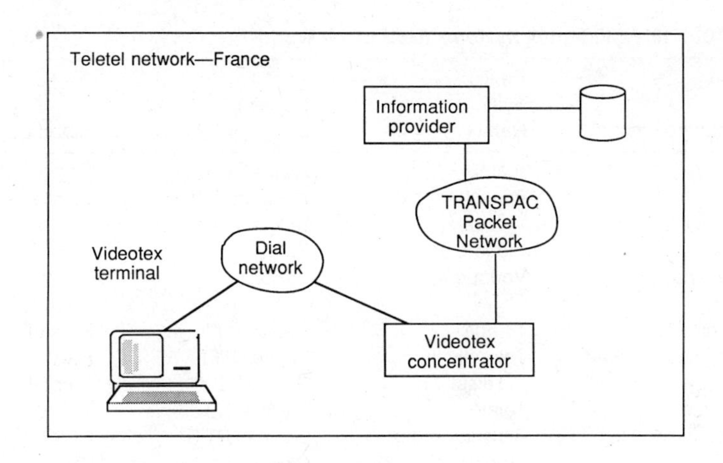

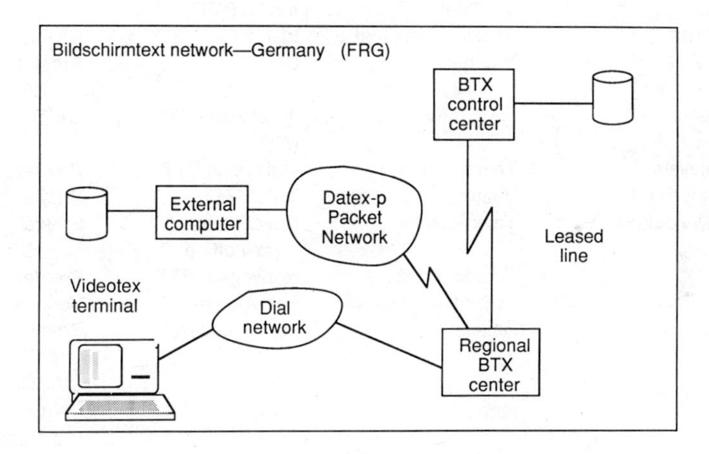

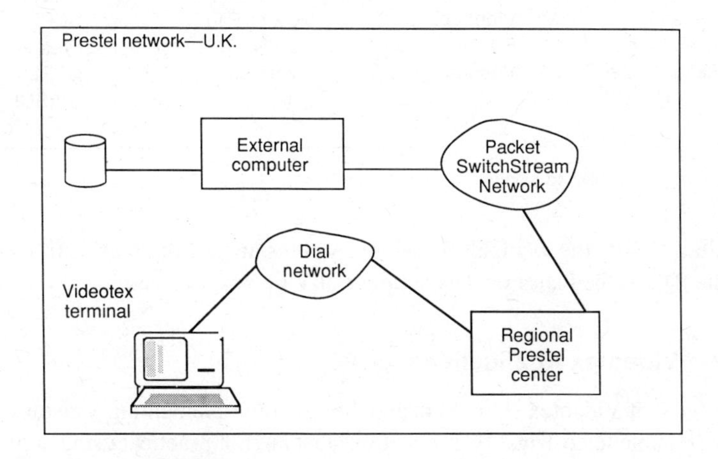

Figure 12.11. National Videotex systems.

TABLE 12.7. Videotex Systems Around the World

County	Public System Name	Network Operator	Standard
Australia	Viatel	Australian PTT	Prestel
Austria	Bildschirmtext	Austrian PTT	Prestel
Belgium	Videotex	Belgian PTT	Prestel
Brazil	Videotexto	Telesp	Teletel
Canada	Various	Various	Telidon NAPLPS
Denmark	Teledata	Danish PTT	Prestel
Finland	Telepalvalu	Finnish PTT	Prestel
	Telset	Sanoma OY	Prestel
France	Teletel	French PTT	Teletel
Germany (FRG)	Bildschirmtext	German PTT	CEPT
Greece	Videotex	Greek PTT	CEPT
India	Videotex	Indian PTT	Teletel
Ireland	Access to several	Irish PTT	CEPT
Italy	Videotel	SIP	Prestel CEPT
Japan	CAPTAIN	Japanese PTT MPT	CAPTAIN
Malaysia	Telita	Malaysian PTT	Prestel
Netherlands	Viditel	Dutch PTT	Prestel
New Zealand	Telematics	New Zealand post office	Prestel
Norway	Teledata	Norwegian PTT	Prestel
Portugal	To be announced	Portuguese PTT	Various
South Africal	Beltel	South African PTT	Prestel CEPT ASCII
Spain	Ibertex	CTNE	Unique standard
Sweden	Telebild	Swedish PTT	Prestel
Switzerland	Videotex	Swiss PTT	CEPT
United Kingdom	Prestel	British Telecom	Prestel
United States	Various	Various	NAPLPS ASCII Prestel

Source: PTT published reports.

Table 12.7 lists the available Videotex systems around the world. In most countries, the PTT is the major or only supplier of Videotex services.

12.4.2. Videotex Applications

The low cost of Videotex equipment and the ease of implementing Videotex systems make it possible to cost-justify applications that are not practical using conventional data processing terminals and communications networks. It is now possible to provide access to information systems for a casual or infrequent user with little or no

DP training. For example, field sales representatives in remote locations can be given a portable terminal. They can telephone the company's computer from home, a hotel room, or their customer's office to inquire on stock availability, place orders, or send messages.

The following is a list of existing Videotex applications followed by additional details on the more popular applications:

- Highly interactive applications
  Tailored insurance quotations
  Order entry

- Rapidly changing information
  Airline seat bookings
  Hotel reservations
  Stock market prices

- High-volume data capture
  Mail order
  Stock control of fast-moving goods

- Large databases
  Bibliographic database
  Time tables
  Directories
  Media

- Applications with high security requirements
  Banking
  Sensitive Databases

Order Entry

Order today and get delivery tomorrow. All intermediate steps become superfluous: filling out order forms, mailing, collection, and entering orders. Each representative can enter an order on his or her TV set. The system checks immediately on validity, credit rating, and if the articles ordered are in stock. Upon confirmation, the orders are processed (e.g., overnight), and the necessary documents are created so that delivery can take place the following morning. Both advance invoicing and regular invoicing are possible.

Transaction Processing

Travel agencies, travel agents, hotels, and information bureaus can use Videotex to inform clients about availability, location, time schedules, and price of services. Without lengthy telephoning, information screens immediately show the various possibilities. After the customer makes a choice, a reservation can be made directly with a simple transaction.

Information Exchange
In many industries, cooperation between enterprises can reduce costs, for instance, between retailers who carry trendy or seasonable articles that vary according to area. An article that is a leftover in one region can still be in demand in another area. From one party to another, supply and demand can be brought together using Videotex.

Electronic Mailbox
An electronic mailbox can save time and telephone expense. Messages can be entered on a terminal and sent to one or more addresses. The message, request for information, or announcement is sent automatically to the electronic mailbox of the addressee. As soon as the users turn on their TV or Videotex terminals, they receive a message to check their mailbox.

Videotex Conferencing
Videotex conferencing is a new facility that allows one user to display frames on the terminals of one or more users. It allows a group of users, from different locations, to look at the same information simultaneously. One person in the group is responsible for accessing the frames and is called a *controller*. The controller may transfer control to another member of the group who then takes over as controller for the session.

12.4.3. Videotex Applications in the United Kingdom

Since the launch of Prestel in September 1979, public awareness of the service and its penetration into the home market have been a slow uphill climb. Prestel has over 1000 information providers (IPs) that offer information services. The main product areas are shown in Table 12.8.

TABLE 12.8. Main Product Areas of Videotex in the United Kingdom

Agriculture	Weather information for farmers, agricultural advice, financial services
Education	Student information, administration
Electronic shopping	Catalogs, order entry, mail order
Government	Land registry, cultural events, data centers, local government information
Home banking	Inquiries, information, self-service
Home computing	Calculator, interest rates, budgeting
Insurance	Agents' information and inquiries, quotes
Investors' services	Credit, loans, stock information
Manufacturing	Spare parts, material reporting, planning, order entry, inventory
Media	Newspapers, classified ads, advertising, billboarding
Transport	Airline, railway, shipping information
Travel booking services	Airline availability, hotel and ticket booking, holiday promotion

Source: Prestel, British Telecom, London, United Kingdom.

One advantage of Videotex over the more traditional media is that readership can be measured with the kind of accuracy that pollsters and market analysts dream about. Each time a user reads or accesses a frame of information, the count is incremented. Prestel produces a monthly listing for each IP showing the most highly accessed IPs.

Micronet was the first IP to break the 1 million access barrier in August 1983. Since that date it has been at the top of the polls with more than 6 million accesses each month. Micronet was launched in March 1983 as a cooperative project between British Telecom (BT) and East Midlands Allied Press (EMAP).

Electronic Home Shopping

Teletran is an electronic retail credit card transaction system established by a joint venture between British Telecom (BT) and Cresta. It provides the limit authorization of the credit card, records the sale, and passes the information to BT computers. Later, transaction details are forwarded to the credit card companies.

Telecard is a Prestel offering for ordering supermarket goods for home delivery. Moves, a distribution company, will deliver food, wine, and household goods to customers in Westminster and the London area. Orders can be placed 24 hours a day for deliveries to 9 p.m. on weekdays and 4 p.m. on Saturdays. Orders over $50 are delivered free.

12.4.4. Videotex Applications in France

Teletel in France is one of the most successful Videotex systems in the world with over 2.5 million *Minitel* terminals in use at year end 1986. The French call it *Le Phenomene Minitel*. The French PTT, in most cases, provides the *Minitel* terminal free to the telephone subscriber.

Teletel handled 15 million calls per month over the PTT telephone system in 1986. Half the calls were to the PTT's electronic telephone directory. About 30% of the calls were to services run by entrepreneurs offering messaging and other applications. An additional 20% of the calls were used for business. Minitel's success is largely the result of a KISS philosophy—"Keep It Simple Stupid." Minitel believes that people use Videotex to electronically chat and for quick answers to specific questions. The system is simple and includes an inexpensive black-and-white display. The PTT has given away over 2.0 of the 2.5 million terminals and expects to have 6.9 million installed throughout France by 1990.

Electronic Directory

The French PTT has a program to replace the White Pages directory. For example, there were 104 publishings and 22.8 million telephone books distributed in France in 1982.

Electronic directories were defined and developed by the French PTT and are supplied to telephone subscribers. At present, the service is provided free to all requesting subscribers. The major advantages are the savings in printing and distribution costs for the paper directories and the ability to update directories on a daily basis.

12.5. VIDEO CONFERENCING

Video conferencing enables groups of persons in specially equipped studios or meeting rooms at widely dispersed locations to speak and see one another over TV monitors and audio equipment. By using video conferencing instead of traveling, expenses can be reduced for accommodations, travel, and executive time.

Two types of video conference are used, full-motion video and freeze frame or slow scan. Full-motion video will scan a scene at 60 times a second, while freeze frame takes a "picture" on demand every 8–60 seconds depending on the resolution required for each frame or scene. Full motion provides continuous motion, while freeze frame takes a picture and transmits it over voice-grade lines to a remote site where it is displayed on a TV monitor as it is received. The remote viewers look at a still picture 8–60 seconds later. Freeze-frame systems send a video snapshot of the participants scanned by the video camera. Although freeze frame is good for graphs and charts, it can be distracting in an interactive session if several persons speak in rapid succession.

The cost for full-motion and freeze-frame video is measured in bandwidth or the equivalent voice channels required to transmit the video signal. For example, full-motion color requires the equivalent of 1500 voice channels. Black and white requires 1000 voice channels. Compression techniques are used to reduce the bandwidth or voice channel equivalent circuits. Equipment for full-motion color now exists for compressing the signals into a digital data stream for operation on a T1 or 1.544-Mbps channel or the equivalent of 24 voice channels. Black-and-white full-motion systems

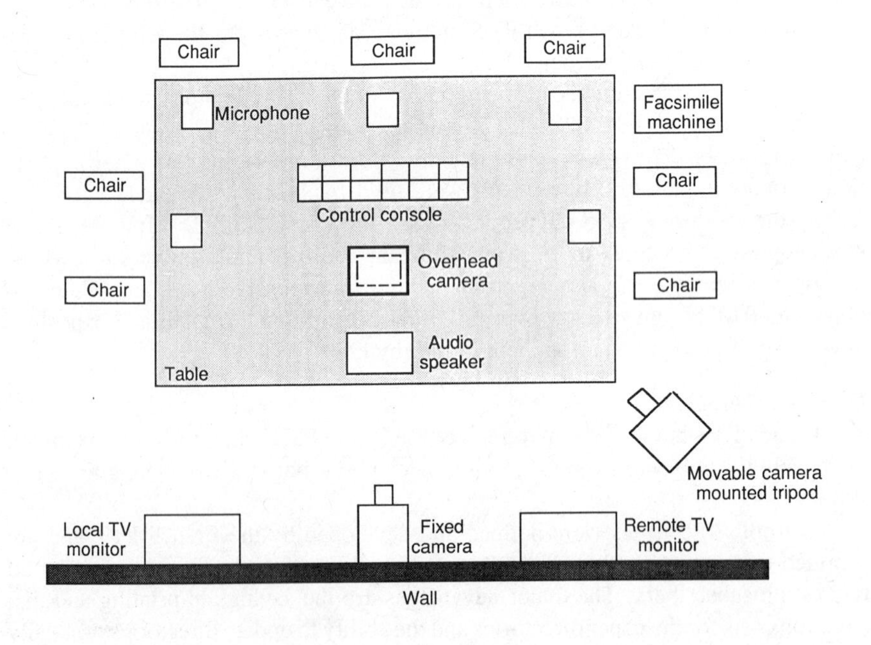

Figure 12.12. Video conferencing center.

are available for operation over a 56-kbps digital link. In comparison, black-and-white freeze-frame systems operate over 4.8- or 9.6-kbps voice-grade lines.

A facsimile machine can be used to support audio and video equipment. Facsimile provides a hard-copy output of documents, graphs, tables, and diagrams used in a meeting for later reference. A separate leased line or telephone voice-grade line is required.

An electronic blackboard, one in each studio, allows a speaker to write on one board and have an electronic image of that writing appear on a remote TV monitor. A blackboard system at the remote site can be used to add or delete elements from the received image, with the changes appearing on the monitors at both sites.

A video conference room layout is shown in Figure 12.12. The layout includes audio, video, and support equipment. Equipment costs start at $75,000 for an audio and video black-and-white system. Transmission costs and meeting room design and equipment are additional.

Video conferencing has been accepted slowly due to the cost of equipment, studio installation, and transmission costs. The ongoing transmission costs are the main deterrent for high-quality full-motion video conferencing. New digital techniques and ISDN will increase the use of video conferencing by dramatically reducing line costs.

13

SATELLITE COMMUNICATIONS

In this chapter we will

- Discuss satellite growth and system components
- Provide information on international satellite services
- Look at direct broadcasting satellites
- Discuss major regional and national satellite programs

Satellites have a number of unique advantages for solving communications needs. They are a growing sector of radio communications and offer the greatest potential for transmission capacity in voice, data, text, and video communications. Communications satellites launched in the past two decades are usually geostationary or appear "parked" above the equator in geosynchronous orbit. That is, the satellite appears stationary to a fixed ground position. To achieve a fixed position in space, satellites travel in the same direction as the earth with an angular velocity of revolution around the earth that is the same as the earth's angular rotational velocity. To keep up with the earth's fixed position, the satellite must be in a circular orbit at 22,300 miles (35,680 kilometers) above the equator and travel at 6900 miles (11,000 kilometers) per hour in the same direction as the earth's rotation.

The geostationary orbit has advantages for continuous broad coverage. A geosynchronous satellite is permanently in view; fixed antennas can be used, and because of their distance, the satellite is in the line of sight from over 40% of the earth's surface. Three satellites are able to cover most of the globe, except the polar regions.

Communication satellites act as relay stations in that an earth station transmits or beams radio signals to the satellite, which receives the signals, converts them to a lower frequency with additional power, and retransmits or beams back the signals to earth. On the ground, antenna stations or earth stations receive the signals and redistribute them through a frequency division multiplexer or time division multiplexer.

One advantage of satellite signals beamed back to earth is that numerous antennas can pick up the signals, convert them, and use the information. The area that can receive a satellite's signals is called an earth coverage beam and depending on design can encompass the continental United States, Western Europe, or the Atlantic Ocean region.

Satellites are exceptional for TV distribution, video transmission, and bulk data transfers for point-to-point or point-to-multipoint applications. For example, the *International Herald Tribune* is edited and published in Paris and then transmitted and printed simultaneously in London, Zurich, Hong Kong, Singapore, The Hague, Marseille, and Miami. Simultaneous transmission to these remote locations can be easily accomplished by INTELSAT's satellites, one covering the Atlantic Ocean region, another in the Pacific Ocean region, and a third in the Indian Ocean region. Another example is TV programming, where "live by satellite" can bring programs into a home from another continent shortly after they happen.

It takes time for the transmitted signals to travel up to the satellite and back down to earth. The time the signal takes to reach the satellite is the distance traveled, 22,300 miles, divided by the speed of light or 186,000 miles per second. On a two-way trip, traveling up and down varies from 240 to 279 milliseconds, depending on the location of the ground antenna system. These delays can be bothersome for telephone conversations, highly interactive computer to remote terminal sessions, or in some transaction processing applications.

Satellites are used to support four types of communications services:

- Trunk telephone circuits between common carriers or PTTs
- TV program exchanges to a CATV network controller for further distribution or studio to studio for rebroadcasting
- Direct broadcasting of TV programs
- Specialized business services such as high-speed data exchange, video conferencing, and bulk data transfer

Trunk telephone services require large earth stations with antenna diameters above 10 meters. These are expensive and must handle high traffic flows to be cost effective. For this reason, they are associated with international telephone gateway exchanges and are relatively few in number or about one or two in most developed countries.

TV program exchange services or *TV relay* requires smaller earth stations with antenna diameters below 10 meters. They are also expensive and must handle sufficient traffic flows to be cost effective. Program exchange is primarily used to feed cable television networks with TV programs. There are about 20–40 earth stations in most developed countries.

Direct broadcasting services require a small antenna with diameters below 1 meter. They are not expensive and are installed on customer premises. Their services will compete directly with cable television networks fed by the PTT's TV program exchange.

The specialized business services require smaller earth stations with antenna diameters of about 3 meters. These are less expensive, but the traffic flows are totally unpredictable and are shared by a small number of business users. There are two approaches: an earth station on the customer's premises or an earth station on the PTT's premises connected to the customer by the PTT's existing local distribution network. Specialized business services include telephone leased lines, high-speed data transmission, and video conferencing.

Trunk telephone traffic is easy to predict, but demand for TV programs and specialized business services is unpredictable. TV programs will be in competition with cable TV networks fed by low-power satellites and direct TV broadcasting from high-power satellites requiring small antennas. The volume and user locations for specialized business services such as video conferencing are unpredictable.

The growth of national satellite communications in the United States is the result of independent private carriers. As in most developed countries, the operation of the long distance terrestrial network is more profitable than local distribution, and the profit from long lines compensates for the loss of local loops. Satellite communications are now able to compete on the profitable long distance services. As a result, the tariffs of terrestrial services have been restructured and are more cost oriented. Satellite communications in the United States has grown as the bypass technology for the long distance terrestrial networks.

In most European countries, satellite communications are operated by the same PTTs that operate the terrestrial network. The PTT has the telecommunications monopoly, and competition is limited to the earth station equipment and not allowed for the provision of the supplier's network. Therefore, satellite communications will compete with terrestrial media, e.g., coaxial or fiber optic cables, on a strict cost basis. Satellite circuits are integrated into the national network where cost-justified. Satellites will be used to speed up the availability of ISDN for business users because they can provide end-to-end digital connections to any location covered by the satellite's beam.

Developing countries use satellites to bridge long distances and inaccessible areas at a cost independent of distance and geography. In many of the developing countries, particularly Africa, interconnecting roads between cities are nonexistent. Microwave towers required for repeater stations are difficult to install and maintain as their signals cross barren terrain or jungles. Therefore, for rapid development, satellites achieve a quicker area coverage.

One example is Algeria in Northern Africa, which recently established a national network in 1 year, while a similar microwave network could have taken 5–10 years. This type of justification is not found in Europe. Terrestrial networks are smaller in average length and include advanced services such as Teletex, Videotex, and Telefax.

13.1. SATELLITE SYSTEM COMPONENTS

A basic satellite communications system is shown in Figure 13.1. It consists of two earth stations and a satellite. One earth station transmits and receives signals from a satellite located 22,300 miles above the equator. The satellite detects the transmitted signal, amplifies it, changes its frequency or format, and retransmits it to earth station B.

A circuit is a transmission path from earth station A to the satellite to earth station B and back again. A channel is a path from one earth station to the satellite and back to the earth station. Satellite tariffs are normally charged by half circuits, that is, the transmission from one earth station to the satellite and return to the same earth station.

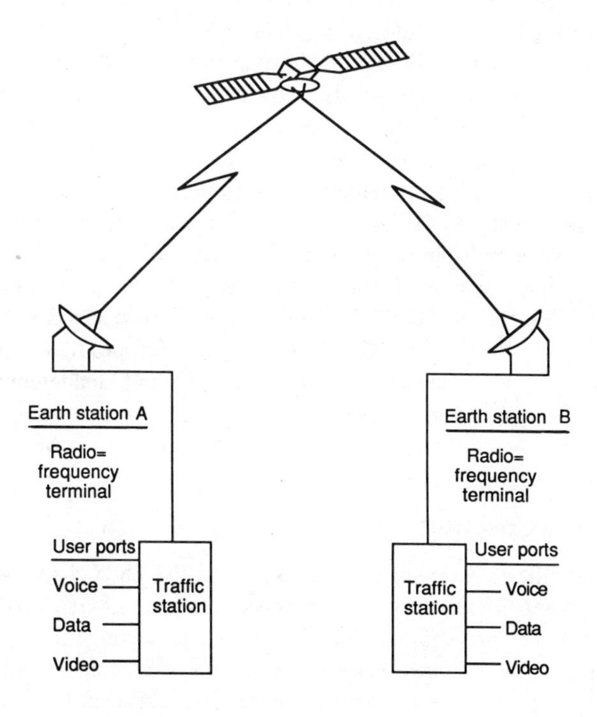

Figure 13.1. Major satellite system components.

13.1.1. Satellite Trends

As communications requirements for the user increase, follow-on generations of satellites are designed to provide higher capacity rather than putting more satellites in orbit. One of the main reasons for this is that each country has a limited amount of space in which to "park" satellites above the equator. The FCC now has a mandatory spacing of 2 degrees in the space allocated for the United States, down from 4 degrees in 1983. Closer satellite spacing allows more satellites in orbit, but additional precision is required by earth stations to differentiate between signals from other satellites. At 22,300 miles, a 2-degree spacing results in a satellite separation of about 778 miles. The increased capability is provided by increased power on the satellite, requiring increased launch weight. Increased power on the satellite means that smaller antennas are required on the ground. Smaller antennas result in lower cost and greater usage.

For example, the development of higher-frequency satellites operating in the 12- to 14-gigahertz (GHz) band allows small earth stations on the ground at frequencies that do not interfere with existing microwave terrestrial frequencies.

13.1.2. Multiple Spot Beams and Shaped Beams

Satellite antennas are directional and aim or point their signal to a small area of the earth. Some international satellites, e.g., INTELSAT, carry two earth-coverage antennas and two narrower-angle antennas to span an ocean and to illuminate specific areas. Domestic satellites are designed to cover a country such as Canada, India, Indonesia, or Japan. The reflectors on an antenna are shaped so that the signal contours on earth follow the shape of the country to be covered.

Beam shaping will continue to be developed to permit a close match of satellite antenna coverage pattern and the land areas to be illuminated. In addition, there is a trend toward the use of multiple spot beams for point-to-point and area coverage. One example of a multiple beam application is the French Telecom 1 Satellite, which is designed to cover France, its neighbors, and six French overseas departments. The major beam covers all of France and most of Western Europe. In addition, smaller spot beams cover the Caribbean islands of Martinique and Guadeloupe as well as French Guinea, Saint Pierre, and Miquelon in South America. Another beam picks up La Reunion and Mayotte, two small islands off the coast of southeast Africa.

13.1.3. Satellite Spectrum

All operating satellites use one or two frequency bands. The C band has an uplink (earth to satellite) of 6 GHz or 6 billion cycles per second and 4 GHz on the downlink (satellite to earth). The problem with the C band is that existing microwave systems on the earth in the United States, Europe, and Japan operate at 4 and 6 GHz. This presents a serious problem for satellites at these frequencies because at certain elevation angles, severe interference can occur. This difficulty can be eliminated if satellites operate at higher frequencies.

Another operating frequency band is the Ku band, with uplinks in the 12-GHz range and downlinks in the 14-GHz range. For higher-capacity satellite systems, there will be a continuing shift to the 12–14 GHz band in order to obtain additional transmission bandwidth and because of increased congestion in the geostationary area. New designs in the Ka band, e.g., ITALSAT, will work on an extremely high frequency with uplinks in the 20-GHz range and downlinks in the 30-GHz range.

Each operating band has its advantages and disadvantages. The C band is useful for wide area coverage, but signals in this band are relatively weak. The earth's atmosphere is transparent to signals in the C band, and signals can pass through heavy fog and rain without interference. However, the satellite's signals interfere with existing microwave networks which also operate in the 4–6 GHz range. Since microwave networks are predominant in large cities, C band satellite earth stations are installed in remote locations, miles from the center of cities. This results in additional expensive cabling or microwave systems to connect from the earth station to the user's site. Ku band properties are opposite to C band in that their signals are immune to microwave frequencies but are sensitive to atmospheric conditions such as heavy fog and thunderstorms. Depending on the satellite system, satellite communication links have error rates of less than one error in 10 million bits. However, under heavy rain, the error rate can deteriorate to one error in 500 bits.

13.1.4. Multiple Access

Each satellite has a receiver and transmitter pair called a transponder. Transponders affect the amount of information or bandwidth (capacity) that can be shared by multiple earth stations. Transponders receive the signals, amplify them, convert them to another frequency, and retransmit them back to earth. For example, one of the 12 Weststar transponders has a bandwidth of 36 MHz. The capacity of this bandwidth allows either one-color TV program with sound, 1200 voice-grade channels, or a data rate of 50 Mbps.

To improve and make full use of this capacity, multiple access is used to interconnect simultaneous communications links from a number of earth stations through one communications satellite. The advantage of multiple access is that a dedicated transponder is not required, but many users can use the transponder on a sharing basis. There are three methods of sharing a transponder by multiple earth stations:

1. Frequency division multiple access (FDMA)
2. Time division multiple access (TDMA)
3. Code division multiple access (CDMA)

In frequency division multiple access (FDMA), each earth station is assigned a different portion of the transponder's frequency band. Each user has a specific uplink and downlink carrier frequency within the bandwidth of the transponder. FDMA requires a reservation controller so that requests for channel capacity can be made through dynamic assignment of the transmitting frequency. The satellite transponder is sensitive to all the assigned frequencies. By using this method, each earth station is assigned a different frequency and tunes into its frequency much like an FM radio receiver.

In time division multiple access (TDMA), each transmitter is assigned a time slot, and the satellite transponder handles incoming signals in serial. To operate efficiently, TDMA requires accurate timing for each station in the network to compensate for different distances from the satellite. In addition, a dynamic reservation system is required to allocate a particular time slot to handle each earth station's specific requirements. TDMA is mainly used for digital transmission since blocks or bursts of data can be transmitted at specific times and synchronized with other stations sending out bursts of data. In TDMA, the entire capacity or bandwidth of a transponder is used by any number of earth stations, each from multiple customers, for a short period of time. This period can be a few microseconds if all earth stations are synchronized and controlled.

Messages are coded and transmitted simultaneously in code division multiple access (CDMA). At each remote earth station, input data is combined with a high-speed cyclic sequence to produce a broadband or wide frequency spectrum output. Since a special code is used for each message, the satellite transponder can decipher and retransmit the coded message to the proper address. At the central receiving station, the process is reversed to obtain the original data stream. This form of multiple access is primarily used by the military. As communications security becomes more critical, this type of multiple access could be used.

13.1.5. Advantages of Satellites

In addition to voice and low-speed data, a wide range of communications services can be provided by satellite. Many of these services would not be possible without the wide bandwidth, reduced transmission costs, and wide geographical coverage provided by satellites, for example the following:

1. Distance insensitive: Earth stations can be installed throughout the United States or a region consisting of several countries to receive the same information simultaneously.
2. Direct terminal-to-terminal connections do not require access to the telephone network. Ku band and direct broadcasting satellites allow small dish antennas (1–3 meters in diameter) that can be mounted on customer premises.
3. TDMA allows interactive two-way satellite networks consisting of several thousand terminals or PCs linked to a central computer by small earth stations. Many customers can share the same facilities.
4. Multibeam design allows satellite coverage for specifically shaped areas or country contours, e.g., northern Africa and the Middle East (ARABSAT), France and its overseas departments, Indonesia and its 13,000 islands, Japan's elongated shape, and the broad area of the United States.

13.1.6. Portable Earth Stations

One of the most significant products introduced to enhance satellite usage is the portable earth station.

Telesystems, a division of COMSAT, has developed the TCS-9000 portable earth station that can be carried in two suitcase-type containers by one person. The TCS-9000 can provide long distance telephone service, telex, and data communications to remote locations. It is designed for simplicity and can be set up in 15 minutes. There are a number of custom options including a 56-kbps rate for high-speed data, a telephone expansion unit for up to six remote telephones, and a PC interface.

13.2. INTELSAT

INTELSAT or the International Telecommunications Satellites Organization has over 110 member countries that own and operate satellites used by most of the world for international communications. INTELSAT currently has a network of 16 satellites serving 170 countries and territories and carries two-thirds of the world's international telephone calls and virtually all international television. Some countries use INTELSAT's satellites for communications within their own borders as well.

The average phone call via INTELSAT is first switched through the PTT's local and central trunk exchanges and next through their international exchange where the equipment automatically selects the overseas circuit that will carry the phone call. After the satellite circuit is selected, the call is routed to the appropriate earth station.

At the remote end, the receiving earth station is capable of receiving all the signals being transmitted by the satellite but selects only those signals that apply to its country. Information may include telephone calls, telex messages, telegrams, and television programs.

INTELSAT provides international telecommunications capacity to country PTTs, who, in turn, sell this capacity in the form of telephone calls, circuits, telegram facilities, and satellite television channels. In addition, a number of countries use INTELSAT for communications within their borders as well as for international traffic. These countries use spare capacity that is leased to them on INTELSAT satellites for their domestic communications. INTELSAT provides satellite capacity for national domestic services in the following countries: Algeria, Australia, Brazil, Chile, Colombia, Denmark (Greenland), France, India, Mexico, Niger, Nigeria, Norway, Oman, Peru, Saudi Arabia, Spain, Sudan, and Zaire. Gabon was one of the first four nations to purchase a transponder for domestic services to establish a national network for telephone, TV, and data. Other countries planning to use INTELSAT satellites for their domestic requirements are Argentina, Libya, Malaysia, Morocco, Portugal, Thailand, and Venezuela.

INTELSAT is also working on its next series of satellites, INTELSAT VI. When fully deployed, the VI Series will have a capacity of 100,000 voice circuits and 3 TV channels or 3 billion bits of information per second. INTELSAT VI will be launched by the European Ariane 4 vehicle. 59.25% of Ariane is owned by French shareholders and 19.6% by German (FRG) shareholders. Other shareholders, providing less than 5% of the company's capital, are Belgium, Denmark, Italy, Ireland, the Netherlands, Switzerland, Spain, Sweden, and the United Kingdom.

One of INTELSAT's biggest competitors is underseas fiber optic cables. The TAT-8 link from the United States to France and the United Kingdom will have 40,000 voice channels. The total cost the IRCs will pay to lease a voice grade circuit will be $2400 per year for fiber versus $7320 for a satellite link. See the next section of this chapter for new lower-cost alternatives. When completed, TAT-8's 40,000 channels plus all existing TAT series cables of 11,200 voice channels will total 51,200. The PTAT-1 and PTAT-2 private underseas fiber optic cables will add 24,000 voice channels. Table 13.1 lists the total INTELSAT channel distribution by programs. After the INTELSAT VI program is fully deployed, the Atlantic region will have a significant number of satellite and cable circuits. As a result of the additional capacity, the U.S. carriers and European PTTs will be forced to reduce prices.

As shown in Table 13.2, INTELSAT has a number of standards for earth stations based on the type of traffic to be carried over the satellite network.

13.3. INTELSAT BUSINESS SERVICES

INTELSAT has developed an integrated service to provide a full range of business applications for both domestic and international coverage and connectivity. INTELSAT Business Services (IBS) uses the INTELSAT International Satellite Network. The applications cover digital voice and data, audio and video conferencing, facsimile, and remote distribution.

IBS leased services are available for full-time (24 hours a day, 7 days a week), part time (7-day fixed data time slot), or occasional use (reserved time). Low to medium

TABLE 13.1. INTELSAT Channel Distribution

Program	Year	Voice Channels		TV Channel
INTELSAT I	1965	240	or	1
INTELSAT II	1967	240	or	1
INTELSAT III	1968	1,500		4
INTELSAT IV	1971	4,000		12
INTELSAT IVa	1976	6,000		20
INTELSAT V	1980	12,000		2
INTELSAT VI	1986	30,000		3
(5 satellites)	1987+	500,000+		100+
Total channels		524,000+		Over 143

Source: INTELSAT, Washington, D.C.

speed capacity with data rates of 64–768 kbps are available for data transfer, facsimile, and digital voice. High-speed rates from 1.544 to 8.448 Mbps are available for full-motion color video conferencing, bulk data transfer, and multiplexed services.

Global coverage is provided by global beams in the C band range and spot beams in the Ku band range. Earth station standards are A, B, F1, F2, and F3 for C band and C, E1, E2, and E3 for Ku band transmission and reception.

Several international record carriers (IRCs) and IBS carriers lease transponder space on a monthly basis and offer the IBS services packaged for unique customer requirements between the United States and Europe. Plans exist to add connectivity to Central and South America, the Middle East, Africa, and the Asia/Pacific area. Japan's IRC, Kokusai Denshin Denwa (KDD), offers IBS high-speed digital leased circuits for international communications and video conferencing to Hong Kong, France, the United Kingdom, and the United States.

TABLE 13.2. INTELSAT Earth Station Standards

Earth Station Standard	Antenna Size, Meters	Type of Service	Frequency Band, GHz
A (existing)	30–32	International voice, data, TV, INTELSAT Business Services (IBS)	6/4
A (revised)	15–17	International voice, data, TV, IBS	6/4
B	10–13	International voice, data, TV, IBS	6/4
C (existing)	15–18	International voice, data, TV, IBS	14/11
C (revised)	11–13	International voice, data, TV, IBS	14/11
D1	5	VISTA (international or domestic) Telephony to remote islands	6/4
D2	11	VISTA (international or domestic) Telephony to remote islands	6/4
E1	3.5–4.5	IBS (Ku band)	14/11 & 14/12
E2	5.5	IBS (Ku band)	14/11 & 14/12
E3	8–10	International voice, data, TV, IBS (Ku band)	14/11 & 14/12
F1	4.5–5	IBS (C band)	6/4
F2	7.5–8	IBS (C band)	6/4
F3	9–10	International voice, data, TV, IBS (C band)	6/4

Source: INTELSAT, Washington, D.C.

The following examples of IBS offerings include joint venture agreements, separate offerings, and special licensed IBS carriers:

1. Comsat and TRT Communications have an agreement to establish, own, and operate IBS earth stations in the New York City and Washington, DC areas. The first earth station on Staten Island will transmit and receive digital transmissions (56 kbps to 2.048 Mbps) to and from Europe, North and South America, and Africa. TRT's international service is available from the United States to Belgium, Canada, Denmark, France, Germany (FRG), Ireland, Italy, Mexico, the Netherlands, Norway, Sweden, Switzerland, and the United Kingdom.
2. MCI International has established a network of four earth stations in Washington, DC, New York, Chicago, and Detroit for IBS to Europe and Latin America. Although MCI introduced its service via the Staten Island Teleport Complex to Europe, it will operate its own earth stations in Manhattan and Washington, DC, followed by service in Chicago and Detroit.
3. Overseas Telecommunications Inc. of Alexandria, Virginia, was the first IBS carrier to obtain an operating agreement with Mexico's Secretaria de Communicaciones y Transportes for service to Mexico for U.S. multinational corporations. A 1.544-kbps digital circuit operates between Mexico City and OTI's earth station in Detroit for both voice and data applications. An INTELSAT F2 IBS earth station has been installed in Mexico by OTI on Chrysler Corporation's premises.
4. RCA Global (a unit of GE) and International Record Carrier signed a joint agreement to operate an earth station to offer IBS total digital, transatlantic, and transpacific service. They have correspondence agreements with both British Telecom International and Mercury Communications in the United Kingdom. In addition, International Record Carrier has correspondence agreements with the Federal Republic of Germany and Switzerland.
5. From its headquarters in New York City, FTC Communications, a Pacific Telecom Company, is offering IBS, a satellite service to Europe. The FTC offering includes transmission speeds of 9.6 to 2.048 Mbps with end-to-end digital service for voice, facsimile, data transmission, and video conferencing.

The FTC community earth station specifications conform to INTELSAT E3 antenna standards for Ku band operation. International routing across the Atlantic Ocean is via the INTELSAT Satellite System.

The operating points include satellite paths from the United States to Belgium, France, Germany (FRG), Ireland, the Netherlands, Switzerland, and the United Kingdom. Future expansion into the Pacific area will cover Australia, Japan, and Singapore.

The service will offer dedicated digital leased channels. The monthly charges for the U.S. share or half channel are as follows:

9.6 kbps	$ 2,500
56 kbps	5,000
64 kbps	5,000
128 kbps	9,500
256 kbps	18,000
512 kbps	24,000
768 kbps	28,000
1.544 Mbps	45,000
2.048 Mbps	55,000

Charges do not include hardware, local line extensions, taxes, and installation charges.

Local FTC extensions are via the Manhattan Coaxial Cable System, the Teleport Fiber Optic Network, Nynex's DDS or T1, or copper loops. Figure 13.2 illustrates some typical arrangements for local network extensions. The number of carriers offering IBS services has caused a reduction in the monthly charge for a T1 circuit. The U.S. side is a free-for-all, whereas the European side is restrictive. That is, only the PTTs are allowed to offer earth stations and charge their normal tariffs. There is no competition on the European half circuit except in the United Kingdom, which has two competing carriers, British Telecom International and Mercury Communications.

13.4. U.S. INTERNATIONAL PRIVATE SATELLITE PROVIDERS

The FCC, in its July 25, 1985, International Satellite Decision, granted permission to five applicants to compete with INTELSAT's IBS program as private satellite providers. They are not common carriers. In effect, the companies are private providers of satellite capacity or providers of channel capacity in the sky. Each U.S. customer purchases a direct earth station for transmitting and receiving information. The situation is different in Europe because the earth stations are owned by the PTTs. In addition, the PTTs are reluctant to set up another satellite service that competes with INTELSAT and EUTELSAT since they all own a percentage of each satellite network.

Each U.S. private satellite provider was assigned an orbital position above the equator to allow communications to the Americas and/or Europe. The satellites will be designed, built, launched, and operated in the Ku band range, allowing antennas 2 to 5 meters in diameter. Each satellite carrier must obtain landing rights with at least one foreign country as a step toward final authorization. The proposed system must be coordinated under the INTELSAT consultation process under Article 14(d) of the INTELSAT agreement.

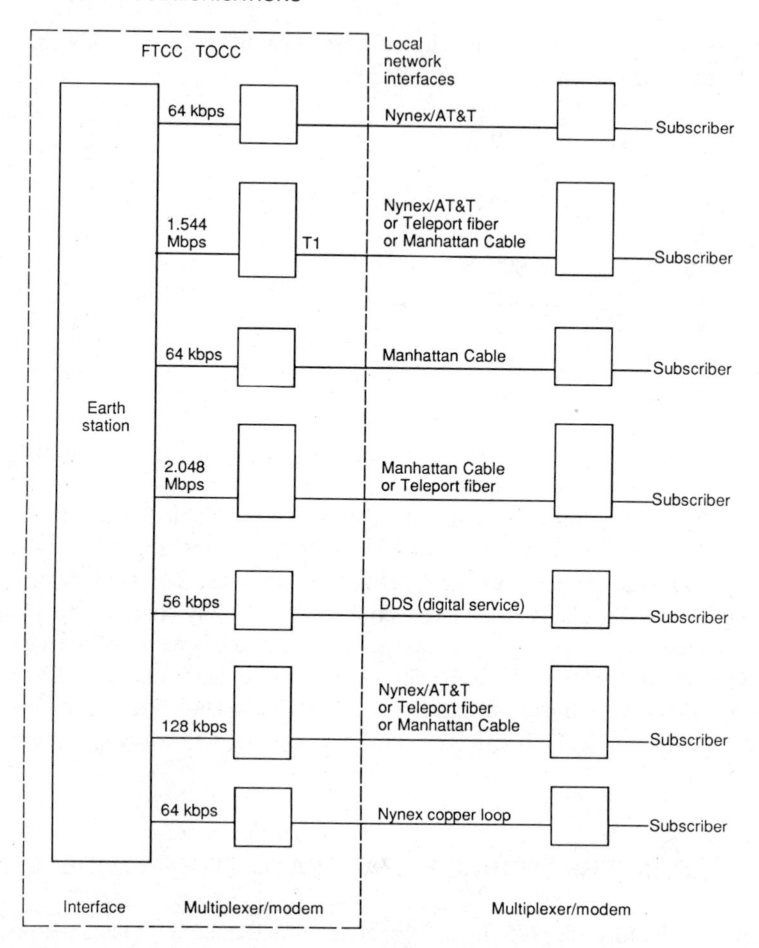

Figure 13.2. Local network extensions.

The U.S. State Department then begins the process of coordinating with INTELSAT to determine if the satellite can be technically coordinated with the INTELSAT system and to ensure that the new system will not cause economic harm to the international cooperative. Of the five original private satellite providers that have an FCC agreement, the following three are currently in business:

1. *Financial Satellite Corp.*, 1250 Connecticut Ave NW, Washington, DC 20036: Financial Satellite Corporation's FINANSAT will be used for private voice and data transmissions for banking and brokerage firms. Brokerage firms will use the service for up-to-the-minute information offered by direct satellite access to exchanges and personal computer networks.
2. *International Satellite, Inc. (ISI)*, 1331 Pennsylvania Avenue NW, Washington, DC 20004: International Satellite Inc. (ISI) was created to develop

new satellite services in the North Atlantic region. There are three major partners in ISI. TRT Communications, Inc. has the principal interest in ISI. TRT provides U.S. domestic and international record and voice communications services and is also a partner in the TAT-8 undersea fiber optic cable consortium and other international communications projects. Satellite Syndicated Systems, Inc. (SSS), a major partner in ISI, provides common carrier distribution of video signals in the United States and to selected points in the Caribbean and Canada. Kansas City Southern Industries, Inc. (KCSI), also a major partner, is a diversified corporation with investments in transportation, data processing, insurance, and telecommunications. KCSI, through subsidiary corporations, operates an extensive private microwave network and has interests in a multistate fiber optic system and cellular radio programs.

Customized communications services to be carried on the ISI system will include the following:

- Video distribution and communications (including TV programming and video conferencing)
- Digital communications (including publishing and energy exploration)
- Audio communications
- "Message" communications such as facsimile and electronic mail and Videotex

3. *Pan American Satellite Corp.*, 460 West 42nd Street, New York, NY 10036: At mid-1986, PanAmSat negotiated successfully with Peru to supply its satellite system for video and high-speed data services. Peru's television networks and multinational corporations are expected to sign on as customers. Other discussions are in different stages with potential correspondents in other South American, Central American, and European countries for private customers such as TV programs.

13.5. INMARSAT

International Maritime Satellite Organization (INMARSAT), based in London, provides satellite and ground support facilities to improve maritime communications for distress and safety services, ship management, public correspondence, and position fixing. Over 48 countries representing most of the world's major maritime nations have signed the INMARSAT's convention or operating agreement. The signatories are generally the PTT administrations.

INMARSAT provides communication services to shipping and offshore industries. INMARSAT and INTELSAT signed an agreement whereby INMARSAT will lease residual space capacity from INTELSAT to provide temporary land-based, nonmaritime, fixed satellite services. This agreement allows government and commercial firms with a need for satellite communications in remote areas of the world to use equipment such as the TCS-9000 Portable Earth Station. Setting up small earth stations is conditioned on there not being adequate terrestrial or INTELSAT satellite facilities in the remote location.

Although the TCS-9000's size and transportability is a breakthrough, there is a reluctance by national PTTs to provide licenses for new private international satellite carriers to bypass local facilities and avoid tariffs by connecting directly with the INMARSAT-INTELSAT systems. In Europe, there is strong resistance to approving portable earth stations except for short-term, emergency applications.

The INMARSAT systems consist of satellites stationed over the Pacific, Atlantic, and Indian Oceans with ground support stations to provide full communications coverage to the major offshore drilling sites and shipping routes of the world. See Table 13.3 for the INMARSAT primary and backup satellite systems.

The three main elements of the INMARSAT system are the following:

- Satellites and ground support facilities
- Coast earth stations (CES) that provide an interface between the space segment and the national and international fixed PTT networks
- Ship earth stations (SES), satellite communications terminals that are purchased or leased by individual ship owners or operators from manufacturers and suppliers of radio equipment

Each of the three ocean regions has its own network coordination station (NCS) operated by Comsat of the United States and Kokusai Denshin Denwa Co. Ltd (KDD) of Japan. The Atlantic region NCS, operated by Comsat, is at Southbury, Connecticut. The Indian and Pacific regions are at Yamaguchi and Ibaraki (Japan), respectively. Each NCS regulates the distribution of available telephone channels to ship and coast earth stations.

INMARSAT Service

Over 5000 vessels worldwide are equipped for satellite communications via the INMARSAT system. They include yachts, fishing boats, container ships, passenger liners, and oil rigs. Each vessel has its own satellite terminal that allows instantaneous access to telecommunication networks 24 hours a day in any weather.

TABLE 13.3. Satellite Location of INMARSAT Systems

Area	Prime Satellite	Backup Satellite
Atlantic Ocean	Marecs B2	INTELSAT-MCS-B
Indian Ocean	INTELSAT-MCS-A	INTELSAT-MCS-C
Pacific Ocean	INTELSAT-MCS-D	Marecs-A

Some of the typical ship-to-shore applications are the following:

Computer-to-computer data links
Engine room monitoring
Weather observations, forecasts, and warnings
Slow-scan TV
Transmission of ships' manifests and accounts
Vessel position reporting
Fleet and national group calls
Berthing and scheduling information
Plans, schematics, charts, and photographs via facsimile
Navigation warnings
Medical advice and assistance

Each vessel or oil rig has a random satellite antenna with a covering to shield the antenna from the elements. The antenna is about 1 meter in diameter and automatically rotates and remains pointed at the satellite regardless of the vessel's direction. The receiver and transmitter are used for telephone, telex, facsimile, data communications, and for instant distress alerts. Below-deck equipment includes a telephone handset, teleprinter, transmitter, and receiver. A list of INMARSAT services is provided in Table 13.4.

TABLE 13.4. INMARSAT Services

Service	Support
Telephone	Ship-to-shore calls are automatic and can be routed to most countries of the world.
Facsimile	Group 2 and 3 Fax is supported.
Telex and cablegram	Service available to telex subscribers throughout the world. Telex channels are assigned to ships by individual coast earth stations.
Marine radio	Completely automatic, including unattended teleprinter operation. Information services include medical and maritime assistance, navigational warnings, and weather forecasts.
Data transfer	Medium speed up to 2400 bps. High-speed (56-kbps) ship-to-shore service is available from the U.S. and the U.K.
Distress access facility	Immediate access to rescue coordination center.
Slow-scan TV	Allows TV pictures over telephone facilities.
Compressed TV	Full motion, full color for video conferencing.

Source: INMARSAT, London, United Kingdom.

13.6. DIRECT BROADCASTING SATELLITES

Direct broadcasting satellites (DBSs) are set up primarily for television program reception and require small dish antenna with diameters less than 1 meter. They are not expensive and can be installed on customer premises or near a home. Some of the major European DBS satellite programs will be discussed.

France/Germany (FRG)

A joint German/French DBS program consisting of two operational satellites, TV-SAT (Germany) and TDF-1 (France), is controlled by a steering committee of four members. The German representatives are from the Federal Ministry for Research and Technology and the Federal Ministry for Post and Telecommunications. The French representatives are from the Centre National d'Etudes Spatiales (CNES) and Telediffusion de France (TDF).

The satellite transmitter will have 230 watts per channel or about 10 times more power than communications satellites. This higher power will allow direct reception by small-dish-type antennas below 1 meter.

Ireland

Two Irish DBS satellites will be built by Hughes Communications for Atlantic Satellites. Hughes Communications, part of General Motors Corp., owns 80% of Atlantic Satellite. Each satellite will have 5 high-powered broadcasting channels and 24 low-powered channels for transatlantic communications or cable TV relay in Western Europe.

Luxembourg

The Societe Europeenne des Satellites (SES) and Arianespace have signed a launch contract for Luxembourg's first television satellite, to be launched by the Ariane 4 launch vehicle in May 1987.

Luxembourg's Astra Vision program will offer cable TV systems in most of Western Europe. Astra Vision will lease 16 channels of its satellite to program providers. Eight channels will be pay TV channels. Although the company would like to be involved in direct broadcasting to viewers' households, the International Telecommunications Union (ITU) regulations bar Astra Vision from doing so because they use telecommunications frequencies.

In addition to the ITU, Astra Vision has had pressure from neighboring country administrators who have tried to block the project. The latter believe that programs and commercials beamed into their countries will destroy their national heritage and present contrary political programming.

United Kingdom

SKY Channel, operated by Satellite Television plc, London, provides programming services over EUTELSAT's ECS-1 satellite. SKY Channel has a contract with British Telecom to use one of its transponders to broadcast radio and TV programs free to European cable TV operators in 15 Western European countries. Over 941 cable TV operators pick off the signals and distribute them to over 6.5 million viewers. SKY Channel provides 17 hours of nonstop radio and TV broadcasting in English per day.

British Satellite Broadcasting—a consortium of five companies—will provide direct broadcasting of TV programs to individual households in the United Kingdom. The United Kingdom government awarded a 15-year franchise for the provision of direct broadcast services with 4 TV channels (one pay TV channel).

Eastern Europe

Eastern European countries are organized into a union called Organisation Internationale de Radiodiffusion et Television (OIRT) with offices in Prague, Czechoslovakia for distributing news. They are delivering television programs, called Intervision, to member nations that include Bulgaria, Czechoslovakia, the German Democratic Republic, Hungary, Poland, Romania, the USSR, and Cuba. The members within the Intervision news exchange are also members of Intersputnik. News items from western broadcasting organizations are transmitted by INTELSAT from London to Moscow and relayed from Moscow, via Intersputnik, to Intervision member countries.

13.7. REGIONAL SATELLITES

Satellites that cover a region or number of countries are set up by governments to create a critical mass of technology, provide area coverage for compatible and common interests, and add specific TV programming. They also provide services to remote countries within the region and reduce their dependency on INTELSAT for all satellite requirements.

Three major regional programs are ARABSAT, EUTELSAT, and Tele-X.

13.7.1. ARABSAT

The Arab Satellite Communications Organization (ARABSAT) is an organization of the League of Arab States, headquartered in Riyadh, Saudi Arabia. ARABSAT has grown into an important telecommunications medium for cultural and socioeconomic activity in the Arab world. Twenty-two member nations from northern Africa through the Middle East are participating in direct TV reception, telephone service, and business communications. ARABSAT is responsible for procuring satellites and launch vehicles and satellite operations. ARABSAT provides data communications among Tunisia, Kuwait, Jordan, and Djibuti. Many member nations are leasing spare transponder capacity for their domestic communications networks.

There are three satellites in the program. The two operational satellites are ARABSAT-1A and ARABSAT-1B (backup), and the third is a "ground" spare. The ARABSAT communications payload offers 500 MHz in the C Band with an uplink from 5.925 to 6.425 GHz and a downlink from 3.7 to 4.2 GHz.

ARABSAT-1A supports up to 8000 voice channels or equivalent telex or data with 7 transponders, while ARABSAT-1B supports TV and radio channels with 7 transponders. Each satellite has 13 transponders.

Three main types of earth stations were developed for ARABSAT:

- Stations with 11-meter diameter antennas for telephone service (trunk routes) in large metropolitan areas
- Stations with 11-meter-diameter antennas for TV reception in large metropolitan areas
- Stations with 2.5-meter-diameter antennas for TV reception in remote areas

13.7.2. EUTELSAT

The international European convention conferred intergovernmental status on the European Telecommunications Satellites organization (EUTELSAT) on September 1, 1985. There are 26 member states—all the countries of Western Europe plus Yugoslavia. EUTELSAT is responsible for the design, construction, and operation of a communications satellite system for international public services provided by the PTT administrations of the member states. EUTELSAT was established in 1977 as a provisional European organization known as Interim EUTELSAT. It had the responsibility of providing two European regional satellites for 20 PTTs, all members of CEPT, to provide TV services, telephone, data transmission, and transponder leasing for cable TV. The EUTELSAT system also covers Eurovision in collaboration with the European Broadcasting Union (EBU) and international business links under the Satellite Multiservice System (SMS).

Financial participation of each member country was based on that member's initial share. With the new convention, the financial contribution was fixed according to the usage requirements of the country. France and the United Kingdom are the largest shareholders with 16.4% each. Other significant shareholders are Italy (11.4%), the Scandinavian countries (11.2%), Benelux (10.4%), and the Federal Republic of Germany (10.8%).

Most of the PTTs plan to make the EUTELSAT capacity available to customers within their nation, primarily for distribution of television programs to cable networks.

ECS-1 is the first communications satellite of EUTELSAT. Under agreement between EUTELSAT and the European Space Agency (ESA), ESA will provide the space segment and control of five satellites for a period of 10 years.

The ECS-1 communications payload offers a full 500-MHz bandwidth with an uplink from 14 to 14.5 GHz and downlink from 10.95 to 11.2 GHz and 11.45 to 11.7 GHz. ECS-1 is based on digital technology for telephone and data services and will use time division multiple access (TDMA). Seven transponders are allocated to telephone service to carry up to 12,000 simultaneous calls per satellite between transit exchanges separated by more than 800 kilometers.

Two transponders will support the TV broadcasting service, allowing two TV channels per beam. The Atlantic spot beam will cover the Azores, Canary Islands, Portugal, and Spain. The western spot beam will cover Iceland, Ireland, the United Kingdom, Norway, Sweden, Benelux, France, northern Italy, Germany, and Austria, and the Eastern spot beam will cover Finland, Italy, Yugoslavia, Greece, Cyprus, and Turkey.

ECS-2 and follow-on satellites will be used for a business service called the Satellite Multiservice System (SMS) that will offer data rates ranging from 2.4 kbps to 2 Mbps. SMS will offer a wide variety of services including teleconferencing, file transfer, remote printing, high-speed facsimile, Teletex, and video services. Two transponders will be allocated to SMS.

ESA will provide EUTELSAT with the fourth satellite of the ECS generation. The fourth in-orbit satellite will serve as a backup and will replace ECS-3 in 1987.

13.7.3. Tele-X

Tele-X is a joint project financed by the Swedish, Norwegian, and Finnish governments. The investment shares are 82% for Sweden, 15% for Norway, and 3% for Finland. Tele-X will provide two types of services: direct broadcasting of TV and radio programs and business services for point-to-point and point-to-multipoint communications.

The operation of Tele-X will be the responsibility of the Nordic Telecommunications Satellite Corporation (NOTELSAT), a company jointly owned by the Swedish and Norwegian PTTs. Both PTTs will operate the earth stations for business services after a successful launch in 1987. The public broadcasting administrations in each country will jointly administer two TV channels.

Business services will provide the user access to switched digital communications links at the following rates:

Application	User Data Rates
Data communications	64 kbps and 2 Mbps
Video conferencing	2 Mbps
Tele-education (one-way video conference)	64 kbps and 2 Mbps
Satellite news	34 Mpbs

The Tele-X communications payload consists of two TV channels and two data and video channels. Two types of services will be available: Direct broadcasting of TV programs and high-speed transmission of data and video from 64 kbps to 140 Mbps.

The data and video transponder will handle one 40-MHz data channel and one 86-MHz data and/or video channel. The 40-MHz transponder will handle 500 channels at 64 kbps or 20 channels at 2 Mbps, while the 86-MHz transponder will handle speeds in the range 2–140 Mbps. The same receiving and transmitting antenna can be used for either application.

13.8. NATIONAL SATELLITE PROGRAMS

Many of the larger nations, in terms of population, have launched satellites for TV broadcasting and business services and, in some cases, to provide initial telecommunications to remote undeveloped regions. Satellites provide additional capacity and complement existing PTT networks.

Except for private communications satellite systems in the United States, satellites are owned by the PTT administration or national broadcasting administrations. This affects the acquisition of earth station equipment. A business customer does not purchase satellite equipment but rents it from the PTT. The following is a scenario that could apply when a business rents or purchases a transponder for international transatlantic traffic between the United States and Europe. On the U.S. side, the earth station equipment can be purchased, rented, or leased. Europe is a different situation. The PTT cannot be bypassed. It would not be unusual for the customer to purchase earth station equipment for its own use and be required to donate it to the PTT, who in turn would rent the same equipment back to the customer.

13.8.1. Australia: AUSSAT Program

AUSSAT, the Australian national communication satellite system, consists of three in-orbit satellites and a network of earth stations. The project is managed by AUSSAT Proprietary Ltd. AUSSAT-1 was launched aboard Space Shuttle Discovery in August 1985 and is now fully operational in its geostationary orbit. AUSSAT-2 was launched by the Space Shuttle Atlantis in late 1985. The Ariane missile launched AUSSAT's third satellite in 1986.

AUSSAT covers Australia and Papua New Guinea. The Papua New Guinea beam also includes New Britain and New Ireland. Australia is covered in its entirety by two identical national Australian beams. AUSSAT Proprietary Ltd. has established the coverage areas to ensure service to sparsely populated regions as well as urban users. Individual spot beams provide service to western Australia, central Australia, northeast Australia, and southeast Australia. All satellites will operate in the Ku band range and offer DBS television, voice, data, and video.

There are several classes and sizes of earth stations. The direct-to-home service will require an antenna diameter of less than 1.5 meters. Two-way voice telecommunications are also available to rural portions of Australia.

AUSSAT-3, launched in 1986, has one 30-watt and two 12-watt transponders available for lease to 20 Pacific islands. Services offered are television, telephone, and point-to-multipoint distribution. The 14/12 GHz Ku band satellite's 12-watt transponders will handle 600 voice circuits, while the 30-watt transponder will handle television. AUSSAT-3 will use 4.5-meter antennas and an 8-meter earth station for TV. The projected voice circuit cost per year will be about $3300 each.

13.8.2. Brazil: BRASILSAT Program

The PTT activities in Brazil are almost totally controlled by the Ministry of Telecommunications and a holding company called Telebras, which has 26 state-owned subsidiaries. One of them is EMBRATEL, which is responsible for interstate and international communications and the satellite network.

The Brazilian Satellite Telecommunication System (BRASILSAT) invested a total of $210 million. Brazil is using three transponders on INTELSAT at an annual lease cost of $6 million. In comparison, an annual investment of $16 million will provide 24 transponders in each of the two satellites—already in space.

The 48 transponders will have a capacity of 48 TV channels or 24,000 voice circuits. Data transfer will be used for video conferencing, facsimile, data applications, and simultaneous printing of newspapers and magazines.

13.8.3. Canada: Anik Program

Telesat Canada, a member of Telecom Canada, is the telecommunications carrier for domestic satellite capacity in Canada. Telesat is owned by the government of Canada (50%), Bell Canada (25%), other Telecom Canada telcos (16%), and other common carriers (9%).

Telesat operates the Anik Series of satellites supplying telecommunications services from coast to coast. There are over 125 major earth stations across Canada. The satellite telecommunications network consists of the Anik satellite series. Four Anik satellites are currently in service:

Anik B: operates in the 6/4-GHz and 14/12-GHz bands.
Anik C2: 16 transponders operate in the 14/12-GHz band.
Anik C3: 16 transponders operate in the 14/12-GHz band.
Anik D1: 24 transponders operate in the 6/4-GHz band.

Anik D2, a 24-transponder 6/4-GHz satellite, launched in November 1984, is currently parked in a 2-year storage orbit. Anik C1, a 16-transponder 14/12-GHz satellite, was launched in April 1985 and is in a storage orbit as a spare for C2 and C3. Anik E, the fifth series of satellites, should be operational in 1990.

13.8.4. France: Telecom 1 Program

The French Telecom 1 satellites are medium-sized satellites for PTT use and defense needs. Telecom-1A and -1B were the first two communications satellites launched by Ariane. A third satellite is kept as backup. The Telecom-1 program will perform three main missions in three different frequency ranges:

1. *14/12 GHz:* Dedicated to business telecommunications traffic; intracompany links for video conferencing, computer connection, electronic mail, and telephone
2. *6/4 GHz:* Domestic communications service from France to French territories; telephone links and TV program distribution to La Reunion Island, Mayotte, Saint Pierre and Miquelon, Guadeloupe, Martinique, and French Guiana
3. *8/7 GHz:* Classified military service for the Department of Defense; will provide the French Army or Navy with voice channels for fixed or moving points.

An agreement was signed between the French and German PTTs for the joint use of Telecom-1. Germany will use transponder capacity for its own needs. Another agreement was signed between the French PTT and the EUTELSAT organization for capacity to cover Western Europe.

13.8.5. Germany (FRG): DFS Kopernikus Program

In 1987, the Deutsche Bundespost will install and operate a domestic communications satellite system called Deutsche Fernmelde Satellit Kopernikus or DFS Kopernikus. The overall system will consist of three satellites (two in orbit and one ground spare) and 34 earth stations. DFS will operate in the Ku band (10 transponders) and Ka band (1 transponder). The satellite systems will provide telephone, TV transmission, TV distribution, and new business services for Germany and West Berlin.

Business services will be able to have digital connections at rates for 64 kbps to 2 Mbps for voice, data, and video conferencing. Unused capacity on the system will be used by the Deutsche Bundespost to back up trunk lines in its terrestrial telephone network.

13.8.6. Indonesia: Palapa Program

Indonesia was the first developing nation to have its own satellite. The Palapa Al was built by Hughes and launched by NASA in 1976. The satellite provides Indonesia with a means of communicating to its 150 million persons, 80% of which are in rural areas. Indonesia's territory covers 13,000 islands from western New Guinea to North Sumatra. In addition to the geographic and cultural diversity of the country (250 languages and dialects are spoken), Palapa has helped to develop a greater sense of national consciousness and provide a means of further regional development.

13.8.7. Italy: ITALSAT Program

ITALSAT is an important project of the Italian national space program. The Italian satellite system was designed by Telespazio under contract to the Italian National Council for Research (CNR). Telespazio proposed a test system in the 30/20-GHz Ka band to be launched in late 1989.

The ITALSAT communications satellite will contain three main applications:

1. Digital telephone support with a capacity of 11,000 voice circuits at 30/20 GHz using six 20-watt transponders accessed by time division multiple access (TDMA). Two antennas will cover Italy.
2. Data transmission support at 30/20 GHz using three 20-watt transparent transponders accessed by TDMA.
3. A propagation experiment at 50/40 GHz.

Eventually, ITALSAT will have a 40,000 voice circuit capacity and will be integrated into Italy's national PTT network.

13.8.8. Japan's Satellite Program

Currently, satellites in Japan are administered by the Ministry of Posts and Telecommunications (MPT). The National Space Development Agency of Japan (NASDA) provides launching facilities for national communications satellites. Operation and maintenance are performed by the Telecommunications Satellite Corporation of Japan (TSCJ).

Japan has two broadcasting satellites and two communication satellites in operation. Nippon Telegraph and Telephone (NTT), the national PTT, is a major user of communication satellites. Kokusia Denshin Denwa (KDD), the Japanese international carrier, is not involved in domestic operation and uses INTELSAT for international telecommunications.

Private firms can buy communications satellite facilities through NTT. In addition, under the new liberalization program of April 1985, type 1 common carriers can be private satellite communications firms. Two private companies are approved as telecommunications business operators and own communications satellites. One is Japan Communications Satellite, Inc. (JCSAT), and the other is Space Communications Corporation (SCC). JCSAT's satellite will be developed by Hughes Aircraft and SCC's by Ford Aerospace and Communications Corporation. The two satellite carriers will offer business services in 1989.

NASDA's next communications satellite will bring the total communications satellites in orbit to seven in 1988. The total number of transponders will increase from 16 to more than 100 by year end 1988.

13.8.9. PACSTAR

PACSTAR was created to apply satellite communications technology to the special geography and needs of the Pacific Ocean countries and territories. PACSTAR is a joint venture between the Papua Guinea PTT and Pacific Satellite, Inc. (PSI), a U.S. corporation in which TRT Communications, Inc. has the principal interest. The PTT provides a full range of telecommunications services for Papua New Guinea and other countries in the Pacific region. By using satellites that can operate with low-cost, 2- to 5-meter-diameter earth stations, PACSTAR can provide cost-effective, economic, social, and cultural links among the nations in the Pacific.

Communications services that can be provided by PACSTAR are the following:

- Wide area video and facsimile distribution
- TV programming, video conferencing, and educational programming
- Data communications including computer-to-computer communication, financial services, and database access
- Domestic and regional telephone, audio, and data services

PACSTAR's space segment capacity will use both the C band and Ku band radio spectrum. The Ku band will minimize coordination problems with existing terrestrial radio services in the more densely populated areas. C band is effective in overcoming the effects of heavy rainfall and will be used for domestic telephone, data, and television distribution for the island nations of the Pacific region.

Ku band satellite coverage will include Guam, Hong Kong, Manila, the Philippines, Port Moresby, Seoul, Shanghai, Singapore, Taipei (ROC), Tokyo, Hawaii, and California.

13.8.10. U.K. Satellite Services

SatStream provides a business communications service from British Telecom International (BTI) to most of Europe. In addition, satellite services are offered to Canada and the United States. SatStream provides multinational companies with digital links at speeds up to 64 kbps. The earth station equipment includes a small dish antenna that can be located on or near the customer's premises. The antenna can be linked through the INTELSAT Business Services (IBS) program. This is a precedent for the United Kingdom. Prior to the SatStream offering, multinational enterprises using international satellite communications were required to use British Telecom's local PTT systems before users could be connected. SatStream will be extended to other European nations through EUTELSAT and Telecom 1, the French satellite system.

Mercury Communications Ltd., in conjunction with its parent company Cable and Wireless, offers transatlantic digital leased services to the United States via INTELSAT. Mercury can provide information circuits between 56 kbps and 8 Mbps. Each circuit can be used in a variety of subdivided voice and data channels using multiplexers. The rates are suitable for voice, high-speed data transmission, facsimile, and full-motion video conferencing.

A United Kingdom to Hong Kong leased line service is available for combined data and/or voice circuits at 56 and 64 kbps between customers' premises in both countries. Voice services use encoding techniques at 28.8 and 32 kbps. The remainder of the 56- or 64-kbps circuit can be used for a mixture of lower-speed data circuits. Distribution throughout Hong Kong and the new territories is provided by the Hong Kong Telephone Company, a Cable and Wireless subsidiary.

13.8.11. U.S. Communications Satellites

There are four types of satellite providers in the United States: the satellite operating company, the common carrier, the private satellite provider, and the reseller. The first category is the satellite operating organizations that are responsible for the design, launching, and continuing operation of their satellite systems, e.g., AT&T, Comsat, GTE Spacenet, General Motor's Hughes Communications, RCA (a unit of GE), MCI's SBS, and Western Union. They provide satellites for their own related companies or sell capacity (transponders) to the common carriers. In some cases, such as AT&T, Hughes, and SBS, they are the common carrier and rent directly to the end user.

The second type of provider is the common carrier, which leases or buys capacity (transponders) from the satellite operating organizations to offer services to its end user customer base. In many cases, the common carriers are legally related to the satellite operating organizations, e.g., AT&T-C, GTE, RCA Americom (a unit of GE), and SBS. If the carrier does not have a satellite operating company, the trend is toward a common carrier developing its own satellite system.

Another class of the satellite operating company was created by the FCC in its July 25, 1985 International Satellite Decision. The three remaining applicants created under this decision were listed in the section on international business satellites. They are authorized to provide international private services.

The fourth type of communications provider is the reseller and/or large end user who buys one or more transponders and leases or rents capacity to third parties. Resellers are proliferating in the United States and include Allnet, Argo Communications, Cylix Communications Corporation, Starnet Corporation, ITT/United States Transmission Company, and numerous other companies. Another example is an end user, e.g., Citibank Corp., who went directly to the satellite operating company. Citibank purchased transponders from Western Union Telegraph Company and is marketing a VANS service offering to their clients. Other resellers are IRCs, who lease space or transponders from INTELSAT Business Services (IBS) and coordinate a digital end-to-end international service.

Table 13.5 is a listing of business communications service offerings from six major domestic suppliers.

TABLE 13.5. Major Domestic Satellite Business Offerings

Type of Service	Information on Service
American Satellite Co., Commercial Services	
Private line data	Voice, data, Fax, and video conferencing available. Line speeds of 4800 and 9600 bps, 19.2 and 56 kbps, and 1.544 Mbps supported. Error rates of less than 1 error in 10 million bits (BER).
Voice-grade channel	Supports voice and data rates up to 9600 bps.
Satellite Voice Exchange (SVX)	Effective for voice traffic and tie line applications.
SVX Foreign Exchange	Voice, data, and alternate voice and data supported.
Transponder	36-MHz bandwidth between two or more earth stations for voice, data, and video conferencing.
AT&T Skynet Satellite Services	
Skynet Audio	Audio broadcasting.
Skynet Video	TV broadcasting.
Skynet Transponder	Voice, data, and video supported. Up to 36 MHz is available on a point-to-point basis. Earth station equipment can be located on the customer's premises.
Skynet 1.5	Voice, data, and video supported up to 3 Mbps.
GTE Spacenet Corp.	
Spacenet	Voice, data, and video support. Leader in micro-earth stations. Full or partial transponder service on an hourly or long-term basis.
Skystar	Data network for customer applications. High-speed point-to-point service within satellite coverage area.

(continued)

TABLE 13.5. Major Domestic Satellite Business Offerings

Type of Service	Information on Service
News Express	Satellite news gathering and broadcasting service for TV stations.
Sprint Digital Satellite Network	Switched voice, data, and private line line traffic.
Private Channel Service Network	High-speed channel service capable of transmitting up to 1.544 Mbps between selected cities.
MCI's Satellite Business Systems (SBS)	
Communications Network Services (CNS)	Data, voice, and image supported. Full-duplex operation with rates to 1.544 Mbps for switched and nonswitched applications. Earth satellite stations are located on the customer's premises.
RCA American Communications (a unit of GE)	
Business Network Services	Voice, data, Fax, and video conferencing supported. Transmission speeds up to 1.544 Mbps are offered. Over 6000 receive-only stations are in use for CATV programming. Interstate traffic on a dedicated private line is offered for distances over 700 miles.
Western Union Satellite Service	
Voice-grade channel	Voice, data, and alternate voice and data available.
Wideband channel	Two channels are offered: 48 or 240 kHz.
Space Tel channel	Telephone service is available on a PABX-to-PABX basis between 21 cities.
Transponder	Voice, data, video, and TV is supported. Large users can obtain a full transponder (36 MHz).

Source: Auberback Reports, CCMI/McGraw-Hill, Datapro Telecommunications Guide, and Company published reports.

Domestic satellite carriers offer a company the ability to install its own earth station. However, unlike a domestic receive-only earth station, which may be installed without FCC permit or license, transmitting stations require FCC authorization. Each transmit earth station operator must apply for a construction permit to construct and install the earth station and a license to operate the station. The application submitted by the applicant to the FCC must be submitted by the party that will operate the station, and the "real party of interest" must be disclosed. The license may then be granted if the applicant can demonstrate that it is legally, financially, and technically competent to construct and operate the facility.

13.9. SUMMARY

As satellite systems in the higher-frequency Ku and Ka bands become more prolific, earth station equipment size and cost will be reduced, thus opening up the market to thousands of end users. Small-diameter "dishes" will become standard in the parking lots and on the rooftops of business organizations, hotels, and homes. Dramatic changes will take place when the PTTs integrate satellite systems into their integrated services digital networks (ISDNs). Video and other high-speed information will be standard and priced comparably to a regular telephone call. Small conference rooms can be turned into video conferencing studios, international electronic mail services will place emphasis on high-speed messages, and information around the world will be available within hours. Travel costs can be reduced and long distance face-to-face contact improved with video conferencing. The net result is that work locations will be closer to where people prefer to live.

14

TRENDS THROUGH THE YEAR 2000

In this chapter we will

- Present trends in liberalization
- Discuss major changes in applications, enhanced services, and future networks

Changes in telecommunications products, services, and networking will have a great impact on society through the year 2000. Information will be available more quickly throughout the world and will touch all businesses and service sectors. These changes will be brought about by some liberalization of the PTT monopolies, new products based on international standards, and enhanced services available through new switched digital networks.

14.1. LIBERALIZATION

The greatest benefit to the subscriber in terms of lower prices, availability of customer premises equipment, and enhanced services will come about through liberalization and standardization. Liberalization will remove restrictions for use of equipment and VANSs. Standardization will result in higher product volumes for manufacturers of customer premises equipment. Higher volumes allow greater investment in high-density VSLI chip sets, resulting in products with additional functions at lower cost.

Liberalization of the PTT monopolies will open up markets for equipment and services by private vendors or competing common carriers. The U.K. government's liberalization of equipment and the establishment of two competing common carriers has resulted in cost savings to customers. The combination of liberalization and lower international tariffs to the United States has resulted in additional revenues for British Telecom and the U.K. computer and telecommunications industries. Over 60% of all U.S. multinational corporations use the United Kingdom as their European telecommunications center.

Japan has opened its domestic telecommunications markets for equipment and services by allowing competition with NTT, the domestic telephone company. The government is attempting to revitalize Japan's telecommunications services and encourage innovation.

Today, the individual U.S. subscriber is paying less for long distance calls due to competing long distance carriers (e.g., AT&T, MCI, and U.S. Sprint). Businesses will realize the potential of liberalization with a variety of competing services. If traffic is high enough between two or more points, they can install microwave or satellite networks to bypass or eliminate the local Telcos. Many PTTs in developed nations view these alternatives as a loss of revenue. Since the PTT is one of the largest employers of personnel in most countries, there is no rush to duplicate the U.S. model. However, there are trends in many European countries that will lead to fewer restrictions on equipment and VANSs.

Depending on the type of leased line service (old or new), the Deutsche Bundespost's leased line monthly charges will increase linearly over time to a maximum of nine times the present fixed monthly rate. Clearly this method will deter heavy traffic loading and force some applications to switched digital data networks. Greater revenues from leased lines due to usage and time charges will lead to some liberalization of their use. Under volume charging, electronic document distribution, electronic data interchange, electronic mail and other VANS applications will be allowed on national and international leased lines within the same enterprise. VANSs providers are expected to expand their services to connect enterprises. Another trend toward liberalization will occur through acquisitions, mergers, and joint ventures between multinational companies and foreign PTTs. A joint venture between a VANS provider and PTT usually extends the VANS marketing arm into that PTT's country. Restrictions are then eliminated since both parties share in the revenue and profits.

There will be pressure by businesses, consumer groups, and trade organizations to liberalize through government review of PTT monopolies. Competition usually results in lower prices and additional services to the subscriber. Any positive and significant increase in a PTT's revenue in the United States, the United Kingdom, or Japan will accelerate this trend.

14.2. NEW APPLICATION TRENDS

Future trends for information transfer over public switched data networks will create opportunities for a new range of applications and services that are difficult or impracticable to implement with present facilities. Future telecommunications applications may have one or more of the following characteristics:

- *Traffic:* There is a trend toward shorter messages exchanged at high speeds. Many applications will involve transactions that can be handled in very short message pairs. In addition, on-line file updating of distributed databases can reduce the length of transmitted files. Traffic will move in bursts, and faster response time will be an important requirement.
- *Multiple host access:* A single user work station will access several mainframes in various locations throughout the world.
- *Large dispersed terminal populations:* Terminal-based services will be available for home or business use. For example, electronic funds transfer and credit checking networks could be accessed by over 50,000 terminals.
- *Interenterprise traffic flows:* At present, communications traffic is restricted to specific interest groups within well-defined boundaries. The trend will be toward applications that involve intercommunication between customers, suppliers, VANS providers, and nonrelated businesses.

14.2.1. Fiber Optic Networks

Local fiber optic networks offering network extensions to national links, trunk circuits, and central Telco offices are being installed in many cities.

Save your railroad stock: Railroads offer common carriers a major direct right-of-way between cities. Fiber optic cable can be laid next to railroad tracks. This direction was set by Mercury Communications in the United Kingdom. Initially Mercury was one third owned by British Railway, who allowed it to lay fiber optic cable next to the railroad tracks. The right-of-way saved years of negotiations for Mercury. In another move, Mercury purchased the defunct London Hydraulic Works, which owned an underground system in The City, a financial area in London. Essentially, Mercury purchased the London Hydraulic Works for its right-of-way up to the customer's premises. Mercury is now in the process of laying fiber optic cable to several potential key customer locations.

Today, PTTs measure fiber optic cable in tens of thousands of kilometers. In the future, installed fiber optic cable will be measured by those same PTTs in millions of kilometers.

14.2.2. International OSI

OSI is a worldwide standards effort to interconnect systems from different manufacturers. Future systems may be required to meet this specification and implement the model in order to provide products to public agencies in some countries. In addition, if VANSs are offered to third parties over public data networks, the service provider may be required to meet OSI standards.

Within OSI, provision will be added for the integration of voice, data, text, and image into one work station or computer. Ultimately one work station will edit, store, print, operate, transmit, and receive a compound document. A compound document could include voice annotation, data, text, and image reproduction of letterheads and signatures.

OSI will impact computer product design. Initially, midrange computers and larger mainframes will offer a gateway that will coexist with OSI and their network architecture. Once all layers become standardized and the price of chips (electronic building blocks) drops, vendors will implement the full seven-layer OSI model into their computer products. Existing vendor-supplied network architectures will coexist with OSI through a gateway product, as shown in Figure 14.1.

Although European administrations initiated OSI, most U.S. computer vendors have stated their intention to meet the OSI standard. Japanese vendors now have the key for moving into new accounts, the OSI standard, and will move deeper into the European market, resulting in more competition and lower prices. European vendors, with their smaller national markets and smaller economies of scale, will be more pressed to meet competition.

For example, as of December 1986, 57 U.S. companies (including all major computer manufacturers) were members of COS (the Corporation for Open Systems). COS was formed to provide connectivity between different vendor architectures based on the OSI model. SPAG (Standards Promotion and Applications Group) was formed by 12 European computer manufacturers with a similar goal. Eight SPAG members (Bull, ICL, Nixdorf, Olivetti, Philips, Siemens, Stet, and Thomson) created a Belgian company, SPAG Services S.A., to verify the compatibility of their products to common standards based on OSI. The formation of SPAG Services was in response to COS and POSI, a Japanese group organized for similar purposes.

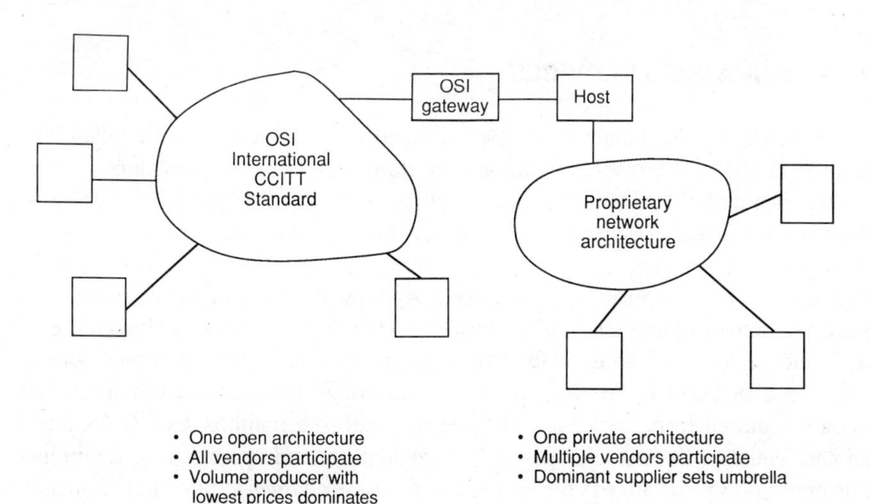

Figure 14.1. Coexisting with OSI.

14.2.3. ISDN

The basic integrated services digital network (ISDN) with two 64-kbps channels will allow the integration of voice and data into one work station. When compared to today's networks, a 64-kbps channel will cost about two times a normal voice call between selected countries. For example, the Federal Republic of Germany has tariffed a 64-kbps ISDN link to the United States at twice the price of a dial-up call.

Lower cost and higher data rates will open up a new range of applications, resulting in the following:

- Greater emphasis on distributed databases
- Image processing and database access
- Industry networks
- Electronic mail and messaging
- Multifunction work station for voice, data, text, and image

ISDNs include higher speeds, faster switching times, and improved quality through end-to-end digital connections. The digital network and information services will have a positive impact on customer premises equipment.

Future products could include a single-line multifunction work station or PC capable of voice and data transfer to multiple networks. Teleconferencing workstations could include text and image within the same device, allowing text (stored in memory) to be merged and transferred along with drawings or sketches over the same communications line. Higher bandwidth will allow one work station to handle compound documents including voice, data, text, and image. Some potential replacement products for existing telecommunications offerings are shown in Figure 14.2.

14.3. ENHANCED SERVICES

In the future, international interenterprise electronic mail and electronic data interchange (EDI) will provide a major opportunity for VANS providers. Vendor compliance with the X.400 Message Handling System will simplify the mechanism for sending messages between different operating systems and separate competing electronic mail services. Another restriction to a universal electronic mail capability is the requirement in most countries that VANS providers offer their services over switched network facilities. Since international VANS providers use leased lines as their backbone networks for both efficiency and quality, switched connections are impractical and costly. To overcome these restrictions today, international subscribers are required to make a long distance dial call to a mailbox facility located in another country. Several changes in PTT regulations and operations will eliminate this problem. One change is the new usage and time charges applied to national and international leased lines.

ISDN is a technological change that will open up national markets for enhanced services. Since ISDN offers a switched digital point-to-point connection, all calls

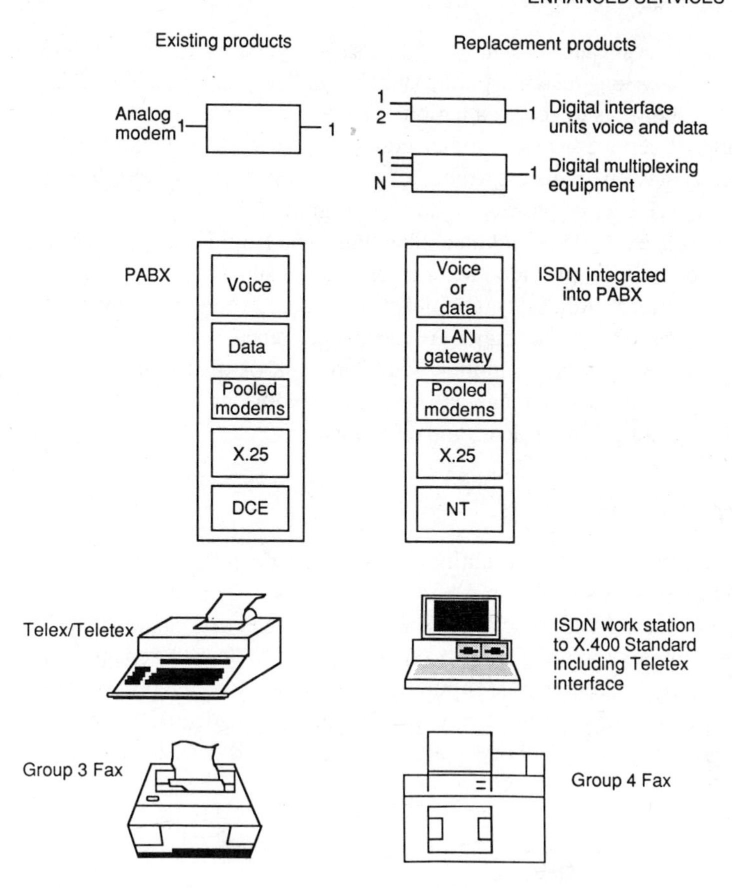

Figure 14.2. ISDN impact on existing products.

(voice, data, text, image, and video) will be billed on usage and time. Switched facilities, therefore, eliminate the PTTs' main objective to VANSs over a leased line. Today, many PTTs do not allow voice and data to coexist on the same leased line. In ISDN, voice and data are designed to coexist on the same wire pair into the customer's premises. Alternatively, PTTs could offer "virtual" private networks over ISDN. These networks would be similar to AT&T software-driven networks.

14.3.1. International Videotex

Accessing a Videotex system in another country is now possible by a long distance call to the national PTT or private system. For example, British Telecom's Prestel Videotex system is offered in the United States. A U.S. user dials into a concentrator that is linked to databases in the United Kingdom. If a user resides in a city where Prestel has installed a concentrator, a local call to the concentrator switches the user directly to London.

How does a user access Videotex frames and databases in other countries? Access is normally by a long distance public switched call to an access city in that country. A preferred method is to access a national Videotex service in your country and have that national service access another country's Videotex service. That is, the PTT provides a gateway for access to another country's Videotex system. An international Videotex service with gateway is shown in Figure 14.3.

A practical example would be in planning a trip from France to Germany (FRG). Using a French PTT-provided Minitel terminal, a subscriber in France would book hotel reservations, train seats, and tickets to special events in Germany and have the tickets debited to his or her major credit card in France.

Another example is a supplier with hundreds of distributors in Europe. The supplier could update one database overnight with new product announcements, promotional campaigns, prices, and discounts.

14.3.2. Teleports

A teleport is a separate legal entity that provides a combination of telecommunications services to customers in a building complex. Teleports can exist separately, serving an entire geographic area. Alternatively, teleports can also be attached to a building complex, creating an environment similar to Shared Tenant Services.

A company can rent, lease, or purchase office space in a teleport building complex in order to have access to very expensive and sophisticated telecommunications services as well as telephone, facsimile, telex, and PTT data networks. Each

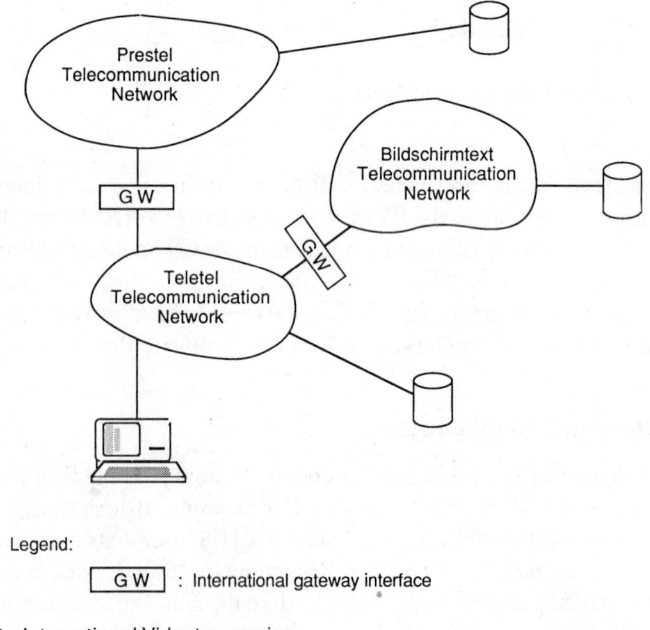

Figure 14.3. International Videotex service.

occupant shares the telecommunications facilities of the teleport. Individually, many companies could not afford to pay for a broad array of standard and enhanced telecom services. The idea is to share telecommunications costs over a number of users.

Over $600 million has been invested in the U.S. teleport business since its inception in the early 1980s. Twenty-two teleports are in operation, and over 20 were in the planning stages at year end 1986.

Teleports provide a long distance gateway, usually a satellite earth station coupled with a local area network. The local area network can be a telephone company local area data channel (LADC), a normal cabling system, or a private or public fiber optic network within a city area.

The telephone company LADC can be very inexpensive for voice and data communications between two locations in the same general vicinity. LADC circuits are hardwired copper circuits. There is no signal amplification of the user equipment at either end, and the lines are not tied to the switched network. LADCs are designed for digital communications between two customer locations in the same serving wire center. For example, an LADC from the teleport located on Staten Island, New York could be connected to other locations in Manhattan.

Teleport Examples

Several key examples of the magnitude of teleports will be presented. The IDB Communications Group Ltd. of Culver City, California has contracted with Teleport Communications to install and operate five satellite earth stations at the teleport on Staten Island. IDB's Technical Operating Center at Staten Island will work with IDB headquarters as a communications and control hub for satellite news gathering. IDB currently owns and operates transmit/receive satellite earth stations in 27 cities. Its system, called the Sports Satellite Interconnect, is used for radio programming.

South Star Communications Inc., the first Teleport developer and operator in southern Florida, is participating in a $20 million teleport complex to be developed near Colorado Springs, Colorado. The Colorado teleport at the Falcon Air Force Base is part of an area master plan called Nova, a mixed-use complex that will include a space technology university, high-tech industrial park, hotel, and shopping center. The complex is called WorldStarWest and will have two 11-meter C band antennas, two 7-meter Ku band antennas, and two INTELSAT Standard B antennas for global satellite communications.

Three major Japanese cities are building teleports for international firms and information processing companies. Yokohama, Tokyo, and Osaka are developing teleports to turn obsolete docks and port areas into new residential and business districts.

Yokohama is investing $8 billion to clear and reclaim 460 acres of land for offices, housing, and recreational use. Development is being carried out by Yokohama Minato Mirai 21 Co. Ltd. (MM21), which is owned by the city, several other government bodies, and about 30 private companies. The teleport will be the centerpiece for advanced telecommunications facilities that will include earth stations and fiber optic networks. In addition to the satellite earth station, the teleport is expected to include a database center offering economic, industrial, scientific, and technical data from Japan and abroad. Computer time sharing services will also be available at the teleport. Yokohama city officials do not expect the teleport to be in full operation until the twenty-first century.

Yokohama MM21 will also include a public facility information and control center that will offer computerized control of heating and ventilation to nearby buildings on a fee basis. In addition, a visual information center will market advanced Videotex services to offices in the surrounding area. Yokohama plans to install four systems, one each for economic, technological, foreign trade, and distribution information, at the teleport.

Tokyo's planned teleport will be part of an extensive redevelopment of the city's seaport area. The teleport will cover about 100 acres of reclaimed land located 3.5 miles from central Tokyo.

The satellite earth station area will consist of six 33-feet-diameter dishes located in a square and surrounded by an office building complex. Data communications, image transmission, high-speed facsimile, and video conferencing services will be supplied by the telecommunications center within the complex. Tokyo government officials do not expect the teleport to begin operations until the twenty-first century.

Osaka's teleport, part of a $10 billion seaport redevelopment program, will be built on a number of reclaimed islands. The teleport will include a satellite earth station consisting of four antennas and a large office park. The emphasis is on information processing firms and the computer facilities of large corporations.

14.4. THROUGH THE YEAR 2000

Teleservices will include new innovations in messaging, Telefax, and Videotex. Messaging enhancements will include inputs by voice command instead of keyboard, network mailboxes and delayed sending, and merged documents with text, letterhead, and signature. Telefax Group 4 on digital networks will provide high-quality copies at the rate of one page every 5 seconds. Videotex will be enhanced with audiovideotex, or associating sound and Videotex with higher-speed transmission and alphaphotographics.

Video conferencing services will include full-motion color as a standard ISDN information service offering. Satellite communications will allow video conferencing with facsimile for high-speed display and document transmission.

Overall, information services for remote access, control, monotoring, and database support will touch every phase of a person's life. A telephone call from a terminal will permit subscribers to send messages, pay bills, shop, and bank without leaving home. Teleaction services for monitoring security alarms, reading utility meters, and adjusting temperature controls will be offered by VANSs for apartments and homes. In summary, by the twenty-first century, a person will have access to information around the world and the capability for outside control of his or her environment.

APPENDIX A

DIRECTORY OF U.S. COMMON CARRIERS

A.1. COMMON CARRIERS FOR DOMESTIC SERVICES

Common carriers provide facilities for the transmission of voice, data, text, video, and Fax over their wholly owned private networks consisting of leased lines, microwave, or satellite facilities.

Allnet Communications Services, Inc.
100 South Wacker Drive
Chicago, IL 60606
Tel.: (312) 443-1444

American Satellite Company
1801 Research Boulevard
Rockville, MD 20850
Tel.:(301) 251-8300

American Telephone and Telegraph Communications, Inc. (AT&T-C)
32 Avenue of the Americas
New York, NY 10801
Tel.: (212) 219-6000

Argo Communications Corporation
145 Huguenot Street
New Rochelle, NY 10801
Tel.: (914) 576-7400

Graphnet, Inc.
329 Alfred Avenue
Teaneck, NJ 07666
Tel.: (201) 837-5100

ITT/United States Transmission Systems, Inc.
100 Plaza Drive
Secaucus, NJ 07096
Tel.: (201) 330-5000

MCI Communications Corp.
1133 19th Street, NW
Washington, D.C. 20036
Tel.: (202) 872-1600

RCA American Communications (a unit of GE)
4 Research Way
Princeton, NJ 08540
Tel.: (609) 987-4000

Satellite Business Systems (MCI Communications)
8283 Greensboro Drive
McLean, VA 22102
Tel.: (703) 749-7000

U.S. Sprint Communications Company
P.O. Box 974
Burlingame, CA 94010
Tel.: (415) 692-5600

Western Union Telegraph Company
One Lake Street
Upper Saddle River, NJ 07458
Tel.: (201) 825-5000

A.2. INTERNATIONAL CARRIERS

International carriers are used for the transmission of voice and data, while international record carriers may carry voice, data, text, video, and Fax services to other nations.

There are three portions to an international leased line connection. The first is the U.S. link to the IRC or IC, the second is the foreign link to the PTT administration, and the third is the foreign connection within the country. The following companies will coordinate the U.S. portion to their gateway cities and interface to the foreign PTT administrations. The subscriber would have to negotiate with the foreign PTT administration for the third portion leading into the local subscriber's premises.

AT&T Communications
412 Mount Kembal Avenue
Morristown, NJ 07960
Tel.: (201) 644-6000

Communications Satellite Corp. (Comsat)
950 L'Enfant Plaza, SW
Washington, DC 20024
Tel.: (202) 863-6000

FTC Communications, Pacific Telecom Company
90 John Street
New York, NY 10038
Tel.: (212) 669-9700

International Record Carrier, Inc.
5 Hanover Square, 12th Floor
New York, NY 10004
Tel.: (212) 809-9797

International Relay, Inc. (IRI)
600 Third Avenue
New York, NY 10016
Tel.: (212) 953-0070

International Telecommunications Satellite Corp. (INTELSAT)
3400 International Drive
Washington, D.C. 20008-3098
Tel.: (202) 944-6800

ITT World Communications, Inc.
67 Broad Street
New York, NY 10004
Tel.: (212) 607-2338

MCI International, Inc.
2 International Drive
Rye Brook, NY 10573
Tel.: (914) 937-3444

RCA Global Communications (a unit of GE)
60 Broad Street
New York, NY 10004
Tel.: (212) 806-7000

Telenet Communications Corp.
8229 Boone Blvd
Vienna, VA 22180
Tel.: (703) 442-1000

TRT Telecommunications Corp.
1331 Pennsylvania Avenue, NW
Washington, D.C. 20004
Tel.: (202) 879-2200

Tymnet McDonnell Douglas
2710 Orchard Parkway
San Jose, CA 95134
Tel.: (408) 946-4900

Western Union International, Inc. (MCI Communications)
One WUI Plaza
New York, NY 10004
Tel.: (212) 701-2000

Western Union Telegraph Company
One Lake Street
Upper Saddle River, NJ 07458
Tel.: (201) 825-5000

APPENDIX B

DIRECTORY OF INTERNATIONAL PTTs

Appendix B includes a listing of international PTTs. If there are two or more listings for a country, the first address is for international traffic or services, and the second represents national or internal services. The exceptions are Canada (two international listings plus the provinces), Indonesia (listings for two main islands), the Philippines (which is served locally by several IRCs), and the United Kingdom (which has two competing common carriers).

For countries that have a packet switched data network (PSDN), the PTT will provide detailed information for both international and national traffic. However, for immediate results on international PSDN service from the United States, contact either ADP Autonet, Telenet, Tymnet, or one of the international record carriers in Appendix A.

A telephone listing and/or a telex number is provided after each telecommunications administration. The telephone country code is enclosed by parentheses.

Afghanistan
Ministry of Communications
The Director of Liaison Office
Kabul, Afghanistan
Telex: 97 MOC AF

Albania
General Direction of Posts & Telecommunications
Dir. of Telecommunications—42 Myslym Shyri Street
Tirane, Albania
Telex: 2174 GENTEL AB

Algeria
Ministry of P&T—General Secretariat
4 Blvd Salah Bouakouir
Alger, Algeria
Telex: 52020 GENTEL ALGER DZ

American Samoa
The Director of Communications
American Samoa Government
Pago Pago, American Samoa
Telex: 782 502 SB

Angola
Empresa Publica de Telecomunicacoes
Rua do 1/0 Congresso, 26, 2/0 Andar
P.O. Box 625
Luanda, Angola
Tel.: (244) 1 31042

Anguilla
Cable & Wirelesss (West Indies) Ltd
P.O. Box 77
The Valley, Anguilla
The West Indies
Tel.: (809) 497

Antigua
Cable and Wireless (West Indies) Ltd
P.O. Box 65
42-44 St. Mary's Street
St. John's
Antigua
Tel.: (809) 46 20840

Argentina
Empresa Nacional De Telecomunicaciones
Rivadavia 618, Piso 1
Buenos Aires 1002
Tel.: (541) 490 913

Ascension
Cable and Wireless Public Limited Co.
Ascension, Ascension
Telex: 215 BOOTH AV

Australia
International Services
Overseas Telecommunications Commission
32-36 Martin Place
Sydney 2000, Australia
Tel.: (61) 2 230 5000

National Services
Telecom Australia
Communications House
199 William Street, Melbourne
Victoria 3000, Australia
Tel.: (61) 3 606 6434

Austria
International Services
Radio Austria A.G.
Renngasse 14
1010 Wien, Austria
Tel.: (43) 222 63 66 51

Bahamas
Bahamas Telecommunications Corp.
Chase Manhattan Building
Oaks Field
P.O. Box N3048
Nassau, Bahamas
Tel.: (809) 323 49 11

Bahrain
Bahrain Telecommunications Company (BATELCO)
P.O. Box 14
Manama, Bahrain
Tel.: (97) 248.212
Telex: 8790 BTCCOM BN

Bangladesh
Telegraph & Telephone Board of Bangladesh
Central Office
Sher-E-Bangla Nagar
Dhaka-7, Bangladesh
Telex: 642020 BTTB BJ

Barbados
Barbados External Telecommunications, Ltd
P.O. Box 32
Bridgetown, Barbados
Tel.:(809) 427 5200

Belgium
Regie des Telegraphes et Telephones
Boulevard de l'Imperatrice 17-19
1000 Bruxelles, Belgium
Tel.: (32) 2 213 3636
Telex: 29280 cddata b

Belize
Cable and Wireless PLC
P.O. Box 414
Belize City, Belize, Central America
Telex: 222 CWADMIN

Bermuda
Cable and Wireless PLC
20 Church Street
P.O. Box HM151
Hamilton HMAX, Bermuda
Tel.: (809) 29 54777

Bolivia
Direccion General de Telecomunicaciones
P.O. Box 4475
Calle Mercado 1115
La Paz, Bolivia
Tel.: (591) 368 788

Botswana
Botswana Telecommunications Corp.
P.O. Box 700
Gaborone, Botswana
Tel.:(267) 353 611

Brazil
Empresa Brasileira de Telecomunicacoes—EMBRATEL
Avenida Presidente Vargas 1012
Sala 982
20071-Rio de Janeiro/RJ, Brazil
Tel.: (55) 21 216-7709

British Virgin Islands
Cable and Wireless (West Indies) Ltd
P.O. Box 440
Road Town, Tortola, British Virgin Islands
Tel.: (809) 49 44444

Brunei
Telecommunications Dept
Bandar Seri Begawan, Brunei
Telex: 2277 JTBHQ BU

Bulgaria
Ministry of Posts & Telecommunications
Sofia C, Bulgaria
Telex: 22846 Gentel BG

Burundi
National Office of Telecommunications
B.P. 60
Bujumbura, Burundi
Telex: 81/82 UU

Cameroon Republic
Ministry of Posts and Telecommunications
Yaounde, Cameroon
Telex: MINPOSTEL 8285KN KN

Canada
International Services
Teleglobe Canada
680 Sherbrooke Street West
Montreal, Quebec, Canada H3A 2S4
Tel.: (514) 289 7272

National Services
Telecom Canada
Room 1890
160 Elgin Street
Ottawa, Ontario, Canada K1G 3J4
Tel.: (613) 560 3000

Worldwide Consulting
Bell Canada International
1 Nicholas Street
Suite 800
Ottawa, Ontario, Canada K1N 9M1
Tel.: (613) 563-1811

Services Within the Provinces
Alberta Government Telephones
10020-100 Street
P.O. Box 2411, Edmonton
Alberta, Canada T5J 2S4
Tel.: (403) 425 2110

Bell Canada
1050 Beaver Hall Hill
Montreal, Quebec, Canada H3C 3G4
Tel.: (514) 870-1511

British Columbia Telephone Company
3777 Kingsway, Burnaby
British Columbia, Canada V5H 3Z7
Tel.: (604) 432 2151

Island Telephone Company
P.O. Box 820, Charlottetown
Prince Edward Island, Canada C1A 7M1
Tel.: (902) 566 5501

Manitoba Telephone System
P.O. Box 6666, Winnipeg
Manitoba, Canada R3C 3V6
Tel.: (204) 941 4111

Maritime Telegraph and Telephone Company Ltd
P.O. Box 880, Halifax
Nova Scotia, Canada B3J 2W3
Tel.: (902) 421 4311

New Brunswick Telephone Company Ltd
One Brunswick Square
P.O. Box 1430, Saint John
New Brunswick, Canada E2L 4K2
Tel.: (506) 694 2340

Newfoundland Telephone Company Ltd
P.O. Box 12088, St. John's
Newfoundland, Canada AlB 4C8
Tel.: (709) 739 2450

Saskatchewan Telecommunications
2121 Saskatchewan Drive
Regina, Saskatchewan, Canada S4P 3Y2
Tel.: (306) 347-3737

Cayman Islands
Cable and Wireless (West Indies) Ltd
P.O. Box 293
Grand Cayman
Cayman Islands, British West Indies
Tel.: (809) 94 97800

Central African Republic
Central African Office of P&T
Immeuble Sica II—Avenue des Martyrs
Bangui, Central African Republic
Telex: DIRTEL 5245 RC

Chad
Office National des Postes et Telecommunications
N'Djamena, Chad
Telex: DIRTEL 5256 KD

Chile
International Services
Empresa Nacional de Telecomunicaciones S.A.
ENTEL Chile
Santa Lucia 360
Santiago, Chile
Tel.: (56) 712 121

National Services
Compania de Telefonos de Chile (CTC) S.A.
San Martin 50
Santiago, Casilla 16-D, Chile
Tel.: (56) 202 102

Subsecretaria de Telecomunicaciones
SUBTEL
Amunategui 139
Santiago, Chile
Tel.: (56) 699 2236

China
International Services
GM—Shenda Telephone Co. Ltd
8F Telecommunication Bldg.
Shennandong Road
Shenzhen, People's Republic of China
Telex: 44383 SDTCL CM CN

Hopei Province
Pktelcom
Beijing Telecommunications Administration
11, Xi Chang An Street
Beijing, People's Republic of China
Tel.: (716) 22211

Kiangsu Province
Shanghai Post and Telecommunication Industrial Corp.
350 North Suzhou Road
Shanghai, People's Republic of China
Tel.: (716) 24 45 00

Colombia
Empresa Nacional De Telecomunicaciones
Calle 23 No. 13-49 Piso 8
Bogota, Colombia
Tel.: (57) 269 4077

Congo
National Office of P&T
Brazzaville, Congo
Telex: OFIPOSTEL 5208 KG

Costa Rica
Radiografica Costarricense
Apartado 54
San Jose, Costa Rica
Tel.: (506) 33 55 55

Cuba
Empresa Telecomunicaciones Internacionales
Zanja y San Francisco
Havana, Cuba
Telex: 511107 EMTEL CU

Cyprus
Cyprus Telecommunications Authority
P.O. Box 4929
Nicosia, Cyprus
Tel.: (357) 21 77111

Czechoslovakia
Federal Ministry of Posts and Telecommunications
International Division—Olsanka 3
130 00 Praha (Prague), Czechoslovakia
Telex: 111488
Tel.: (42) 714-1111

Denmark
General Directorate of Posts and Telegraphs
Favergade 17
DK-1007 Copenhagen, Denmark
Tel.: (45) 252 9111

Eastern Denmark (Copenhagen and Sealand)
Teleregion 1
Roedovrevej 241
DK-2610 Roedovre
Tel.: (45) 1 41 50 55

South Jutland and Funen
Teleregion 2
Banegaardspladsen 2
DK-6000 Kolding
Tel.: (45) 5 52 41 99

Central and Northern Jutland
Teleregion 3
Kannikegade 16
DK-8000 Aarhus C
Tel.: (45) 6 12 07 88

Djibouti
Office des Postes et Telecommunications
Boulevard de la Republique
Djibouti
Telex: 580DJ/DION OPT DJ

Dominican Republic
Dominican Republic Telephone Company
Santa Domingo, Dominican Republic
Tel.: (809) 567-2211

Ecuador
Instituto Ecuatoriano de Telecomunicaciones
Casilla Postal N. 3066
Quito, Ecuador
Telex: 22202 IETGG ED
Tel.: (593) 2-560700

Egypt
National Telecommunications Organization
P.O. Box 2271
Cairo, Egypt
Telex: 92100 RADMSR UN

El Salvador
Administracion Nacional de Telecomunicaciones
Edificio de ANTEL—Centro de Gobierno
San Salvador, El Salvador
Tel.: (503) 227 273

Ethiopia
Ethiopian Telecommunications Authority
P.O. Box 1047
Addis Ababa, Ethiopia
Tel.: (252) 1 150 500

Falkland Islands
Cable and Wireless PLC
Box 179
Port Stanley, Falkland Islands
Telex: 2411

Federal Republic of Germany
Deutsche Bundepost
Fernmeldetechnisches Zentralamt
Kundenberatung fuer Dateldienste
Referat T 21
Postfach 50 00
6100 Darmstadt
Federal Republic of Germany
Tel.: (49) 6151 83-4641
Telex: 419511 ftz d

Fiji
Fiji International Telecommunications Ltd
P.O. Box 59
Suva, Fiji
Tel.: (679) 312 933

Finland
General Directorate of Posts and Telecommunications
P.O. Box 528
00101 Helsinki, Finland
Tel.: (358) 0-1951

France
Direction des Telecommunications des Reseaux Exterieurs
246 Rue de Bercy
75584 Paris Cedex 12, France
Tel.: (33) 43.42.62.00

French Antilles
Guadeloupe
Agence Commerciale des Telecommunications
Tour Secid Plan de la Renovation
B.P. 392
97162 Pointe-a-Pitre Cedex, Guadeloupe
Tel.: (596) 821 800

Martinique
Agence Commerciale des Telecommunications
41 Rue Gabriel Peri
B.P. 697
97200 Fort de France Cedex, Martinique
Tel.: (596) 714 695

French Guiana
Agence Commerciale des Telecommunications
97305 Cayenne Cedex, French Guiana
Tel.: (594) 31 85 14

French Polynesia
Direction de l'Office des P&T
Papeete, French Polynesia
Telex: POSTELEC 530 FP

Gabon
Telecommunications Internationales Gabonaises
B.P. 2261
Libreville, Gabon
Tel.: (241) 74910

Gambia
Gamtel Company Ltd
G.P.O. 387
Banjul, The Gambia
Telex: 2201

Ghana
Posts and Telecommunications Corp.
The Director General
Accra, Ghana
Telex: 2010 ENG HQ ACCRA GH

Gibraltar
Cable and Wireless PLC
P.O. Box 203
Gibraltar
Tel.: 350 75687
Telex: 2202 CWADM

Greece
Hellenic Telecommunications Organization
OTE Ltd Co.
International Communications Department
1 Veranzerou Street
106 77 ATHINAI, Greece
Tel.: (30) 1 363 8099
Telex: 219797

Grenada
Cable and Wireless (West Indies) Ltd
P.O. Box 119
St Georges's, Grenada
Tel.: (809) 440 3382

Guam
International Services
RCA Global Communications (a unit of GE)
P.O. Box EH
Agana, Guam 96910
Tel.: (671) 477 8261

Guatemala
Empresa Guatemalateca de Telecommunications
7A Avenida 12-39 Zone 1
Guatemala City, Guatemala
Tel.: (502) 2 285 898

Guinea
Direction Generale des Telecommunications
Conakry, Guinea
Telex: 607 GE

Guinea Bissau
Secretariado de Estado Dos Correios e Telecomunicacoes
Bissau, Guinea Bissau
Telex: BI

Guyana
Guyana International Telecommunication Corp.
P.O. Box 239
Georgetown, Guyana
Telex: 2216

Haiti
Conseil National de Telecommunications
P.O. Box 2002
Avenue Marie Seanne
Port au Prince, Haiti
Tel.: (509) 24123

Honduras
Empresa Hondurena de Telecomunicaciones (Hondute)
Apartado Postal 1794
Tegucigalpa, DC, Honduras
Telex: 1220 Hondutel HT HO

Hong Kong
International Services
Cable and Wireless HKG, Ltd
P.O. Box 597
Hong Kong
Tel.: (852) 5 862 1111

National Services
Hong Kong Telephone Company Ltd
P.O. Box 479
Hong Kong
Tel.: (852) 5 288 111

Hungary
Posta Kozponti Taviro Hivatal
Post Central Telegraph Office
Varoshaz u. 18
P.O.B. 1
H-1364 Budapest V., Hungary
Tel.: (36) 1 184 811
Telex: 225885 BCTBP H

Iceland
The Director General of P&T
P.O. Box 270
150 Reykjavik, Iceland
Tel.: (354) 1 26000
Telex: 2000 GENTEL IS

India
Overseas Communications Services
Videsh Sanchar Bhavan-Mahatma Gandhi RD, Fort
Bombay 400 023, India
Telex: 11 2429 OCSH IN

Indonesia
Perusahaan Umum Telekomunikasi
Jln. Cisanggarung 2,
Bandung, West Java, Indonesia
Tel.: (62) 22 59100 59400

PT Indosat
Wisma Antara It Merdeka
Selatan 17
Jakarta, Indonesia
Tel.: (62) 21 342770

Iran
Telecommunications Company of Iran (TCI)
DR. Shariati Avenue
Tehran, Iran
Telex: 212799 TCI IR

Iraq
The President of Posts, Telegraphs
and Telephones State Organization
Baghdad, Iraq
Telex: 212002 COMGEN IK

Ireland
Telecom Eireann
6-8 College Green
Dublin 2, Ireland
Tel.: (353) 1 60 22 22
Telex: 32844

Israel
Bezeq: Israel Telecommunications Corp. Ltd
P.O. Box 1088
Jerusalem 81010, Israel
Tel.: (972) 3 335 716

Italy

International Services

Italcable
Via Calabria 46-48
00187 Roma
Tel.: (39) 6 4770 1
Telex: 611146 ITC DG I

Ivory Coast

Intelci
BP 1838
Abidjan 01, Ivory Coast
Tel.: (225) 32 49 85

Jamaica

International Services

Jamaica International Telecommunications Ltd
15 North Street
P.O. Box 138
Kingston, Jamaica
Tel.: (809) 92 22953

Japan

International Services

Kokusai Denshin Denwa Co. Ltd (KDD)
3-2 Nishi-Shinjuku 2-chome,
Shinjuku-ku, Tokyo 163, Japan
Tel.: (81) 03 347 7000

National Services

Nippon Telegraph and Telephone Corp. (NTT)
1-6 Uchisaiwaicho 1-chome,
Chiyoda-ku Tokyo, 100, Japan
Tel.: (81) 03 509 5111

Jordan

Telecommunication Corporation
Amman, Jordan
Tel.: (962) 6 73 111

Kenya

Kenya Posts & Telecommunications Corp.
P.O. Box 30301
Nairobi, Kenya
Telex: 22245 DIRPOSTS KE

Kuwait
Ministry of Communications
P.O. Box 16, Kuwait
Tel.: (965) 819033

La Reunion
Agence Commerciale des Telecommunications
26 Rue de la Compagnie
97489 St Denis de la Reunion Cedex, La Reunion
Tel.: (262) 210 312

Lebanon
Ministere des Postes et Telecommunications
Exploitation et Maintenance—Ave. Sami El Solh
Beyrouth, Lebanon
Telex: MINIPT 23700 LE or DIRGEM 23600 LE

Lesotho
Lesotho Telecommunications Corp.
P.O. Box 1037
Maseru 100, Lesotho
Tel.: (266) 501 24 211

Liberia
Ministry of Posts and Telecommunications
Monrovia, Liberia
Telex: 4551 MINPOSTEL LI

Libya
Directorate General of Posts and Telecommunications
Tripoli, Libya
Telex: 20000 BARIDAM LY

Liechtenstein
Office des Relations Internationales
Regierung des Furstentums Liechtenstein
FL-9490 Vaduz, Liechtenstein
Telex: 77855 IPVDZ FL

Luxembourg
Administration des Postes et Telecommunications
P.O. Box 999
17 Rue de Hollerich
2999 Luxembourg
Tel.: (352) 499 14 27 or (352) 499 17 22
Telex: 3410 ptdt lu

Macao
Director Geral
Companhia de Telecomunicacoes de Macau S.A.R.L.
P.O. Box 868
Macao (via Hong Kong)
South China
Tel.: (853) 55 8448

Madagascar
Societe des Telecommunications Internationales
de la Republique Malagasy
B.P. 763
Antananarivo, Madagascar
Telex: 22398 STIMAD MG

Malawi
The Postmaster General
P.O. Box 537
Blantyre, Malawi
Telex: 4100 POSTGEN MI or 4422 POSTSERV MI

Malaysia
Syarikat Telekom Malaysia
The Telecommunications Company of Malaysia
Bukit Mahkamah
Jalan Raja Chulan
50672 Kuala Lumpur, Malaysia
Tel.: (60) 3 232 9494

Maldive Islands
Maldive Department of Posts and Telecommunications
34 Marine Drive, Maafannu
Male 20-02
Republic of Maldives
Tel.: (960) 2805

Malta
Wireless Telegraphy Branch
Auberge de Castille
Valletta, Malta
Telex: 406 1100 MOD MLT MT

Mauritania
Office des Postes et Telecommunications
de la Republique Islamique de Mauritanie
Nouakchott, Mauritania
Telex: 600 MTN MQ

Mauritius
Overseas Telecommunications Services Corp. Ltd
P.O. Box 65
Port Louis, Mauritius
Telex: 4210

Mexico
Direccion General de Telecomunicaciones
Torre Central de Telecomunicaciones
Eje Central Lazaro Cardenas No 567
Mexico 12 D.F.
Tel.: (52) 5 530 30-60

Mongolia
Ministry of Communications
Ulan Bator, Mongolia
Telex: MH

Morocco
Ministere des P&T
Direction des Telecommunications
Rabat, Morocco
Telex: GENTEL B 31001 or GENTEL C 31948

Mozambique
Telecomunicacoes de Mocambique
Caixa Postal 25
Maputo, Mozambique
Telex: 6509 DGTDM MO

Nepal
Nepal Telecommunications Corp.
Singha Durbar
Kathmandu, Nepal
Telex: NP201 TELCOM NP

Netherlands
Netherland Postal and Telecommunications
P.O. Box 30000
2500 GA The Hague, The Netherlands
Tel.: (31) 70 75 86 11

New Caledonia
Direction de l'Office des P&T
Noumea, New Caledonia
Telex: POSTEL 089 MN

New Zealand
Telecom Corporation of New Zealand Ltd
7-27 Wakefield Quay
Wellington, 1 New Zealand
Tel.: (64) 4 738 444

Nicaragua
Empresa Nicaraguense de Telecomunicaciones
Managua, Nicaragua
Telex: 1548 NU

Niger
Office des Postes et Telecommunications
Direction Generale de l'Office
Niamey, Niger
Telex: 5209 OFPOSTEL NI

Nigeria
Cable and Wireless Ltd
c/o Business Communication (Nigeria) Ltd
17/19 Festival Road
Victoria Island
Lagos, Nigeria
Tel.: (234) 659 966
Telex: 11001

Norway
Norwegian Telecommunications Administration
P.O. 6701
St. Olavs plass
N-0130 Oslo 1, Norway
Tel.: (47) 2 488-990

Oman
General Telecommunication Organization
P.O. Box 3789
Ruwi, Muscat, Oman
Tel.: (968) 701 844

Pakistan
Pakistan Telegraph & Telephone Department
Government of Pakistan
Islamabad, Pakistan
Telex: 5824 DGIBA PK

Panama
Intel
Apartado 9A 659
Panama City, 9A, Panama
Tel.: (507) 64-0271

Papua New Guinea
Posts and Telecommunications Corp.
P.O. Box 1349
Boroko, Papua-New Guinea
Telex: PNGXTEL NE 22254 NE

Paraguay
Adm. Nacional de Telecomunicaciones (Antelco)
Dir. Asuntos Internacionales-Casilla Correo 84
Asuncion, Paraguay
Telex: 412PY DAIANTELCO PY

Peru
Empresa Nacional de Telecomunicaciones
Las Begonias 475-3ER Piso
Lima 27, Peru
Tel.: (51) 421-388

Philippines
International Services
Philippine Long Distance Telephone Company
R. Coujuangco Bldg. Legaspi Village
Makati, Metro Manila, Philippines
Tel.: (63) 816-8121

X.25 Packet Switched Network
Philippine Telegraph and Telephone Co.
Spirit of Communications Center
106 Alvardo Street, Legaspi Village
Makati, Metro Manila, Philippines
Tel.: (63) 818-0511

IRCs with Local Offices
Service via WUI Communications
Eastern Telecommunications Philippines, Inc.
Telecoms Plaza Bldg
306 Sen Gil J. Puyat Ave, Salcedo Village
Makati, Metro Manila, Philippines
Tel.: (63) 815-8920 or 816-0001

Service via RCA Global Communications (a unit of GE)
Philippines Global Communications, Inc.
8755 Paseo de Roxas
Makati, Metro Manila 3117, Philippines
Tel.: (63) 817-0246 or 817-0776

Service via ITT World Communications
Globe Mackay Cable and Radio Corp.
ITT Building
669 United Unions Avenue
Manila, RP 2801, Philippines
Tel.: (63) 521-3550 through 3554

Poland
Glowny Urzad Telekomunikacji Miedzymiastowej
00-686 Warsaw, Barbary 2, Poland
Telex: 813423 GUTM PL

Portugal
Companhia Portuguesa Radio Marconi (CPRM)
Praca Marques de Pombal, 15-7
Lisbon 1200, Portugal
Tel.: (351) 1 370 051

Puerto Rico
National Services
Puerto Rico Telephone Company
1500 Roosevelt Avenue
San Juan, Puerto Rico
Tel.: (809) 792-8480

IRCs with Local Offices
ITT World Communications, Inc.
665 Ponce de Leon
San Juan, Puerto Rico 00904
Tel.: (809) 721-2523

RCA Global Communications, Inc. (a unit of GE)
P.O. Box 3388
Old San Juan, Puerto Rico 00904
Tel.: (809) 724-2574

Qatar
Qatar National Telephone Service
P.O. Box 14
Doha, Qatar
Tel.: (974) 400 222

Romania
Direction Generale des P&T—Service du Trafic
B-Dul Dinicu Golescu 38
77 113—Bucuresti, Romania
Telex: 11372 DGPTC R

Rwanda
Dirgentel
B.P. 1332
Kigali, Rwanda
Tel.: (250) 6777

Saudi Arabia
International Services
Cable and Wireless
Saudi Arabia Ltd
P.O. Box 6196
Riyadh 11442
Saudi Arabia
Tel.: (966) 1 465 7092

National Services
Ministry of Post, Telegraph and Telephone
Riyadh 11112
Saudi Arabia
Tel.: (966) 1 463 1152

Senegal
Societe Nationale des Telecommunications
Internationales du Senegal
B.P. 69
6, rue Wagane Diouf
Dakar, Senegal
Tel.: (221) 23 10 23

Seychelles
Cable and Wireless PLC
P.O. Box 4
Mahe, Seychelles
Tel.: (248) 22 221

Sierra Leone
Sierra Leone External Telecommunications Ltd
P. O. Box 80
Freetown, Sierra Leone
Tel.: (232) 22801
Telex: 3210

Singapore
Telecom Singapore
31 Exeter Road
Singapore 0923, Republic of Singapore
Tel.: (65) 734 3344

Solomon Islands
Solomon Island International Communication Ltd
P.O. Box 148
Honiara
Solomon Islands, South West Pacific
Tel.: (677) 215 76

Somali
Telecommunications Department
Mogadishu, Somali
Telex: SM

South Africa
Telecommunications Commercial
Private Bag X74
Pretoria 0001
South Africa
Tel.: (27) 12 293 1156

South Korea
Data Communications Corp. of Korea
Korea Stock Exchange Building
33 Yeoeido-Dong, Yeongdungpo-Ku
Seoul, South Korea
Tel.: (82) 2-784-0381

South Yemen
Cable and Wireless PLC
P.O. Box 168
Sanaa, Yemen Arab Republic
Tel.: (967) 2 231800
Telex: 2224

Spain
International Telephone Services
TELEFONICA
Departamento Comercial Negocios
Ventas Internacional
Paseo de Recoletos, 37
Madrid 28001, Spain
Tel.: (34) 1 410 00 38

Sri Lanka
Overseas Telecommunications Service
P.O. Box 235—OTS Building—Duke Street
Colombo 1, Sri Lanka
Telex: 21110 MUX COLOMBO CE

Sudan
Sudan Telecommunications Corp.
Ministry of Transport and Communications
Khartoum, Sudan
Telex: 299 GENTEL KM SD

Suriname
Telecommunicatiebedrijf Suriname
Heiligenweg Paramaribo Republiek—P.O. Box 1839
Paramaribo, Suriname
Telex: LATEL SN 131 SN

Swaziland
Posts and Telecommunications
P.O. Box 125
Mbabane, Swaziland
Telex: 2033 WD

Sweden
Televerket
Marbackagatan 11
Farsta S-123 86, Sweden
Tel.: (46) 8 713 1000

Switzerland
Direction Generale des PTT Suisses
Viktoriastrasse 21
3030 Berne, Switzerland
Tel.: (41) 31 62 11 11

Syria
Minister of PTT
Telecom Direction
Damascus, Syria
Tel.: (963) 11-119855 or 11-11220

Taiwan (Republic of China)
International Telecommunications Administration
31, Ai-kuo East Road
Taipei, Taiwan (Republic of China)
Tel.: (886) 2 344 3770

Tanzania
Tanzania Posts and Telecommunications Corp.
P.O. Box 9070
Dar-es-Salaam, Tanzania
Tel.: (251) 51 31 155

Thailand
Communications Authority of Thailand
New Road
Bangkok 5, Thailand
Tel.: (66) 2 233 1050

Trinidad and Tobago
Textel
P.O. Box 3
Corner of Edwards Street and Independence Square
Port of Spain
Trinidad, West Indies
Tel.: (809) 62 54431

Tunisia
Direction Generale des Telecommunications
3 Rue Bis D'Angleterre
Tunis, Tunisia
Telex: GENTEL 12350 TN or GENTEL 13360 TN

Turkey
Directorate General of PTT
Department of Telegraphs and Telephones
Ankara, Turkey
Telex: (607) 42400
Tel.: (90) 41 125252

Uganda
Uganda Posts and Telecommunications Corp.
P.O. Box 7171
Kampala, Uganda
Tel.: (256) 41 30 540

United Arab Emirates
Business Communications (U.A.E.) PVT Ltd
Emirates Telecommunications Corporation
P.O. Box 2534
Abu Dhabi
United Arab Emirates
Tel.: (971) 2 720646

United Kingdom
International Services
British Telecom International
Room 723
Holborn Center
120 Holborn
London EC1N 2TE, United Kingdom
Tel.: (44) 1 936 2743

Mercury Communications Ltd
82-83 Blackfriars Road
London SE1 8HA, United Kingdom
Tel.: (44) 1 633-9577

International Packet Switching
Packet SwitchStream (British Telecom)
G07 Lutyens House 1-6 Finsbury Circus
London EC2M 7LY, United Kingdom
Telex: 883040 G

International and National Services
Mercury Communications, Limited
90 Longacre
London WC2E 9NP, United Kingdom
Tel.: (44) 1 836-2449

Upper Volta
Direction Generale de l'Office des P&T
Ouagadoudou, Upper Volta
Telex: 5200 UV

Uruguay
Administracion Nacional de Telecomunicaciones
Casilla de Correo 1477/Paraguay
2431 Montevideo, Uruguay
Tel.: (598) 23 53 72

USSR
Ministere des P&T de l'URSS
7 Rue Gorki
Moskva K-375, USSR
Telex: 412961 MTG SU

Vanuatu
Vanuatu Posts and Telecommunications Department
Port Villa, Vanuatu
Tel.: (678) 2000

Venezuela
Compania Anonima Nacional de Telefonos de Venezuela
Avenida Libertador
Edificio C.A.N.T.V.
Caracas 101, Venezuela
Tel.: (58) 02 782-5346 or 02 500-1111
Telex: 21380

Vietnam
Direction Generale des P&T
18, Rue Nguyen Du
Ha Noi, Vietnam
Telex: VT

Virgin Islands (United States)
ITT Telecommunications Operating Company, Inc.
P.O. Box 3768
St. Thomas 00801, U.S. Virgin Islands
Tel.: (809) 774-6750

Western Samoa
Western Samoa Post Office Department
Apia, Western Samoa
Tel.: (685) 23 456

Yemen
Yemen Telecommunications Corp.
P.O. Box 1256—Tawahi
Aden, Yemen
Telex: 2300 YEMENTEL AD

Yugoslavia
PTT Zajednica
Community of Yugoslav PTT
Palmoticeva, 2
YU—11000 Beograd, Yugoslavia
Tel.: (38) 011 338 921
Telex: 11421 YU GENTEL YU

Zaire
Exploitation des Telecommunications
B.P. 13798
Kinshasa-1, Zaire
Telex: 21360 GENTEL ZR

Zambia
Posts and Telecommunications Corp.
P.O. Box 71660
Broadway Road
Ndola Copperbelt Province, Zambia
Tel.: (260) 26 5201

Zimbabwe
Posts and Telecommunications Corp.
P.O. Box 8061—Causeway
Harare, Zimbabwe
Tel.: (263) 791 711

APPENDIX C

BIBLIOGRAPHY

Auerbach Report, "Electronic Office, Management and Technology," Auerbach Publishers, Inc., Pennsauken, NJ, 1986.

Bergerud, Marly, and Gonzalez, Jean, *Word/Information Processing Concepts*, John Wiley & Sons, Inc., New York, 1981.

Bleazard, G. B., *Why Packet Switching*, NCC Publications, Manchester, United Kingdom, 1979.

Business Week, "For Videotex, the Big Time Is Still a Long Way Off," Jan. 14, 1985.

Business Week, "Why the French Are in Love with Videotex," Jan. 20, 1986.

CCITT, "VIIIth Plenary Assembly—Document 66, Study Group VII—Report R 38," Geneva, 1984.

CCITT Recommendation D.1, "Recommendations for the Lease of International Telecommunications Circuits for Private Service," CCITT, Geneva, 1984.

CCITT Recommendation D.3, "Special Conditions for the Lease of Intercontinental Telecommunication Circuits for Private Service," CCITT, Geneva, 1984.

CCITT Recommendation D.6, "General Principles for the Lease of International (Continental and Intercontinental) Private Telecommunication Circuits," CCITT, Geneva, 1984.

CCITT Recommendation G.711, "Pulse Code Modulation (PCM) of Voice Frequencies," *Yellow Book*, Vol. III, Fascicle III.3, CCITT, Geneva, 1981.

CEPT Information Technologies and Telecommunications Task Force, "Study of the Introduction of Integrated Services Digital Networks with the Community," Scientific Control Center (SCS), Bonn, the Federal Republic of Germany and Scicon Limited, CEPT, London, 1985.

CEPT Special Group ISDN (GSI), "1982 Report on Integrated Services Digital Network Studies—ISDN in Europe," Stockholm, Nov. 1982, CEPT, Geneva, 1984.

Cheong, V. E., and Hirschheim, R. A., *Local Area Networks Issues, Products and Development*, John Wiley & Sons, Inc., Chichester, United Kingdom, 1983.

Communications Europe, "Cellular Now, ISDN Later," Jan. 1986.

Computer Networks and ISDN Systems, "ISDN and Value-Added Services in Public and Private Networks," Vol. 10, 1985.

Cypser, R. J., *Communications Architecture for Distributed Systems*, Addison-Wesley, Reading, MA, 1978.

Data Communications, "Vendors Want Electronic-Mail Blueprint. Will X.400 Be It?," Jan. 1986.

Data Communications Concepts, GC21-5169-4, International Business Machines Corporation, Rochester, MN, 1983.

Datamation, "Videotex: Into the Cruel World," Sept. 15, 1984.

Datapro, "Telecommunications Guide," Datapro Research Corporation, Delran, NJ, 1986.

Davies, D. W., Barber, D. L. A., Price, W. L., and Solomonides, C. M., *Computer Networks and Their Protocols*, John Wiley and Sons, Inc., Chichester, United Kingdom, 1979.

Deasington, Richard, *A Practical Guide to Computer Communications and Networking*, Ellis Horword Limited, Chichester, United Kingdom, 1984.

Doll, Dixon R., *Data Communications*, John Wiley & Sons, Inc., New York, 1978.

Duc, Nguyen Q., and Chew, Eng K., "Evolution Towards Integrated Services Digital Networks," *Telecommunications Journal of Australia*, Vol. 34, No. 2, 1984.

The Economist, "The World on the Line," Nov. 23, 1985.

EDP Industry Report, "Large Scale Computer Census," Vol. 21, No. 4, 1985.

EDP Industry Report, "Medium-Scale Computer Census and Small-Scale Computer Census," Vol. 21, No. 5, 6, 1985.

EDP Industry Report, "Personal Computer Census," Vol. 21, No. 8, 1985.

Eurodata Foundation, "Data Communications in Europe, 1983–1991, A Management Summary," London, 1984.

Eurodata Foundation Yearbook 1985, "Revision of Tariffs," Eurodata Foundation, London, 1985.

Eurodata Foundation Yearbook 1986, "Data and Text Communications Services in Europe," Eurodata Foundation, London, 1986.

Financial Times Business Information, Limited, "Fibre Optics in the UK and USA," London, 1985.

Flint, David C., *The Data Ring Main, an Introduction to Local Area Networks*, Wiley-Heyden Publication, Chichester, United Kingdom, 1983.

Frantzen, V., "Packet-Switched Data Communications Services in the ISDN," in *Proceedings of the 6th International Conference of Computer Communications*, Publishing Services, IEEE, New York, 1982.

Green, James H., *Local Area Networks, A User's Guide for Business Professionals*, Scott, Foresman and Company, Glenview, IL, 1985.

Griesinger, Frank K., *How to Cut Costs and Improve Service on your Telephone, Telex, TWX and Other Communications*, McGraw-Hill Book Company, New York, 1977.

IBM Systems Journal, "Teletex—a Worldwide Link Among Office Systems for Electronic Document Distribution," Vol. 22, No. 1/2, 1983.

ICC Policy Statements on Telecommunications and Transborder Data Flows, "The Liberalization of Telecommunication Services—Needs and Limits," The International Chamber of Commerce, Paris, 1982.

ICC Policy Statements on Telecommunications and Transborder Data Flows, "An International Programme for Homologation/Certification of Equipment Attached to Telecommunications Networks," The International Chamber of Commerce, Paris, 1983.

ICC Policy Statements on Telecommunications and Transborder Data Flows, "Information Flows—Analysis of Issues for Business," The International Chamber of Commerce, Paris, 1984.

ICC Policy Statements on Telecommunications and Transborder Data Flows, "International Private Leased Circuits—the Business User's View," The International Chamber of Commerce, Paris, 1984.

Information from the Deutsche Bundespost, FTZL 16-14 Order No. 179E, Federal Republic of Germany, March 1983.

Institute for the Future and Arlen Communication Inc., "1985 Videotex—Teleservices Directory," Menlo Park, CA and Bethesda, MD, 1984.

International Herald Tribune "Electronic Route to Being Your Own Travel Agent," Paris, France, August 16, 1985.

International Standards Organization (ISO) 7498, "Information Processing System—OSI Basic Reference Model," ISO, Geneva, 1985.

International Telecommunications Rates, AT&T Communications, Morris Plains, NJ, 1984.

Jay, Frank (ed. in chief), and Goetz, J. A. (chairman SCC 10), *IEEE Standard Dictionary of Electrical and Electronic Terms*, Wiley-Interscience, New York, 1984.

Kee, Richard, and Lewin, David, *ISDN: The Commercial Benefits*, Ovum Ltd., London, 1986.

MAPTEK Europe, "PTTs in Northern Europe Maintain Tight Control over Equipment Supply," Vol. 84, No. 56, Quantum Science Corporation, London, 1985.

Martin, James, *Telecommunications and the Computer*, 2nd ed., Prentice-Hall, Inc., Englewood Cliffs, NJ, 1976.

Martin, James, *Communications Satellite Systems*, Prentice-Hall, Inc., Englewood Cliffs, NJ, 1978.

Meijers, Anton, and Peeters, Paul, *Computer Network Architectures, Pitman Books Limited, London, 1982.*

Meyer, Carl H., and Matyas, Stephen M., *Cryptography: A New Dimension in Computer Data Security*, John Wiley & Sons, Inc., New York, 1982.

Nora, Simon, and Minc, Alain, *The Computerization of Society: A Report to the President of France*, M.I.T. Press, Cambridge, MA, 1980.

Organization for Economic Cooperation and Development (OECD), "Telecommunications Pressures and Policies for Change," Paris, 1983.

Public Data Networks in the Nordic Countries: Denmark, Finland, Norway, Sweden, Televerket, Stockholm, March 1978.

Quantum Science Corporation, *Conference Report: Volume IV, Strategic and Tactical Paths to 1990 and the Race Beyond*, London, June 11–13, 1985.

Rada, Juan F., and Pipe, G. Russell (eds., *Communications Regulation and International Business*, Elsevier Science Publishers, B.V., Amsterdam, 1984.

Roberts, Steven (consultant ed.), *International Directory of Telecommunications*, Longman Group Limited, Harlow, Essex, United Kingdom, 1984.

Rosner, Roy D., *Packet Switching, Tomorrow's Communications Today*, Lifetime Learning Publications, Belmont, CA, 1982.

Rullo, Thomas A. (ed.), *Advances in Data Communications Management*, Vol. 1, Heydon & Son, Inc., Philadelphia, 1980.

Satellite Communications, "Jumping the Videoconferencing Hurdles," Jan. 1986.

Satellite Communications, "Getting that License," Feb. 1986.

Schwartz, Mischa, *Computer-Communication Network Design and Analysis*, Prentice-Hall, Inc., Englewood Cliffs, NJ, 1977.

Siemens Telecom Report, "Services in the ISDN," Vol. 6, Munich, April 1983.

Siemens Telecom Report, "Integrated Services Digital Network (ISDN)," Vol. 8, Munich, April 1985.

Sippl, Charles J., *Data Communications Dictionary*, Van Nostrand Reinhold Company, New York, 1976.

Tanenbaum, Andrew S., *Computer Networks*, Prentice-Hall International, Inc., Englewood Cliffs, NJ, 1981.

Tapscott, Don, *Office Automation: A User-Driven Method*, Plenum Press, New York, 1984.

Tariff F.C.C. Number 4, AT&T Switched Digital Service, U.S. Government Printing Office, Washington, D.C., April 1985.

Tariff F.C.C. Number 9, AT&T Private Lines Services—Interoffice Channels, U.S. Government Printing Office, Washington, D.C., April 1985.

Tariff F.C.C. Number 10, AT&T Mileage Information and Administrative Matters, U.S. Government Printing Office, Washington, D.C., April 1985.

Tariff F.C.C. Number 11, AT&T Private Line Local Channel Services, U.S. Government Printing Office, Washington, D.C., April 1985.

Telecommunications, "Harmonization of New Telecommunications Services in Europe," Oct. 1983.

"Telecommunications Policies in Seventeen Countries: Prospects for Future Competitive Access," *Document NTIA-CR 83-24*, U.S. Government Printing Office, Washington, D.C., 1983.

Telecommunications Policy, "Emerging Economic Constraints on Transborder Data Flows," Butterworth & Company Limited, London, 1985.

Teleteknik, "The Benefits and Costs of Competition in International Communications and the Impact on Small Administrations," English ed., No. 1, 1985.

Transnational Data and Communications Report, "Trade Barriers to Telecommunications, Data and Information Services," Vol. VI, No. 6, Transnational Data Reporting Service, Inc., Washington, D.C., 1983.

Transnational Data and Communications Report, "Transborder (Private) Data Flow and the International Airlines," Vol. VII, No. 5 & 6, Transnational Data Reporting Service, Inc., Washington, D.C., 1984.

Transnational Data and Communications Report, "TDF Barriers to Services Trade," Vol. VIII, No. 4, Transnational Data Reporting Service, Inc., Washington, D.C., 1985.

Transnational Data and Communications Report, "ISDN Plans in Thirty Countries," Transnational Data Reporting Service, Inc., Vol. VIX, No. 8, Washington, D.C., 1986.

Tydeman, John, Lipinski, Hubert, Adler, Richard P., Nyhan, Michel, and Zwimpfer, Laurence, *Teletext and Videotex in the United States*, McGraw-Hill Publications, New York, 1983.

Tymnet International, "Worldwide Information System on Telecommunications," San Diego, 1985.

Vocabulary for Data Processing, Telecommunications and Office Systems, International Business Machines Corporation, Poughkeepsie, 1981.

"The World of International Communications," AT&T International Department, Morris Plains, NJ, 1984.

World Telecommunications, "Latin America, Africa, Asia, and the Pacific, 1971," Vol. 4, Arthur D. Little, Cambridge, MA, 1971.

APPENDIX D

GLOSSARY OF TERMS

Access method
A technique for moving data between main storage and input/output devices. For example, a host computer resident program that provides a communications capability for application programs. It interfaces between the application program and the communications controller or front-end processor.

Acoustic coupler
A device that converts electrical signals into audio signals and vice versa and enables data transmission over the public switched telephone network with a regular telephone handset.

Addressing
The means for a sending or control station to select the unit destination for a message.

Alternate routing
A technique using an alternate path to transmit data when the usual path is not available. Normally used between nodes in public and private networks.

Amplitude modulation
The process by which a continuous high-frequency wave or carrier is caused to vary in amplitude by the action of another wave carrying information. The variation of a carrier signal's strength (amplitude) is a function of the information signal.

Analog facsimile
A system for the transmission of images by converting optical data derived from scanning an original into an electrical analog signal for transmission over voice-grade lines. A typical analog transmitter scans an entire original (characters, graphic

symbols, letterheads, and white space). Each picture element of the original is represented by an analog electrical signal. The analog signals create a continuous electrical waveform that is used to drive a printer or reproduce a series of picture elements resembling the original.

Analog signals

A representation of the value of a variable by a physical quantity that is considered to be continuous. The information content of an analog signal conveyed by the value or magnitude is directionally proportional to the characteristics of the signal such as amplitude, phase, or the duration of a pulse. To extract the signal, it is necessary to compare the value or magnitude of the signal to a standard.

Analog-to-digital converter

A device that converts analog signals to digital signals. Used for converting signals from the public switched telephone network back to digital form. *See* Digital-to-analog converter.

Analog transmission

Transmission of a variable signal with a range of values in signal magnitude between a maximum and minimum value as contrasted with a binary transmission, which has two discrete values.

Archiving

The storage of backup files for a specified period of time. An archival system should have an index of archived files such as name, date, media, file type, length of file, and reference number.

ASCII (American Standard Code for Information Interchange)

A code standard for digital communications using a coded character set consisting of 7 bit-coded characters (8 bits including parity). Used for information exchange among data processing systems, data communications systems, and associated equipment. The ASCII set consists of control characters and graphic characters.

ASR (automatic send/receive)

A communicating device that operates in an automatic send/receive mode of transmission.

Asynchronous mode

Unexpected, unpredictable, or without a regular time relationship. In data transfer, the transmission of one character at a time, preceded by one or two start bits, followed by a stop bit in order to ensure correct receipt at the remote end. Contrast with synchronous data transfer.

Asynchronous transmission

Data transmission in which each data character is individually synchronized on a communication channel by using one or two start bits and a stop bit. Also called start-stop transmission.

AT&T-C

American Telephone & Telegraph Communications, Inc. is a long distance common carrier providing voice, data, video, facsimile, and other services within the United States and to Canada, Mexico, and overseas locations.

Attenuation

A decrease in the magnitude of current, voltage, or power of a signal in transmission between points. It may be expressed in decibels or nepers.

Audio processing

A term used to describe the control, transmission, storage, and recreation of sound, including the human voice.

Autodial

Used in place of a manual telephone dial or keypad in order to automatically dial a number or a group of prerecorded telephone numbers. A modem or X.21 feature on circuit switched data networks and packet switched data networks.

Automatic answer

A machine feature that permits a station to respond to a call it receives over a switched line without any operator action. Provides unattended operation.

Automatic call or autocall

A machine feature connecting to an external device that permits a station to initiate a connection to another station over a switched telephone line without operator intervention.

Background processing

In a computer system programmed for foreground and background processing, critical events are assigned to the foreground partition for immediate processing. Lower-priority or less critical tasks are assigned to the background partition and are processed in the remaining time available after foreground processing.

Bandwidth

A group of consecutive frequencies constituting a band that exists between limiting frequencies. For data transmission, the wider the band, the greater the number of bits that can be sent over a communications channel in a given time. The difference between the two limiting (high and low) frequencies of a band is expressed in hertz. On a digital circuit, the term is called the data rate or line transmission capacity.

Baseband signaling

In the process of modulation, the frequency band occupied by the aggregate of the transmitted signals when first used to modulate a carrier. The term is commonly applied to cases where the ratio of the upper to the lower limit of the frequency band is large compared to 1.

Batch communication

The sending of information or data from one work station in a communications network to a host for processing, or host to work station without intervening data responses from the receiving unit. Contrast with interactive and transaction processing.

Baud

A unit of signaling speed equal to the number of discrete conditions or signal events per second. For example, 1 baud equals ½ dot cycle per second in Morse code or 1 bit per second in a train of binary signals. Telex and teletypewriter communications line speeds are usually rated in baud.

Baud rate
The number of information bits that can be transmitted in 1 second is the baud rate. By definition, the reciprocal of the time of the shortest signal element in a character. For example, 1 baud equals 1 bit per second in a series of binary signals.

Bidirectional
Operating in two directions. Can be simultaneous or one way at a time.

Binary synchronous communications (BSC)
Communications using binary synchronous line discipline. A uniform procedure, using a standard set of control characters and control character sequences for synchronous transmission of binary coded data between stations.

Bis
Appended to a CCITT network interface standard and identifies a second version of the standard. For example, X.21 bis. *See also* Ter.

Bit
A contraction of *binary digit*. It is the smallest element of a computer code with a value of either one or zero. One electrical pulse in a group of pulses or a pulse train.

Bit-oriented protocol
A protocol technique using field position for communications control. Includes ADCCP, HDLC, and SDLC. Contrast with byte-oriented protocol.

Bit rate
The speed at which serialized data is transmitted. Usually expressed in bits per second (bps), kbps (1000 bps), or Mbps (million bps).

BOC
Part of the original Bell System, called Bell Operating Companies. After divestiture, most of the 24 Bell Operating Companies were grouped into seven separate RBOCs (Regional Bell Operating Companies). *See* RBOC.

BPS (bits per second)
In serial transmission, the instantaneous bit speed with which a device or channel transmits a bit.

Broadband
A technique in which a number of channels are provided over a communications medium (such as a LAN cable), usually by frequency division multiplexing (FDM). Information is transmitted across the available bandwidth.

Buffer
A separate device or storage area in a computer or peripheral device used for temporary storage of data in order to compensate for a difference in instantaneous data flow rate from one device to another.

Bypass
A new networking term that allows access to an Interexchange carrier without using the facilities of a local carrier. The network provider in this case uses privately owned or leased transmission media (e.g., microwave antennas or satellite ground support stations) to complete calls between sites without using the facilities of intra/inter-LATA carriers.

Byte
A small group of bits (usually eight) handled as a unit. Interchangeable with characters, e.g., an 8 bit-byte equals one character.

Byte-oriented protocol
A protocol technique using defined characters (bytes) from a code set for communications control. Examples are ASCII, BSC, and DDCMP (DECnet protocol). Contrast with bit-oriented protocol.

Carrier
A continuous known reference frequency that can be modulated or impressed with a second (information carrying) signal to carry the information.

CCIR
The International Radio Consultative Committee is a main committee of the ITU (International Telecommunication Union). CCIR conducts studies, issues opinions concerning technical questions, and sets frequencies for radio, television, satellite, and other services that transmit and/or receive through the air.

CCITT
The International Telegraph and Telephone Consultative Committee (CCITT) is a main committee of the ITU. It is an organization of common carriers or PTT administrations that meet periodically to define telecommunications standards that they will mutually adopt.

CEPT (European Conference of Postal and Telecommunication Administrations)
CEPT is a European organization of postal and telecommunication administrations (PTTs) from 26 countries. Its principal objective is to ensure closer relations between member organizations and to use more effectively their technical potential in preparing joint proposals for submission to international postal and telecommunications organizations.

Certification
A formal statement made on the successful completion of a verification process by an independent government licensing authority.

Character
A letter, number, or symbol such as an asterisk, comma, or percent sign used for data or control in communication devices.

Circuit switching
A switching technique where the connection is made between the calling party and the called party prior to the start of communications.

Clocking
A method of controlling the number of data bits sent on a data communications line in a given period.

Closed user group
A number of subscribers to a public switched data communication service who have established with the carrier a user community having the ability to communicate with one another. Access is not available to other users of the same communications service.

Coaxial cable
A cable consisting of one conductor, usually a small tube of electrically conducting material surrounding a central conductor held in place by insulators, which is used to transmit telephone, telegraph, and television signals of high frequency. Coaxial cable is commonly used for antenna systems from TV to satellite systems and between a work station and its controller or computer.

Code
The representation of information in a form differing from its original form but that can be understood by both sender and receiver.

Common carrier
An organization licensed to provide telecommunications facilities to the public. Usually referred to as PTT in countries other than North America.

Communication
The exchange of information between two or more parties using a common set of signs or symbols and a protocol. Communication tools include voice, writing, telephone, and electronic mail.

Communications adapter
A hardware feature that enables a computer system to be attached to a data communications network. Interfaces the computer to a modem, data sending unit, multiplexing equipment, direct high-speed connections, or LANs.

Compatibility
The ability of a machine to accept or process data prepared by another machine without conversion or modification of the data.

Conditioning
Adding equipment to a communication channel to improve the data transmission qualities of the channel. C- and D-type conditioning can be used for data transfer on private leased lines in North America. In other countries that provide a private leased line service, conditioning is included in the monthly charge for a quality-grade line.

Connect Time
Elapsed time during which a terminal or communicating work station is connected to and functioning as a station in a computer system.

Connectivity
The ability to attach a range of devices to a work station, computer system, or network.

Contention
A condition on a communication channel when two stations attempt to use the same channel simultaneously.

CPS (characters per second)
A measure of the speed of an output device, e.g., 300 CPS.

Crosstalk
Undesired signals, such as voice signals, transferred to another cable in close proximity or in a conduit or duct.

CSDN (circuit switched data network)
Circuit switched data networks use digital switching techniques for data transmission. Their networks implement fast call setup and call clearing on a point-to-point connection.

CSMA/CD
Carrier sense multiple access with collision detection. An access protocol used in a variety of local area networks (LANs). Each station can sense the presence of carrier signals for other stations and thus avoid transmitting a packet that would result in a collision.

Data circuit termination equipment (DCE)
Equipment installed at the user's premises that provides all functions required to establish, maintain, and terminate a connection. DCE includes signal conversion and coding between the data terminal equipment (DTE) and the line.

Data communications
The transmission of data between systems and/or devices over a communications line.

Data compression
A technique to save storage space by eliminating gaps, empty fields, redundancies, or unnecessary characters. This reduces the length of records or blocks and saves transmission time.

Data link
Equipment, procedures, or protocols used for sending data over a communication line.

Data stream
All data transmitted over a data link in a single read or write operation.

Data terminal equipment (DTE)
A terminal, work station, PC, or computer system that is used to send or receive data and attaches to DCE (data circuit terminating equipment).

Datel
A measured use service for international voice and data transmission. Charges are based on the amount of traffic over the line. U.S. connections are provided by the international record carriers from their U.S. gateway cities to the overseas PTT's PSTN (public switched telephone network). A line speed of 1200 bps is considered the normal data rate, although additional speeds of 2400 and 4800 bps are possible.

DCE
See Data circuit terminating equipment.

DDCMP (Digital Data Communications Message Protocol)
A Digital Equipment Corporation discipline for transmitting data between stations in a point-to-point or multipoint data communications system for parallel, serial synchronous, or serial asynchronous data transfer.

DDS (DATAPHONE Digital Service)
AT&T-Communications provides DDS for interstate private line digital communications between major metropolitan cities. It offers two-way simultaneous (duplex) transmission of digital signals at synchronous speeds from 2400 bps to 56 kbps employing end-to-end digital technology.

Decibel
A unit of signal strength, such as a signal on a data communications channel. Usually expressed in dB.

Demodulation
Removing the data signal from the carrier signal to produce the original data signal. Opposite of modulation.

Destination system
A remote system or target system that has resources (data files, peripheral devices, application programs, or system facilities). In a network product offering that supports resource sharing, the destination system is programmed to respond to remote requests.

Developed country
One of the major countries of the world with an organized structure and utilities throughout the country. Some of the developed countries are Canada; all of Europe, including Czechoslovakia, Hungary, Poland, and Western Europe; Japan; Australia; New Zealand; the United States; and the USSR.

Developing countries
Countries that do not have an organized infrastructure and utilities throughout the country. They do not have a living standard as high as developed countries.

Dictionary
A stored vocabulary used for spelling verification and automatic hyphenation. Also, a file of data references for location of data in a database (data dictionary).

Digital data
Data represented by on and off (ones and zeros) conditions called bits.

Digital data network
A network that carries data represented in binary form. Digital data networks include DDS, T1, CSDNs (circuit switched data networks), and PSDNs (packet switched data networks).

Digital signal
The information content of the digital signal is concerned with discrete states of the signal such as the presence or absence of a voltage or a contact in an open or closed position. A digital signal is made up of a stream of on and off (1s and 0s) discrete pulses and has meaning based on the possible combinations of the discrete states of the signal.

Digital-to-analog converter
A device that converts digital signals to analog signals for transmission via an analog network, e.g., the public switched telephone network.

Disk
Rigid, random-access, high-capacity, magnetic storage medium. Disks may be removable (cartridges providing off-line storage) or nonremovable. Capacities range from 1 to well over 1200 Mb per disk.

Diskette
Magnetic-coated mylar disk enclosed in a protective envelope. Same as floppy disk.

Display station
An input output device containing a display screen on which data is displayed and includes an attached keyboard for entering data and commands.

Distortion
An undesirable change in a data communications waveform caused by transients as a result of modulation. The unwanted change in a waveform that occurs between two points in a transmission system.

Distribution list
Allows letters and memos to be sent to a group of persons. The capability of an operating system to store lists of persons and distribute mail to those persons.

Document
A collection of one or more lines of text that can be named and stored as a separate entity.

Document assembly
The ability to assemble new documents from previously recorded text or to join a document with variable information (such as names and addresses) to create a number of nearly identical documents but each addressed to a different person.

Dot matrix
A method of generating display characters in which each character is formed by a grid or matrix pattern of dots. For example, the CEPT Videotex standard uses a 10 × 12 dot matrix to form a character or symbol.

Down converter
A portion of the satellite receiver usually located at the antenna site. The receiver changes the high-frequency microwave signal to a lower frequency.

Downlink
The satellite-to-earth transmission channel.

Duplex
Pertains to simultaneous, two-way, independent transmission. Same as full duplex or FDX. Contrast with half duplex or HDX.

EBCDIC (External Binary Coded Decimal Interchange Code)
An 8-bit code (8 data bits) used to represent data and control characters.

Electronic mail
Electronic transmission of text information within an establishment, enterprise, or interenterprise. Includes telex, computer-based message systems, communicating word processors, and facsimile.

Emulation
The imitation of one system's modes of operation by another system so that they can communicate. Also applies to terminal protocol imitation.
End-to-end protocol
A protocol between two DTEs that are communicating via a virtual circuit over a PSTN, leased line, CSDN, or PSDN.

End user
Ultimate source or destination of information. An end user may be an operator or application program.
Enter
To input data from the screen to the computer by pressing a key. This key may be labeled ENTER, SEND, EXECUTE, etc.
Equal access
Access to the public switched network that is equal in quality and scope to that provided to AT&T Communications. Allows equivalent quality lines to competing common carriers.
European standard paper sizes

	Inches	Millimeters
A0	33.1 × 46.8	841 × 1189
A1	23.4 × 33.1	594 × 841
A2	16.5 × 23.4	420 × 594
A3	11.7 × 16.5	297 × 420
A4	8.3 × 11.7	210 × 297
A5	5.8 × 8.3	148 × 210

Facsimile (FAX) device
A machine employed to store or relay image data via telephone or transmission lines. The CCITT has classified Fax devices into four groups according to operating time for one A4 page (210 × 297 mm):

Group 1: analog devices, 4–6 minutes of operation time
Group 2: analog devices, 2–3 minutes of operation time
Group 3: digital devices, operation time less than 1 minute
Group 4: to be standardized later

FCC
Federal Communications Commission, Washington, D.C. Responsible for interstate regulations and tariffs of common carriers.
File
A collection of records; an organized collection of information directed toward some purpose. Also, a logical unit of data within a storage facility, e.g., a disk or diskette, that is accessible by an application program.

Floppy disk
A removable diskette or floppy disk as distinguished from a nonremovable rigid disk.

Foreground processing
The execution of a computer program that has top priority and preempts the use of computer facilities. Compare with background processing.

Format
The predetermined arrangement of text or data on a data medium. A format statement may include parameters such as fields, lines, page numbers, margin and tab settings, decimal tab settings, centering instructions, paragraph indentations, line spacing, and pitch size.

Four-wire circuit
A circuit that uses four wires (two for transmitting data and two for receiving data) for data transmission. Four wire is normally used for duplex transmission. However, newer modem developments allow duplex operation on two wires.

Frequency division multiplexing (FDM)
A system in which the available transmission frequency range is divided into subchannels for use by several devices.

Frequency modulation
Modifying the frequency of a carrier signal so it can carry data signals.

Front-end processor
A processor that can relieve a host computer of certain communications tasks such as line control, message handling, code conversion, error control, and communications application support functions.

Full duplex (FDX)
See Duplex

Gain
A term used to indicate the level of a signal, expressed in decibels (dB). The gain of an antenna is usually specified by the vendor.

Gateway
Communications product that provides a bridge between two different network protocols. Acts as a translator to change one network protocol to the attached network protocol and vice versa.

Giga (G)
A prefix for 1 billion times a specific unit.

Graphics
Graphics are designed to convey data relationships so as to optimize the user's powers of assimilation. Graphics are based on coded data in contrast to images, which are based on noncoded data.

Half duplex (HDX)
Permitting data communications in opposite directions on a communications link, one direction at a time.

Hardware
Physical equipment used in data or word processing, as opposed to software, programs, procedures, rules, and documentation. Contrast with software.

Harm
Includes electrical hazards to telecommunication agency personnel, damage to telecommunication agency equipment, malfunctions of telecommunication agency billing equipment, and degradation of service to persons other than the users of the subject terminal equipment or their calling or called party.

HDX
See Half duplex.

Hertz (Hz)
The unit of frequency, one cycle per second. Also called cycles per second.

Hierarchical communication
Usually terminal-to-host communications. The host controls all the resources of the network. Each nonhost node must establish a session with the host to access the network.

High-level data link control (HDLC)
Bit-oriented data link control procedure standardized by the International Standards Organization (ISO). In general terms, HDLC formats and protocols are very similar to IBM's SDLC formats and protocols.

Homologation
The process of proving to PTT administrations that the hardware or software attachment product to be connected to their network will not cause harm to the network or to other users on the network.

Host
The controlling or highest-level system in a data communications configuration.

Host computer
A computer that, in addition to providing a local service, acts as a central processor for a communications network.

IC
Interexchange carrier. A company that provides long distance carrier signals between carriers.

Integrated adapter
An interface that is implemented under the covers (integrated) of a product.

Integrated office
An office that permits all office personnel to perform, on a single logically unified computer network, any combination of operations for word, data, voice, and image processing.

Intelligent terminal
Refers to a terminal with some logic capability or a programmable terminal that may perform some of the functions of a computer.

Interactive operation
On-line conversation between a terminal user and a computer application. Normally, a two-way interaction with responses at both ends.

Interenterprise
Between two different legal entities. One enterprise consists of all its establishments, divisions, and subsidiaries.

Interface
A shared boundary. An interface may refer to a hardware component for linking two devices, an area of storage or registers that is accessed by two or more computer programs, or an application programming interface shared by two computer programs.

Inter-LATA
Between LATAs. *See* LATA.

Interoffice channel (IOC)
A connection between interexchange carrier central offices.

Interrate center channel
A connection that originates and terminates in different rate centers of the same LATA. *See* LATA.

Intraenterprise
Within the same enterprise. Usually between two establishments or units within the same overall corporate entity.

Intra-LATA
Within the same LATA. *See* LATA.

Intrarate center channel
A connection that originates and terminates in the same rate center.

ISO (International Standards Organization)
An organization of national standards bodies and liaison members promoting the development of standards to facilitate the international exchange of goods and services and to develop mutual cooperation in the spheres of intellectual, scientific, technological, and economic activity.

Kilo (k)
A prefix for 1000 times a specific unit.

LAN (local area network)
See Local area network.

LAP-B
Link access protocol-balanced is a subset of the official International Standards Organization HDLC protocol supporting balanced transmission. LAP-B is compatible with HDLC and is indicated in the CCITT X.25 specifications as the preferred access protocol.

LATA
Local access and transport area. A geographical boundary loosely based on a standard metropolitan statistical area as defined by the U.S. Census Bureau.

Leased line

A communications channel reserved for the private and sole use of the leasing customer. Also called a nonswitched line.

LEC

Local exchange carrier. A carrier serving only intra-LATA communications.

Letter size

Common U.S. letter paper sheet size (8½ × 11″).

Local area network (LAN)

A system for gathering or exchanging information within a limited geographical area, usually within a building or group of buildings in close proximity.

Logical channel

Logical channels exist between the packet terminal and the network node and identify the permanent or switched virtual circuit. Permanent virtual circuits are permanently assigned to logical channels at network subscription time. Switched virtual circuits are assigned to logical channels dynamically when the call is placed.

Memory

A term for main storage for any logic controlled device.

Menu

List of alternative operator actions supplied for operator selection.

Menu-driven

Interaction with the system is controlled by menus as opposed to commands. Initiated by a program or function key.

Message

A combination of characters and symbols transmitted from one point to another or from one user to another.

Message switching

The technique of receiving a message, storing it until the proper outgoing line is available, and then retransmitting it without using a direct connection between incoming and outgoing lines.

Microwave

A term used to signify radio waves in the frequency range of about 1000 megahertz (MHz) or of the gigahertz region.

Modem

See MOdulator-DEModulator.

Modulation

Altering a carrier signal to carry data signals on a communications channel.

MOdulator-DEModulator (modem)

A device that converts the digital output from a computer to an analog signal that can be transmitted over a telephone line and, in a like unit at a remote location, converts the analog signal back to digital at the receiving end for input to a computer.

Multidrop network

A network configuration in which there are one or more intermediate nodes on the communications path between a central node and endpoint node. A multidrop line has more than one physical communication line ending. Commonly referred to as multipoint operation.

Multileaving remote job entry (MRJE)

Usually a software system function that allows the submission and receipt of remote jobs to a host system over a BSC or SDLC communications line with concurrent two-direction operation.

Multiplexer

A hardware device for attaching several communications devices onto one link for simultaneous message transmission and receipt.

Multiplexing

The division of a communications line into two or more separate channels, either by separating it into independent frequency bands (frequency division multiplexing) or by assigning the same channel to different users at different times (time division multiplexing).

Multipoint line

A connection established among data stations for data transmission in three or more locations on the same channel using a common protocol. The connection may include switching facilities. Contrast with point-to-point line.

Network

An interconnected group of nodes and signal paths connecting input and/or output devices to a system.

Node

In a network, a point where one or more functional units interconnect transmission lines. For example, a node can be a user device or a communications processor. Nodes perform network routing and direct information toward its final destination.

Nonswitched line

A private line connection between systems or devices that does not have to be made by dialing. Usually leased from the common carrier or PTT. Also called leased line.

OCC

Other common carrier. An interexchange carrier other than AT&T Communications.

Octet

A group of 8 bits; synonymous with byte or character.

Office

A place in which services, clerical work, professional duties, or similar activities are carried out.

Office automation

The use of computers to perform normal office functions with greater efficiency and speed.

Official approval

Granting permission to connect to a PTT or Telco network. It may be called registration, certification, homologation, agreement to attach, or type approval.

On-line systems

A system in which input data enters the host directly from the point of origin and/or output data is transmitted directly to where it is used.

OSI (open systems interconnection)

ISO's (International Standards Organization) seven-layer reference model for interchange of information between computer systems having incompatible architectures.

PABX (private automatic branch exchange)

An automatic telephone exchange for internal and external telephone calls for voice and data lines within an organization and to public and/or private networks. *See also* PBX (private branch exchange).

Packet

A packet is the basic transmission unit on a virtual circuit across an X.25 network and consists of a packet header with a routing code and a data field.

Packet assembler/disassembler (PAD)

A packet mode DTE interfacing non-X.25 terminals (e.g., start-stop devices) to an X.25 network. CCITT Recommendations X.3, X.28, and X.29 define standards that several carriers have adopted relating to PAD protocols.

Packet switching

A method of transmission in which a message is divided into fixed length packets within the network. Compared to message switching systems, which route a message in its entirety.

Packet switching node

The X.25 network node to which a DTE is connected via a public or private leased line network access link.

Parallel transmission

Transmitting all bits of a character simultaneously.

Parity check

Checking a data character for an odd or even number of 1 bit. Used for error detection.

Password

A unique string of characters that an operator must supply to gain access to a system or program.

PBX (private branch exchange)

A manual or automatic telephone exchange for internal and external telephone calls for voice and data lines within an organization or to public and private networks. *See also* PABX (private automatic branch exchange).

Peer-to-peer communication
All nodes access network applications and resources. For example, in a PC network, one PC can send a file to another PC without the need for another processor to control the communications session. Any peer can initiate a session. Contrast to host-controlled system.

Permanent virtual circuit (PVC)
A permanent virtual circuit represents a point-to-point, nonswitched circuit across a packet switching (X.25) network, over which only data, reset, interrupt, and flow control packets can flow.

Personal calendar
A function that allows users to create, print, modify, and inquire into personal appointment schedules.

Phase modulation
Altering the phase of a carrier signal to convey data signals.

Pixel
The smallest directly addressable graphic unit. Also called picture element or pel.

Point to point
A channel connected between only two stations. A point-to-point channel can be a switched or nonswitched line or circuit.

Point-to-point line
A communication line that serves as a direct link between two locations. Contrast to multipoint line.

Polling
An ordered procedure for inviting stations on a channel to transmit data.

Presubscription
The ability to choose an interexchange carrier.

Primary station
On a point-to-point channel, the station that gains control of the channel first and on a multipoint channel, the station controlling communications. *See* Secondary station.

Principal
Term often used in office automation to refer to professional and knowledge workers.

Private line
The channel and channel equipment furnished to users as a unit for their exclusive use without interexchange switching arrangements.

Professional
An office worker, generally a specialist trained to gather and present information for a specific discipline such as engineering, finance, marketing, legal, accounting, personnel, customer service, or program development.

Protocol
A mutually agreed upon formal set of conventions governing the format and control of information exchange between two intelligent machines.

Protocol converter
Translation unit that permits a terminal using one communications language to "talk" to a terminal data network or host processor that uses another.

PSDN (packet switched data network)
A public or private network using packet switching techniques for store and forward of messages in standardized packets for end-to-end connectivity between terminals and/or computers. Store and forwarding is rapid, typically a fraction of a second as the packets pass from node to node in a network. Flow control in the network is designed to ensure that storage does not become overloaded while still maintaining line loadings as close to maximum as possible.

PSTN
Public switched telephone network.

PTT
An abbreviated term that represents the post, telephone, and telegraph administrations in the world. Does not apply to Canada and the United States. The national agencies providing public communication services. Private companies may also provide such services in some countries, normally under close government regulation.

Public data network (PDN)
A shared resource data network that offers data communications services to public subscribers.

Public utility commission
A state organization responsible for regulations and tariffs of common carriers operating within the state.

Queue
A line or list of items in a system waiting for service, for example, documents waiting to be printed out.

Rate center
A geographical subarea of a LATA. *See* LATA.

RBOC
Regional Bell Operating Company. One of seven companies formed in the breakup of AT&T.

Real time
Response to requests for service on demand, in contrast to time sharing, in which all requests for service are responded to on a round-robin basis in a predetermined time sequence.

Recognized private operating agency (RPOA)
A CCITT term for a nongovernmental organization that provides a network service, e.g., AT&T and British Telecom.

Remote processing
The use of a computer for processing data transmitted from remote locations.

Resolution
The perceived clarity of an image displayed on the screen or produced by a hard-copy printer. Printer resolution is expressed as the number of lines per unit of length. Image is measured in points per unit of area or pels per inch.

Response time
The time it takes for a data communications system to respond to a request. For example, if you enter a data field on a terminal keyboard, response time begins when you press the last key and ends when the first character of your answer is displayed at the terminal.

Ring
A type of topology (shape) of a local area network (LAN).

Ring network
A network topology with decentralized control in which messages are passed around the ring from node to node until they reach their destination or address.

RJE
Remote job entry.

RS-232-C
A technical specification published by the Electronics Industries Association (U.S.) that establishes a set of unique electrical and physical interface requirements between communicating equipment, e.g., modems.

Satellite earth station
Another name for a satellite receiving system or ground support equipment.

Satellite receiver
The portion of a satellite television system that includes the down converter and user-operated controls.

Scanner
A digitizing scanner, generally used for graphics that already exists in hard copy. It scans an entire page and defines the graphics in digital form to be stored by the computer.

SDLC (synchronous data link control)
An IBM protocol for managing synchronous, code-transparent, serial-by-bit information transfer over a link connection. Transmission exchanges may be duplex or half duplex over switched or nonswitched links. The configuration of the link connection may be point to point, multipoint, or loop.

Secondary station
The station that receives control from a primary station (e.g., on a multipoint channel). *See* Primary Station.

Serial transmission
Transmitting each bit of a data character separately (and sequentially) over one communications path.

Session
A logical connection established between two devices to allow them to communicate with each other. It is also the period of time during which programs or devices can communicate with each other.

Shared logic system
A multiterminal system where each terminal shares the word processing power, storage, and peripherals of a central computer.

Shared resource system
A multiterminal word processing system configuration whereby a number of intelligent, independent work stations share common peripheral devices such as printers, OCR readers, and disk files.

Signal-to-noise ratio
A specific ratio of the value of the signal amplitude to the noise amplitude. A number that expresses the relative strength of the desired signal to that of background noise. An insufficient signal-to-noise ratio results in static in a radio or snow in television. The number is often expressed in decibels (dB).

Simplex channel
A channel that can transmit data in only one direction.

SNA/remote job entry (SRJE)
Refers to the submission and receipt of a remote job communicating over an IBM Systems Network Architecture (SNA) synchronous data link control line or equivalent.

Software
Computer programs, procedures, rules, and associated documentation concerned with the operation of a computer system. Contrast with hardware.

Spelling aid
A program to aid the operator when a word not contained in the dictionary occurs, by suggesting the correct spelling of the word.

SRJE
See SNA/remote job entry.

Star network
A network configuration in which there is only one path between a central or controlling node and each endpoint node.

Start-stop
See Asynchronous transmission.

Station
A system or device that can send or receive data over a communications line. Normally called work station.

Stop bit
In a serial transmission, the last bit that provides subsequent detection of the start bit of the next asynchronous transmission.

Storage
The part of a computer system that stores information for subsequent use or retrieval.

Store and Forward
The handling of data (often messages) in a network by storing it until retransmission at a later time.

Switched access private line
A private line with one end at a user's premises and the other end accessing or accessed by a switched network.

Switched line
A connection between two stations that is established by dialing.

Switched virtual circuit (SVC)
A switched virtual circuit is a dynamically established virtual circuit, analogous to a switched (dialed) physical circuit. In addition to the protocols for transferring data that are available to a PVC (permanent virtual circuit), additional protocols are required to allow for call setup and call clearing of the switched virtual circuit. Also called virtual circuit.

Synchronous data link control (SDLC)
A discipline for the management of information transfer over a data communications channel. *See* SDLC.

Synchronous mode
A mode of data transmission in which all bits and characters are sent in uniform intervals of time or integral fixed constant multiples of the base bit and/or character times.

Synchronous transmission
Sending all bits of a character and all characters of a transmission at a constant rate with defined bit timing.

T1
T1 is a digital transmission system between Telco central offices designed to operate at 1.544 Mbps. T1 digital services for both voice and data are now available with local access for certain cities in Canada and the United States.

Tariff
The established monthly rates for data communications services.

Telco
The former Regional Bell Operating Companies, GTE Telephone, and other independent telephone operating companies.

Teleconferencing
The process of conferring between multiple groups in separate geographic areas by using telecommunications.

Teleprocessing
Processing data that is received from or transmitted to a remote location via communication channels.

Teletex
A high-speed public electronic mail service that will enable the communication of text between Teletex compatible terminals on a memory-to-memory basis. It is a follow-on replacement for telex.

Teletype

A Teletype Corporation trademark for a TTY (teletypewriter) device that uses asynchronous, start-stop transmission with keyboard, printer, and paper tape reader-/punch.

Telex

TELeprinter EXchange. An automatic exchange service originally provided by Western Union in the United States and now available throughout the world.

Ter

Appended to a CCITT network interface standard, it identifies a third version of the standard. *See also* Bis.

Terminal

A device usually equipped with a keyboard and display device, capable of sending and receiving information over a link. Also a point in a system or network at which data can either enter or leave.

Text

In an office context, text consists of strings of alphameric characters arranged to convey information.

Time division multiplexer (TDM)

A transmission system in which characters or bits belonging to different messages are transmitted at successive times on the same channel.

Throughput

The total useful information processed or communicated over a given period of time.

Token passing

A network access method that uses a distinctive character sequence as a symbol (token), which is passed from node to node, indicating when to begin transmission. Any node can remove the token, begin transmission, and replace the token when it is finished.

Token passing ring

Form of a LAN (local area network) in which all stations are connected in a closed ring by point-to-point links. Access to the transmission medium (e.g., twisted pair, coax, fiber optics) is governed by possession of a special bit pattern (free token) circulating on the ring.

Topology

The shape used in describing networks, e.g., ring, loop, hierarchical, or star.

Translate

To transform data from one language or coded character set to another.

Transponder

A receiver/transmitter combination aboard a satellite that receives a signal and retransmits it at a different carrier frequency. Transponders are used in communications satellites for retransmitting signals between earth stations.

Tributary station

A station on a multipoint channel receiving control from a control station.

Trunk line
A telephone communication channel between two PTT or Telco switching systems.

Unattended operation
A mode of operation of a system where certain functions (communication, printing) may be performed without operator assistance.

Uplink
The earth-to-satellite transmission channel.

VANS (value added network service)
A communications network using existing common carrier networks and providing such additional features as message switching, remote computing, remote database, electronic data interchange, document distribution, and protocol conversion. There are two types of VANS: a VANS provider that provides networking, processing, and storage and a VANS service provider that provides only processing and storage—no networking.

VAS (value added service)
An outdated term, now considered a VANS service provider. That is, a value added service supplied over a public or private network. A VANS service provider would not provide any networking capability. *See* VANS.

Verification
The process of determining through either protocol specification analysis or physical testing whether equipment conforms with a set of standard specifications or protocols.

Video
In television, a term pertaining to the bandwidth and spectrum position of the signal resulting from television scanning.

Videotex
A particular type of on-line, real-time public information service that uses the PSTN to access databases for the purpose of retrieving information, conducting order entry transactions, or posting an electronic message. It incorporates graphics that transmit pages (frames) of text and/or graphics.

Virtual call
See Switched virtual circuit.

Virtual circuit
A virtual circuit is a logical connection between two pieces of DTE that enables them to exchange messages according to a standard communication procedure with network preservation of sequence order. For example, in Tymnet's network, blocks are transmitted through the network over full-duplex virtual circuits. These circuits are established at log-on time and exist for the duration of the interaction between the terminal and processor. The circuits are not physical but exist as table definitions in the software associating an input channel with the appropriate output channel at each switching point.

Voice mail system
A message system that allows users to create, store, and send voice messages to other users with immediate or delayed delivery.

Voice-grade channel
A channel suitable for the transmission of speech, analog or digital data, or facsimile, generally using a frequency range of 300–3400 hertz (Hz).

V.24
A CCITT recommendation defining an interface to analog data communications facilities (such as Datel). V.24 is the CCITT counterpart to the EIA RS-232-C specification.

Wideband Channel
A facility that the common carriers provide for transferring data at speeds from 19.2 kbps to 2 Mbps.

Word
A character string or bit string considered as an entity.

Word processor
Equipment used to prepare business correspondence by entering and temporarily storing text for subsequent revision (editing) and printing. A word processor may handle groups of characters such as words, lines, paragraphs, or pages.

Work station
A keyboard/CRT that can be a nonintelligent or intelligent terminal, PC, or word processor. It may consist of a display, keyboard, storage (disk, diskette, or magnetic tape), and printer that together allow the operator to perform a variety of accounting, information processing, and word processing tasks.

X.21
A CCITT recommendation defining the interface between a terminal or computer and a CSDN (circuit switched data network).

X.25 network
A common-carrier-supplied service providing packet switched data transmission conforming to the CCITT X.25 Recommendation.

INDEX